D0515099

Eukaryotic DNA Replication

Frontiers in Molecular Biology

SERIES EDITORS

B. D. Hames

Department of Biochemistry and Molecular Biology
University of Leeds, Leeds LS2 9JT, UK

AND

D. M. Glover

Cancer Research Campaign Laboratories
Department of Anatomy and Physiology
University of Dundee, Dundee DD1 4HN, UK

TITLES IN THE SERIES

Eukaryotic DNA Replication

EDITED BY

J. Julian Blow

Imperial Cancer Research Fund
Clare Hall Laboratories
Hertfordshire, UK

MOHAVE

MOHAVE COMMUNITY COLLEGE LIBRARY

IRL PRESS
at
OXFORD UNIVERSITY PRESS
Oxford New York Tokyo

Oxford University Press, Walton Street, Oxford OX2 6DP

Oxford   New York
Athens   Auckland   Bangkok   Bombay
Calcutta   Cape Town   Dar es Salaam   Delhi
Florence   Hong Kong   Istanbul   Karachi
Kuala Lumpur   Madras   Madrid   Melbourne
Mexico City   Nairobi   Paris   Singapore
Taipei   Tokyo   Toronto
and associated companies in
Berlin   Ibadan

Oxford is a trade mark of Oxford University Press

Published in the United States
by Oxford University Press Inc., New York

© Oxford University Press, 1996

All rights reserved. No part of this publication may be
reproduced, stored in a retrieval system, or transmitted, in any
form or by any means, without the prior permission in writing of Oxford
University Press. Within the UK, exceptions are allowed in respect of any
fair dealing for the purpose of research or private study, or criticism or
review, as permitted under the Copyright, Designs and Patents Act, 1988, or
in the case of reprographic reproduction in accordance with the terms of
licences issued by the Copyright Licensing Agency. Enquiries concerning
reproduction outside those terms and in other countries should be sent to
the Rights Department, Oxford University Press, at the address above.

This book is sold subject to the condition that it shall not,
by way of trade or otherwise, be lent, re-sold, hired out, or otherwise
circulated without the publisher's prior consent in any form of binding
or cover other than that in which it is published and without a similar
condition including this condition being imposed
on the subsequent purchaser.

A catalogue record for this book is available from the British Library

Library of Congress Cataloging in Publication Data
(Data available)
ISBN 0 19 963586 2 (Hbk)
ISBN 0 19 963585 4 (Pbk)

Typeset by Footnote Graphics, Warminster, Wilts
Printed in Great Britain by Bath Press, Bath

Preface

Research into eukaryotic DNA replication is currently in an exciting phase of development. One of the early successes of molecular biology was to explain the way that genetic information, in the form of the self-complementary double helix of DNA, could be replicated as it passed from generation to generation. This work, largely carried out in *E. coli*, revealed the basic mechanism of DNA replication by which DNA polymerases translocate along and duplicate the two parental strands. Much of the basic enzymology of prokaryotic replication has been described in the excellent book entitled *DNA Replication* by Arthur Kornberg and Tania Baker.

For many years, the study of DNA replication in eukaryotes was a poor relation of the much more extensive prokaryotic field. This has now changed, at least in part due to the recent surge in our understanding of the way that the eukaryotic cell cycle is regulated. The eukaryotic cell cycle is quite distinct from that of prokaryotes, as it temporally separates chromosome replication and segregation, possibly as a consequence of its large chromosomes which need to be replicated using multiple replication origins. One particularly gratifying aspect of the recent increase in our understanding of eukaryotic replication is how it has unified work from a range of different model systems, the most important of which are covered in this book. Considering these systems together, we can now for the first time catch a glimpse of an integrated view of eukaryotic DNA replication and the way it is coordinated within the cell division cycle.

Much of the basic enzymology of the eukaryotic replication fork has close homologies with its *E. coli* counterpart. However, as outlined in Chapter 1 by Peter Burgers, there are some crucial differences, including the use by eukaryotes of two or three different DNA polymerases. Proteins have been identified that provide most of the major functions required at the replication fork, and a picture is beginning to emerge of how they work together to create a coordinated replication machine.

The biochemical analysis and reconstitution of SV40 DNA replication, covered by Daniel Herendeen and Tom Kelly in Chapter 2, has provided much of our knowledge of eukaryotic replication. SV40 viral DNA replication requires only a single virally encoded protein, the T antigen, all other replicative enzymes being provided by the host cell. The recent reconstitution of SV40 DNA replication using purified proteins is a major landmark in the study of eukaryotic DNA replication. Although this has identified many of the proteins involved in DNA replication, it is unlikely to tell us much about the normal cell cycle regulation of chromosomal DNA replication, whose control is mainly, though not completely, usurped by the virus.

In large part due to its genetic tractability, DNA replication in the budding yeast

S. cerevisiae is probably better understood than in any other eukaryote. In Chapter 3, York Marahrens and Bruce Stillman describe the thorough dissection of replication origins in this organism. This work has culminated in the identification of proteins that specifically interact with these replication origins and that function in the initiation process. The major limitation to this work has so far been the inability to re-create origin-dependent initiation in yeast cell-free systems; however, the recent availability of purified proteins thought to play important roles in initiation may soon resolve this problem.

Analysis of DNA replication in higher eukaryotes suggests that it may be more complex than in yeast. In Chapter 4, Mel DePamphilis describes metazoan replication origins, which appear to occupy much larger regions of DNA than do their counterparts in *S. cerevisiae*. A thorough understanding of these replication origins is crucial for the identification of replication proteins that interact with them. In addition, the complexity of metazoan origins may also depend on physical interactions with structural components of the nucleus. This latter topic is dealt with in more detail in Chapter 5, in which Peter Cook and colleagues challenge the notion, derived from work with dilute protein solutions, that DNA polymerases track along the DNA as it is replicated. Instead they discuss the evidence suggesting that the DNA polymerases themselves remain fixed to the nuclear substructure, whilst the DNA reels through these fixed sites as it is replicated.

The last three chapters deal with the way that DNA replication is coordinated with progression through the cell division cycle. Chapter 6, by Julian Blow, describes the control of DNA replication in *Xenopus* egg extracts, currently the only eukaryotic cell-free system that supports efficient chromosome replication *in vitro*. DNA replication in this system is dependent on template DNA being assembled into interphase nuclei. This process also regulates the licensing system, which ensures that chromosomal DNA is replicated precisely once in each cell cycle. In Chapter 7, Etienne Schwob and Kim Nasmyth review the cell cycle control of DNA replication in *S. cerevisiae*. This largely depends on a cascade of different cyclin-dependent kinases (CDKs), comprising the catalytic subunit (Cdc28) complexed with at least nine different cyclin partners. This kinase cascade both induces transcription of genes necessary for DNA replication and also triggers the replication process itself, whilst ensuring the regulated progression from G1 into S phase. Although DNA replication in the fission yeast *S. pombe* is less well understood, it provides an interesting contrast to the picture in *S. cerevisiae*, as is described by Susan Forsburg in Chapter 8. In particular *S. pombe* has identified the importance of cyclin-dependent kinases in the dependency relationships that ensure the strict alternation of DNA replication and mitosis in the cell cycle.

Our understanding of eukaryotic DNA replication is set to develop rapidly over the next few years, and although many surprises are likely to be encountered, the basic foundations of the field appear to be firmly established. These foundations, as presented in this book, should be able to form the basis on which researchers in the field, as well as students, teachers, and researchers in other disciplines, can build a dynamic understanding of this fast-moving subject.

I would like to express my gratitude to my wife, Margret, who has given me real assistance in assembling this book and bringing all the different strands together for publication.

South Mimms   J. J. B.
June 1996

Contents

The plate referred to on page 13 follows page 14.

4 Replication origins in metazoan chromosomes   96

MELVIN L. DEPAMPHILIS

7 Cell cycle control of DNA replication in *Saccharomyces cerevisiae*

ETIENNE SCHWOB and KIM NASMYTH

8 Regulation of S phase in the fission yeast *Schizosaccharomyces pombe*

SUSAN L. FORSBURG

Contributors

J. JULIAN BLOW
Imperial Cancer Research Fund, Clare Hall Laboratories, Blanche Lane, South Mimms, Herts EN6 3LD, UK.

PETER M. J. BURGERS
Department of Biochemistry and Molecular Biophysics, Washington University School of Medicine, 660 S. Euclid Ave. – Box 8231, St Louis, MO 63110, USA.

PETER R. COOK
CRC Nuclear Structure and Function Research Group, Sir William Dunn School of Pathology, University of Oxford, South Parks Road, Oxford OX1 3RE, UK.

MELVIN L. DEPAMPHILIS
National Institute of Child Health and Human Development, Bldg. 6/Rm. 416, National Institutes of Health, Bethesda, MD 20892–2753, USA.

SUSAN L. FORSBURG
Molecular Biology and Virology Laboratory, The Salk Institute for Biological Studies, 10010 North Torrey Pines Road, La Jolla, CA 92037, USA.

DANIEL HERENDEEN
Johns Hopkins University, Molecular Biology and Genetics, School of Medicine, 725N. Wolfe Street, Baltimore, MD 21205, USA.

PAVEL HOZÁK
Institute of Experimental Medicine, Academy of Sciences of the Czech Republic, Vídeňská 1083, 142 20 Prague 4, Czech Republic.

DEAN A. JACKSON
CRC Nuclear Structure and Function Research Group, Sir William Dunn School of Pathology, University of Oxford, South Parks Road, Oxford OX1 3RE, UK.

THOMAS J. KELLY
Johns Hopkins University, Molecular Biology and Genetics, School of Medicine, 725N. Wolfe Street, Baltimore, MD 21205, USA.

YORK MARAHRENS
Whitehead Institute for Biomedical Research and Department of Biology, Massachusetts Institute of Technology, Cambridge, MA 02142, USA.

KIM NASMYTH
IMP, Research Institute of Molecular Pathology, Dr Bohr-Gasse 7, A-1030 Wien, Austria.

ETIENNE SCHWOB
Institute de Génétique Moléculaire de Montpellier – UMR 5535, CNRS, 1919 route de Mende, BP 5051, 34033 Montpellier, France.

BRUCE STILLMAN
Cold Spring Harbor Laboratory, PO Box 100, New York, NY 11724, USA.

Abbreviations

6-DMAP	6-dimethylaminopurine; a protein kinase inhibitor
Abf1p	ARS binding factor 1; binds the B3 domain of certain replication origins in *S. cerevisiae*
ACS	ARS consensus sequence
ADA	adenosine deaminase
APE	amplification promoting element
ARS	autonomously replication sequence; *cis*-acting DNA sequence required for extra-chromosomal replication
CAK	CDK activating kinase
cdc	cell division cycle
Cdc2	cyclin-dependent kinase (Cdk1) with major roles in cell cycle regulation
CDC4	*S. cerevisiae* gene required after START but prior to iDS
CDC6	*S. cerevisiae* gene required for DNA synthesis; sequence homology with *S. pombe* cdc18
CDC7	*S. cerevisiae* protein kinase; mutants arrest after START but prior to iDS
Cdc10	component of the MBF (DSC1) transcription complex in *S. pombe*
Cdc13	*S. pombe* B-type cyclin
cdc18	*S. pombe* gene required for iDS, and major target of the cdc10 transcription system; sequence homology to *S. cerevisiae* CDC6
cdc19	*S. pombe* MCM2 homologue; identical to nda1
cdc21	*S. pombe* MCM4 homologue
CDC28	*S. cerevisiae* cdc2 homologue
Cdc34	*S. cerevisiae* ubiquitin-conjugating enzyme; mutants arrest after START but prior to iDS
CDC46	identical to *S. cerevisiae* MCM5
CDK	cyclin-dependent kinase
Cig	*S. pombe* cyclins with potential role in G1
CKI	CDK inhibitor
Clb	*S. cerevisiae* B-type cyclins
Cln	cyclin proteins in *S. cerevisiae* with roles at START
Dbf4	*S. cerevisiae* protein required for the initiation of DNA replication and which interacts with Cdc7
DHFR	dihydrofolate reductase
DMI	densely methylated island
DSC1	alternative name for MBF
DUE	DNA unwinding element
E2F	metazoan transcription factor, activates expression of genes involved in DNA synthesis

EBV	Epstein–Barr virus
Far1	*S. cerevisiae* CKI induced by mating pheromone
FEN-1	nuclease involved in Okazaki-fragment processing; identical to MF1
G1	gap period in the cell cycle after mitosis and before S phase
G2	gap period in the cell cycle after S phase and before mitosis
HMR E	transcriptional silencing locus that regulates mating type in *S. cerevisiae*, and also functions as a replicator
iDNA	initiator DNA; small DNA fragments of approximately 10–25 nucleotides at the 5′ end of Okazaki fragments
iDS	initiation of DNA synthesis
IR	inverted repeat
LCR	locus control region; controls gene expression in the human β-globin locus
MBF	transcription complex acting at START on MCB-containing promoters in yeast; also known as DSC1. Consists of Swi6 and Mbp1 in *S. cerevisiae*; in *S. pombe* probably consists of cdc10 and a res protein
Mbp1	component of the MBF transcription complex in *S. cerevisiae*
MCB	promoter element recognized by MBF (DSC1) in yeast
MCM	*S. cerevisiae* mini chromosome maintenance genes; components of licensing factor (RLF) in *Xenopus*
MF1	maturation factor 1, identical to FEN-1
M phase	mitotic period of the cell cycle
MPF	mitosis (or maturation) promoting factor; a complex between cdc2 and cyclin B
nda1	*S. pombe* MCM2 homologue; identical to cdc19
nda4	*S. pombe* MCM5 homologue
OBR	origin of bidirectional replication
ORC	origin recognition complex; binds the ACS in *S. cerevisiae*
ORE	origin recognition element
ori	origin of DNA replication; the site on DNA where replication initiates
oriC	*E. coli* origin of replication
p48	48 kDa primase subunit of human polymerase α-primase
p58	58 kDa subunit of human polymerase α-primase
p68	68 kDa subunit (B subunit) of human polymerase α-primase
p180	180 kDa polymerase subunit of human polymerase α-primase
PCR	polymerase chain reaction
PCNA	proliferating cell nuclear antigen, a processivity factor for DNA polymerase δ
PK	protein kinase
Pol1/I	*S. cerevisiae* DNA polymerase α
Pol2/II	*S. cerevisiae* DNA polymerase ε
Pol3/III	*S. cerevisiae* DNA polymerase δ
PP2A	protein phosphatase 2A
puc1	*S. pombe* cyclin with sequence similarity to *S. cerevisiae* CLNs
PV	papillomavirus

Rap1p	repressor/activator protein 1; binds to HMR E and to telomeres
RB	retinoblastoma protein, a tumour suppressor
RCC1	chromatin associated protein; involved in regulation of chromosome condensation
Res1/2	putative partners of cdc10 in the MBF (DSC1) transcription complex of *S. pombe*
REV3	gene encoding a putative *S. cerevisiae* DNA polymerase
RF-A	alternative name for RP-A
RF-C	replication factor C; DNA polymerase accessory factor
RLF	replication licensing factor; ensures that DNA is replicated only once in each cell cycle
RLF-M	component of the replication licensing system; a complex containing MCMs 2, 3, and 5
RP-A	eukaryotic single-stranded DNA binding protein involved in DNA replication
rum1	*S. pombe* CKI; overexpression leads to repeated rounds of DNA replication in the absence of mitosis
SBF	transcription complex consisting of Swi4 and Swi6, acting at START on SCB-containing promoters in *S. cerevisiae*
SCB	promoter element recognized by SBF in *S. cerevisiae*
Sic1	CKI which negatively regulates S-phase entry in *S. cerevisiae*
Site II	27 bp recognition sequence for T antigen at the SV40 origin
SPB	spindle pole body; microtubule organizing centre in yeast, equivalent to the higher eukaryotic centriole
SPF	S-phase promoting factor; consists of cdk2-cyclin E/cyclin A-cdc2 in *Xenopus*, and Cdc28-clb in *S. cerevisiae*
S phase	period of chromosomal DNA replication in the cell cycle
SSB	single-stranded DNA binding protein
ssDNA	single-stranded DNA
START	point of commitment to the cell division cycle in yeast
SV40	simian virus 40
Swi4	component of the SBF transcription complex in *S. cerevisiae*
Swi6	component of the SBF and MBF (DSC1) transcription complexes in *S. cerevisiae*
UAS	upstream activator sequence

1 | Enzymology of the replication fork

PETER M. J. BURGERS

1. Introduction

The assembly of the replisome is an ordered process mediated by both protein–DNA and protein–protein interactions. The initial steps in this process, origin recognition by the initiator proteins and localized melting of the origin into a replication bubble, will be discussed in Chapters 2 and 3 of this volume. This chapter focuses on the elongation process of DNA replication and the activities of the proteins at the fork. It combines our knowledge of individual replication proteins from mammalian systems and their application to SV40 DNA replication with genetic and biochemical studies of the analogous proteins in yeast. Comprehensive reviews detailing each of these fields have been published (1–10). With the exception of the putative DNA helicase that unwinds the DNA, all enzymes and factors that are predicted to function at a chromosomal replication fork from studies with model systems have been identified in these two eukaryotes, and most of their genes have been cloned. The eukaryotic replication fork contains both DNA polymerase α and DNA polymerase δ and their accessory factors. Although not evident from viral replication studies, DNA polymerase ε may also participate at the chromosomal DNA replication fork. Finally, recent progress on the maturation of Okazaki fragments initiated by ribonuclease H1 and a 5'–3' exonuclease will be discussed.

2. Biogenesis of the replication fork

In all organisms in which replication origins have been clearly defined, the corresponding initiator proteins that bind to these sites have also been identified. With the exception of the yeast origin recognition complex which so far fails to demonstrate a DNA unwinding activity, partial unwinding of the DNA is mediated by the initiator protein and facilitated by local destabilization of the DNA, e.g., by supercoiling, or by the presence of the single-stranded DNA binding protein (Fig. 1) (11–14).

Further opening of the helix requires the action of a DNA helicase, which is loaded onto the emerging replication bubble by accessory proteins. Additional protein–protein interactions with the initiator complex also promote helicase loading. For

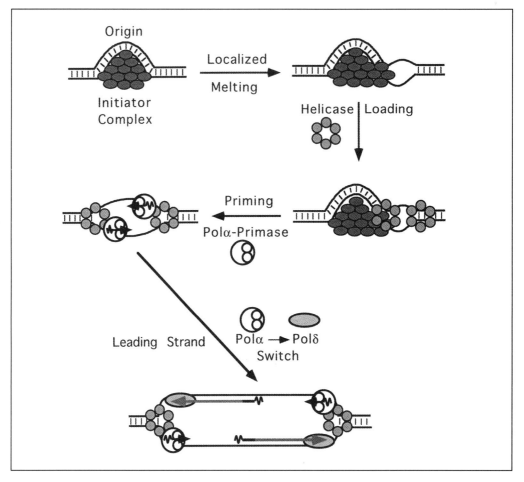

Fig. 1 Early steps in the assembly of the replication fork. For the sake of clarity, the single-stranded DNA binding protein has been omitted. The subunit structure of the cellular DNA helicase is not known, but is proposed to be hexameric by analogy with other systems. The fate of the initiator complex after helicase loading varies between different systems. Its omission from this figure after loading of DNA polymerase α–primase is for the sake of clarity, and is not meant to imply that dissociation of the initiator complex has occurred. See also legend to Fig. 3.

instance, loading of the dnaB helicase at the *E. coli* origin, oriC, requires dnaC but is also aided by direct interactions between dnaB and the initiator dnaA (15, 16). On the other hand, loading of dnaB at the lambda origin requires λ-P, which in turn interacts with the λ-O initiator protein (17, 18). Alternatively, in some prokaryotic and eukaryotic viral systems, the helicase is a separate domain of the initiator protein (reviewed in 12). However, only in the SV40 and papilloma virus systems is there sufficient evidence that the helicase associated with the initiator protein also participates in the propagation of the replication fork, and it is not known whether for instance the herpes simplex virus initiator helicase fulfils a similar function (19–23). It is likely that the cellular DNA helicase responsible for initial expansion of the replication bubble remains as an active component of the replication fork. This helicase could also be instrumental in setting up the fork through protein–protein interactions with other factors. However, additional DNA helicases may have auxiliary or even essential functions at the fork.

Subsequent to bubble formation, RNA priming is carried out by the primase subunit of DNA polymerase α–primase (Polα) (Fig. 1). Direct interactions between T antigen and Polα promote loading of this enzyme at the unwound SV40 origin (24–26). It is the synthesis of a first primer on each side of the bubble which sets up the formation of the replication fork. In SV40, extensive DNA synthesis by Polα beyond that of the primer is inhibited by the binding of replication factor C (RF-C) at the primer terminus and loading of the proliferating cell nuclear antigen (PCNA), followed by DNA polymerase δ (Polδ) (20, 27, 28). In this viral replication system Polδ holoenzyme carries out efficient continuous elongation of the leading strand. In contrast, Polα and Polδ holoenzyme cycle on the lagging strand, Okazaki fragments being primed by Polα and elongated by Polδ (Fig. 2a). No role for DNA polymerase ε (Polε) appears to exist in this viral replication system. Chromosomal forks, however, may show a different composition. One possible scenario is depicted in Fig. 2b with lagging strand synthesis being performed by a Polα–Polε cycle. The rationale for this hypothetical arrangement of enzymes is discussed below under the specific DNA polymerases.

In vitro model replication studies have shown that the leading and lagging strands of the fork are replicated coordinately. Most likely, coordinate replication is controlled through direct interactions between the leading and lagging strand replication machineries. Such a direct interaction is not easily visualized when one considers the forks shown in Fig. 2 as the lagging strand machinery travels in a direction opposite to that of the leading strand machinery. To explain coordinate replication of both strands in the bacteriophage T4 system, Alberts and co-workers have proposed a model in which, by looping of the lagging strand, both strands can be replicated in the same direction, allowing protein–protein interaction between the complexes to be maintained (29). Such a hypothetical fork structure for the eukaryotic replisome containing Polα, Polδ, and Polε, as well as their accessory factors is shown in Fig. 3. As the lagging strand polymerase (in this case depicted as Polε) elongates the Okazaki fragment, it retains contact with the leading strand replication complex, thus extending the Okazaki fragment as a loop. When the Okazaki fragment is completed, the loop is released allowing the polymerase to reattach to the new Okazaki fragment initiated by Polα.

Proper propagation of the fork requires that the stress of positive supercoiling induced by helicase unwinding is relieved by either DNA topoisomerase I (topo I) or topoisomerase II (topo II) (reviewed in 30). This follows from *in vitro* SV40 studies and from genetic studies in yeast (31, 32). Fork propagation in yeast is supported by either topo I (in a *top2^{ts}* strain at the restrictive temperature) or by topo II (in a *Δtop1* strain), but DNA synthesis is shut down if both activities are absent (32). In contrast, decatenation of daughter molecules uniquely requires DNA topo II (31).

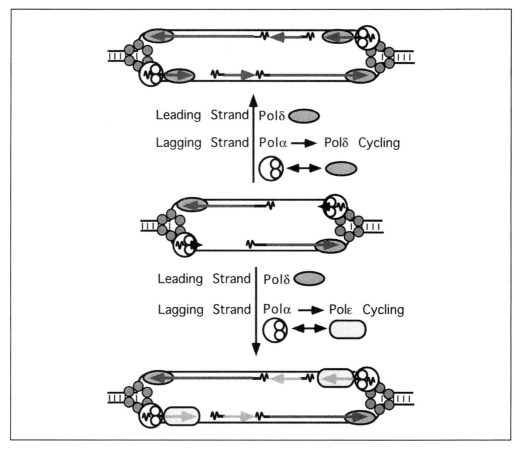

Fig. 2 Assembly of the lagging strand replication apparatus. Two contrasting models are shown, involving either (a) DNA polymerase δ (top) or (b) DNA polymerase ε (bottom) in Okazaki fragment synthesis. The top model represents the SV40 DNA replication fork.

3. DNA helicases

Identification of the DNA helicase(s) active at the eukaryotic DNA replication fork remains a major challenge towards its reconstitution. There are numerous DNA helicases in the eukaryotic cell, but with the exception of a few DNA helicases involved in DNA repair, the functions of most of these enzymes in DNA metabolism remain obscure (for recent reviews see 33–35).

Recent biochemical studies with the *E. coli* rep and uvrD DNA helicases indicate that the aggregation state of an active DNA helicase on the DNA should be that of a dimer or higher (36–38). The replication forks from bacteriophages T7 and T4, from *E. coli* oriC, and from SV40 have in common that the helicase is actually hexameric in structure (39–48). Although there are no compelling biochemical data indicating that a helicase which separates the DNA at the replication forks needs to be hexameric in structure, it would not be unreasonable to assume that the eukaryotic replica-

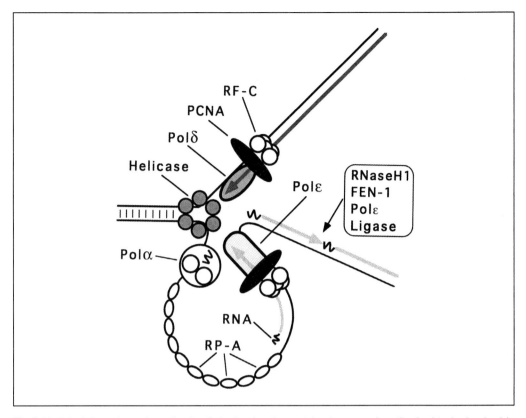

Fig. 3 Model of the eukaryotic replication fork showing the proteins known or hypothesized to be involved in fork elongation and in the maturation of Okazaki fragments. In the model shown in this figure, DNA polymerase ε is the likely enzyme to mature Okazaki fragments. iRNA (wiggly line) and DNA synthesized by Polα-primase is in black, DNA synthesized by Polδ in dark gray, and DNA synthesized by Polε in light gray.

tional helicase would also be a hexamer, as depicted in Figs 1 and 2. Unfortunately, there are no distinctive criteria that would aid in the identification of replicational helicases. Sequence comparison studies of a large number of helicases show that beyond the conserved ATP binding site, there is only minimal sequence homology among the many DNA and RNA helicases, making it difficult to predict such an enzyme with a high degree of accuracy from the primary amino acid sequence (49). Similarly, although they have some motifs in common, the hexameric replicational helicases do not stand out as a clearly distinct class on their own (50). One experimental approach to the replicational helicases has been based on the assumption that they might form stable complexes with other replication proteins, and therefore copurify more or less extensively with these enzymes. Using this approach, DNA helicases that copurify with Polα, Polδ, and replication factor C have been identified and purified (51–54). The significance of these interactions depends on the ability to show coupled helicase–polymerase action in appropriate assay systems, e.g., rolling circle DNA replication (42). Much needed progress in this field may come from further genetic investigations in yeast, and from biochemical reconstitution studies of model replication forks.

4. DNA polymerases

Comprehensive reviews of DNA polymerases α, δ, and ε have been published (3, 5–10, 55). The properties of these enzymes will only be briefly summarized in this chapter. Rather, particular attention will be given to those properties important for our understanding of their function in the replication process. Yeast Polα, Polδ, and Polε were previously designated as PolI, PolIII, and PolII, respectively, but the Greek letter nomenclature in existence for mammalian DNA polymerases has been adopted to reflect the conservation of structure and function between these two organisms (56). In addition to these three essential DNA polymerases, the yeast nucleus also contains DNA polymerase β and a putative DNA polymerase encoded by the *REV3* gene (57–59). Although the phenotypes of deletion mutants indicate that these two DNA polymerases are not involved in DNA replication, the *REV3* DNA polymerase may be incorporated into a mutagenic replisome to bypass sites of DNA damage that form blocks for the regular replicational DNA polymerases. The observation that the *REV1* gene, which interacts genetically with *REV3*, shows sequence similarity to *E. coli umuC* supports this idea (60). UmuC is proposed to be incorporated into a mutagenic form of DNA polymerase III holoenzyme during SOS DNA replication in *E. coli* (61).

4.1 DNA polymerase α–primase

In contrast to viral systems, e.g., bacteriophages T4 and T7, or herpes simplex virus, in which the DNA primase activity is intimately associated with the DNA helicase activity, the cellular primase is found in a tight four-subunit complex with DNA polymerase α (Table 1) (62–69). In this complex the p48 subunit specifies the primase activity and the p58 subunit probably complexes the primase to the polymerase subunit (70, 71). The p180 subunit carries the polymerase activity (72, 73). Although the precise role of the B subunit is not known, several lines of evidence indicate that it is required at an early step. The human B subunit interacts directly with the SV40 T antigen, indicating a role in Polα loading, and the phenotype of yeast *pol12^{ts}* mutants also indicates that this subunit acts at a very early stage of DNA replication (26, 74).

Polα does not contain a 3′–5′ (proofreading) exonuclease activity. Interestingly, there is significant sequence similarity (13–15%) between the exonuclease domains of Polδ and Polε and a region of both human and yeast Polα between amino acids 550 and 750. As with other proofreading polymerases this region is positioned upstream of the polymerase domains. However, with the exception of exonuclease motif II, the motifs in this domain lack some critical acidic amino acids which from mutational studies with other proofreading DNA polymerases are essential for exonuclease activity (6, 75–78). One interpretation of these comparisons would be that Polα lost its proofreading activity during the evolution of the cell. Inactivation

Table 1 Proteins and genes required for fork propagation and termination[a]

Factor[b]	Subunits[b]	Gene[b]	Essential[b]	Function or remarks
RP-A	70	RFA1	Yes	DNA binding
(RF-A)	30	RFA2	Yes	Cell cycle-dependent phosphorylation
(SSB)	14	RFA3	Yes	
Polα	167	POL1	Yes	Polymerase
	79	POL12	Yes	Initiation of replication
	62	PRI2	Yes	Polymerase–primase linkage
	48	PRI1	Yes	DNA primase
Polδ	125	POL3	Yes	Polymerase–exonuclease
	55	–	–	
Polε	256	POL2	Yes	Polymerase–exonuclease
	79	DPB2	Yes	
	23	DPB3	No	NTP binding motif
	29	–	–	
PCNA	29	POL30	Yes	Clamp for polδ and polε
RFC	97	CDC44	Yes	Binds primer-terminus
(Activator 1)	40	RFC2	Yes	
	40	RFC5	Yes	Putative subunit
	38	RFC3	Yes	ATPase
	36	RFC4	Yes	Forms complex with RFC3
RNase H1[d]	89	–	–	Bulk of initiator RNA removal
Exonuclease	45	YKL510	No	Deletions ts for growth;
(FEN-1)		(RTH1)		terminal RNA removal
DNA ligase	85	CDC9	Yes	Chain closure
TopoI	90	TOP1	No	Swivelase
TopoII	164	TOP2	Yes	Swivelase; decatenase

[a] References to factors and their genes are in the text.
[b] Subunit structure of the yeast factors and sizes derived from the yeast gene sequence are given unless indicated otherwise.
[c] Yeast genes and requirement for cell growth are listed.
[d] Enzyme from human cells.

of the proofreading activity of Polα would have disastrous consequences for maintaining its genetic stability if this DNA polymerase was responsible for bulk DNA replication. Studies of SV40 DNA replication in nuclei and in purified systems strongly indicate that DNA synthesis by Polα is limited to very short stretches of DNA (see below).

Temperature-sensitive mutations in the yeast *POL1* gene encoding the polymerase subunit of Polα show various phenotypes in DNA replication and recombination, and also in telomere maintenance (reviewed in 5). The *pol1-17* mutant exhibits a quick-stop in DNA synthesis upon a shift to the nonpermissive temperature consistent with a continuous requirement for Polα during the elongation state of DNA replication. This allele as well as other *pol1* alleles with a quick-stop phenotype have amino acid changes in or near the conserved domains which are required for polymerase function, and Polα activity from these mutants is defective (5, 79–83). These mutational studies strongly indicate that the polymerase activity of Polα is continuously required during DNA replication and not just during the initiation process.

Under conditions where priming is coupled to DNA synthesis by Polα, RNA primers about 10 nt in length are synthesized during the initiation of SV40 DNA replication (28). In a purified system, elongation of the fork can be carried out by Polα in the presence of replication protein A (RP-A or RF-A, the eukaryotic single-stranded binding protein) and a topoisomerase, i.e., the mono-polymerase system (84). However, this reaction is inhibited by RF-C, allowing the loading of PCNA and the participation of Polδ in this replication system (20). In crude extracts in which PCNA has been neutralized by antibodies, thereby inhibiting the activity of Polδ as described below, elongation of the initial 10 nt RNA primers is arrested after elongation with about 25 nt of DNA (28). These small RNA–DNA fragments are called initiator DNA (iDNA). It is not clear from these studies whether iDNA fragments of this size are also synthesized during regular Okazaki fragment synthesis. However, studies of the elongation reaction in isolated nuclei indicate the prevalence of iDNA fragments as precursors of full-length Okazaki fragments (85). Finally, the maturation of Okazaki fragments by Polα and the appropriate maturation factors is very inefficient when the single-stranded DNA is coated with RP-A, and completely inhibited in the presence of RF-C and PCNA (86). These studies indicate that the role of Polα in Okazaki fragment synthesis is limited to that of the synthesis of a short RNA–DNA primer.

4.2   DNA polymerase δ

DNA polymerase δ is the smallest of the replicational DNA polymerases with a catalytic subunit of 125 kDa and a small subunit of about 50 kDa with unknown function (87, 88). Both the polymerase and 3′–5′-exonuclease activities reside in the large subunit (77, 89). Of all the eukaryotic DNA polymerases, Polδ is the most highly conserved showing 44% sequence identity between yeast and humans in comparison to 31% sequence identity for Polα and 39% sequence identity for Polε (6, 90, 91). This high degree of conservation during evolution may explain the observation that some specific protein–protein contacts, e.g., between the polymerase and PCNA, have been conserved between species (88).

Yeast *POL3 (CDC2)* temperature-sensitive mutants show a terminal phenotype consistent with a role for this enzyme in DNA replication (92, 93). Residual DNA synthesis did occur at the restrictive temperature, but this could be completely eliminated if the cells were shifted up after release from a urea block. An even more striking indication that Polδ carries out bulk chromosomal DNA synthesis in yeast follows from the observation that *POL3* exonuclease-deficient mutants exhibit a 100 to 500-fold increase in spontaneous mutation rates (77, 94). This is comparable to the increase in spontaneous mutation rates observed with *E. coli dnaQ (mutD)* strains, which are deficient in proofreading by the ε subunit of DNA polymerase III holoenzymes, if one considers that *dnaQ* mutants are also deficient for mismatch repair, whereas *POL3* exo⁻ mutants are not (94, 95).

The enzymatic properties of Polδ have been reviewed (3, 6, 9). Polδ and Polε are both insensitive to a potent inhibitor of Polα, N^2-(*p-n*-butylphenyl)-dGTP

(BuPhdGTP). This inhibitor has been used in permeable cells or isolated nuclei systems to probe the role of Polα versus Polδ and/or Polε (85, 96). Unfortunately, no inhibitor which clearly distinguishes Polδ from Polε is available. The most distinctive enzymatic property of Polδ is its conversion from a distributive to highly processive enzyme when complexed with PCNA so that a long stretch of nucleotides is synthesized each time it binds to primer DNA (see below). This is commonly measured by the stimulation of Polδ on a poly(dA) template sparsely primed with oligo(dT). The catalytic subunit of yeast Polδ is sufficient for interaction with PCNA (97). In contrast, a form of mouse Polδ lacking the small subunit is not stimulated by PCNA (98). Similarly, a form of *Drosophila melanogaster* Polδ consisting of a single 120 kDa polypeptide is not stimulated by PCNA, whereas the two-subunit enzyme is (99, 100). Although this might suggest a divergence of function from yeast to higher eukaryotes, it could also indicate that both the catalytic and the 50 kDa subunit of Polδ contribute to binding to PCNA, and that the interaction between the mouse or *Drosophila* catalytic subunit and PCNA was not strong enough to give a positive result in the particular assay used. An alternative explanation for these observations is that the single subunit forms of Polδ are actually proteolytic products of Polε, which has been shown to produce a distinct 122 kDa form upon limited proteolysis (101). This could well be the case with the mouse enzyme, but seems to be unlikely for the *Drosophila* enzyme because its template preference properties match those of a δ-type polymerase much more than those of an ε-type polymerase (98, 99).

4.3 DNA polymerase ε

The yeast and human enzymes have a dissimilar subunit structure. Human Polε is a two-subunit enzyme with a catalytic subunit of 225 kDa and a 50 kDa subunit of unknown function (91, 101). Yeast Polε (Pol2) is a four-subunit enzyme (Table 1). In addition to a catalytic subunit of 256 kDa there are three accessory subunits ranging in size from 79 to 29 kDa (Table 1) (8, 102). As with Pol, the exonuclease activity of Polε is contained in the polymerase subunit (78).

Several lines of evidence indicate that Polε carries out DNA repair synthesis. The human enzyme was originally isolated as a factor required for conservative DNA synthesis in a reconstituted system utilizing UV-irradiated permeabilized human diploid fibroblasts (103). In yeast, base-excision repair synthesis in nuclear extracts of *pol1* (Polα) or *pol3* (Polδ) temperature-sensitive mutants was only minimally affected at the restrictive temperature. However, an extract from a *pol2^{ts}* (Polε) mutant was defective at the restrictive temperature and could be complemented by the addition of purified yeast Polε (104). However, the indication of a role for Polε in DNA repair does not exclude an additional role in DNA replication.

Temperature-sensitive *pol2* mutants arrest with a dumbbell shape phenotype typical for a defect in S-phase progression, and fail to synthesize DNA upon a shift to the nonpermissive temperature (93, 105). In addition to its essential polymerization function, the *POL2* gene is also required for checkpoint control in the S phase of the yeast cell cycle (181). During the S phase of the cell cycle, Polε appears to act as a sensor of DNA replication, transmitting inhibitory signals to prevent inappropriate transcriptional and cell cycle functions. These two essential functions of Polε are on separable domains, the polymerase function residing on the N-terminal domain and the checkpoint function on the C-terminal domain. This latter domain is unique to the ε polymerases.

The *DPB2* gene encoding the second largest subunit of yeast Polε is essential for growth and a temperature-sensitive *dpb2* mutant is partially defective in DNA synthesis at the restrictive temperature, indicating that the protein is required for normal yeast chromosomal replication (106). The *DPB3* gene encoding the third subunit of Polε is not essential, but a *dpb3* deletion mutant exhibits an increased spontaneous mutation rate, suggesting that the protein is required to maintain fidelity of chromosomal replication or repair (107). *Pol2* exonuclease deficient exo⁻ mutants show a 10- to 40-fold increase in spontaneous mutation rates and Polε purified from one such exo⁻ strain has an at least 100-fold decrease in exonuclease activity (94, 108). The mutator phenotype of *pol2 exo⁻* is much less severe than that of *pol3 exo⁻*, perhaps suggesting that the primary role of Polε is not in DNA replication (78, 94, 108). However, other considerations, discussed below, strongly indicate a role for Polε in the replication process.

Polε, unlike Polδ, has the capacity to synthesize processively on poly(dA)-oligo(dT), obscuring any possible interaction with PCNA. This difference in processivity and the ability of Polε to carry out DNA synthesis at higher salt and Mg concentrations than Polδ most clearly distinguishes these two enzymes, as inhibitors such as aphidicolin, BuPhdGTP, or carbonyldiphosphonate do not decisively differentiate between the two species (reviewed in 4, 9). However, assays on natural template systems coated with *E. coli* SSB (single-stranded DNA binding protein) or RP-A have shown that Polε does interact with PCNA and RF-C to increase its processivity of action (109–113). Likewise, Polε can be substituted for Polδ in the *in vitro* SV40 DNA replication assay, but this reaction is quite inefficient and probably not of physiological relevance (111).

4.4   Fidelity mechanisms and the fork

It is important to establish the protein distribution at the replication fork in order to understand which factors control the fidelity of DNA replication in the eukaryotic cell. In *E. coli*, both strands are replicated by the DNA polymerase III core enzyme and, therefore, no large differences should exist in replicational fidelity. However, in eukaryotic cells, due to the participation of two or three DNA polymerases and the asymmetric distribution of these enzymes, a distinct asymmetry in the fidelity of DNA replication may exist. A moderate asymmetry has been revealed by Kunkel and co-workers using *in vitro* SV40 assays (114). Because SV40 DNA replication is bidirectional, a specific mutational target can be replicated either as the leading strand or as the lagging strand by placing the origin on either side of this target. Inclusion of a large excess of one dNTP is predicted to induce replication errors due to misincorporation of that dNTP, thereby allowing the replication error to be assigned to the appropriate strand. Error rates for template G.dTTP and template-C.dTTP mispairs increase by several fold if the template is replicated as the lagging strand rather than the leading strand (114). On the other hand, error rates for template-(G,T,A).dGTP mismatches vary by less than two-fold between leading and lagging strands (115). This analysis indicates that the contribution of mutagenic Polα to Okazaki fragment synthesis must be minimal (116). Perhaps, most of the iDNA is replaced by nick translation during Okazaki fragment maturation (Section 6).

In yeast, Sugino and co-workers have established an epistasis relationship between the proofreading functions of Polδ and Polε and the mismatch repair system. In yeast, the *PMS1* gene is required for mismatch repair (117). In these epistasis studies, spontaneous mutation rates in haploid or (partial) diploid strains containing the single *pol2exo⁻*, *pol3exo⁻*, or *pms1* mutation were compared with those of double mutants. The rationale for these experiments has been described (94). Briefly, suppose that X and Y represent two genes, mutations in each of which produces a spontaneous mutator phenotype. In general, synergism in the double mutant X⁻Y⁻, i.e., a mutation rate greater than the sum of mutation rates in X⁻ plus Y⁻, is due (i) to competition between X and Y for the same or overlapping sets of errors or (ii) to action in series of the X and Y pathways. In the latter case, a multiplicative relationship between X and Y should exist in the double mutant X⁻Y⁻. In principle, these two synergistic pathways can be distinguished by using strains which are heterozygous diploid for X or Y, provided that the mutant is partially dominant over wild type. In an X⁺X⁻Y⁻ strain multiplicity between the X⁺X⁻ and Y⁻ single mutants should still remain if X and Y act in series, but synergism in the X⁺X⁻Y⁻ strain should be reduced if X and Y compete for the same errors (94). Haploid cells containing both the *pol3exo⁻* and *pms1*, or *pol3exo⁻* and *pol2exo⁻* mutations failed to grow, but the double mutations could be maintained in homozygous diploids (Table 2) (94, 108). A clear multiplicative relationship exists between *POL3* and *PMS1*, indicating that proofreading by Polδ and mismatch repair are consecutive pathways. This multiplicative relationship is maintained in diploids heterozygous for *POL3*. A similar relationship was observed for *POL2* and *PMS1*, again indicating that the proofreading function of Polε and mismatch repair act in series. In principle, it is possible that Polε acts after mismatch repair, e.g. in long-patch damage repair, but the phenotype of *pol2ᵗˢ* mutants and the epistatic relationship between *pol3exo⁻* and *pol2exo⁻* mutants indicates that Polε acts before mismatch repair during DNA replication, as for example is shown in Fig. 2b. As the proofreading deficient mutants of Polδ and Polε are still expected to replicate with relatively high fidelity, the efficiency and fidelity of mismatch repair by either of these mutant polymerases would not be severely compromised if repair tracts remain small. A more complicated relationship exists between *POL2* and *POL3*. Synergism is observed in a diploid homozygous for the *pol3exo⁻ pol2exo⁻* double mutant and this synergism is reduced when a haploid *pol2exo* strain is made hemizygous for *POL3* by transformation with a *pol3exo⁻* containing centromere plasmid (Table 2). One conclusion from this epistasis analysis is that the proofreading functions of

Table 2 Spontaneous mutation rates in yeast mutator strains[a]

Haploid cells	pol2	pol3	pms1	pol2/pol3	pol3/pms1	pol2/pms1
Reversion rate	10	240	150	Lethal	Lethal	1800
Forward rate	12	130	40			520

Haploid plus plasmid	pol2	POL3/pol3	POL3/pol3 pol2
Forward rates	86	57	180

Diploid cells	pol2/pol2	pol3/pol3	pms1/pms1	pol2/pol2 pol3/pol3	pol3/POL3	pol3/POL3 pms1/pms1	pol3/pol3 pms1/pms1
Reversion rates	10	490	250	1900	6	2400	38 000

[a] Data adapted from (78, 94, 108); POL2 and POL3 indicate the wild-type genes; pol2, pol3, and pms1 indicate pol2exo⁻, pol3exo⁻, and Δpms1, respectively. Mutant combinations pol2/pol3 and pol3/pms1 cannot be obtained in haploids and are therefore presumed to be lethal combinations. Reversion rates of his7-2 to auxotrophy and forward mutation rates of URA3 to ura3 (detected as 5-fluoroorotic acid resistance) were measured relative to the isogenic wild-type strains (=1).

Polδ and Polε compete for an overlapping set of replication errors, produced for instance by the error-prone Polα. Clearly, competition for the replication errors made by Polδ must be limited if it exists at all. Otherwise, if Polε could proofread for Polδ, the *pol3exo⁻* mutant would not display such a strong mutator phenotype. More likely, the proofreading activities of Polδ and Polε are able to correct only those insertion errors made by their own polymerase activity and, therefore, double mutants would be expected to show additivity. The observed synergism could be due to partial inactivation or saturation of the mismatch repair system in the *pol3exo⁻ pol2exo⁻* double mutant. In *E. coli*, mismatch repair is saturated in an *E. coli mutD* strain deficient for proofreading by the ε subunit of DNA polymerase III holoenzyme (95). If this is the case then *pol3exo⁻ pol2exo⁻ pms1* triple mutants should not be much stronger mutators than *pol3exo⁻ pms1* double mutants. In any case, these studies provide additional suggestive evidence that the activity of Polε is required during chromosomal DNA replication. The more modest mutator phenotype exhibited by *pol2exo⁻* mutants than *pol3exo⁻* mutants could be attributed to a minor but essential role for Polε in bulk DNA replication or to an intrinsic higher nucleotide insertion fidelity of this enzyme. Consistent with this latter interpretation, the *in vitro* fidelity of Polε is several fold higher than that of Polδ (116).

5. DNA polymerase accessory proteins

In the last several years a large number of polymerase accessory factors have been identified by biochemical studies. These fall in general in three broad classes, single-stranded DNA binding proteins, primer–template recognition factors, and processivity factors. Two primer recognition factors for Polα, primer recognition protein C1C2 and alpha-associated factor, will not be discussed here (118, 119). As they do not appear to be required for SV40 DNA replication or have been defined in yeast, their role in DNA replication remains uncertain.

5.1 Replication protein A

The eukaryotic single-stranded DNA binding protein, RF-A or RP-A, consists of three subunits, each of which is essential in yeast for cell growth (Table 1) (120–122). RP-A is required for both the initiation and elongation of SV40 DNA replication (123–125). The yeast factor efficiently substitutes for human RP-A in the initiation phase, but only poorly in the elongation phase (126). The 70 kDa subunit binds to single-stranded DNA (124). Phosphorylation of the 34 kDa subunit in a cell cycle regulated manner indicates a regulatory role for this subunit (126, 127). More-over, unphosphorylated p34 subunit is phosphorylated by a DNA-dependent protein kinase during the initiation of SV40 DNA replication (128). RP-A coats approxi-mately 30 nt of single-stranded DNA with moderate cooperativity (129). As with most SSBs, binding to polypyrimidines is strongest. RP-A stimulates the activity of Polδ or Polε rather nonspecifically as other SSBs can substitute for it (110, 111, 125). However, the interaction with Polα is specific (25, 125).

5.2 PCNA

PCNA has been independently identified as a human immunoreactive proliferation-dependent nuclear antigen (from which it derives its current name), as a protein whose synthesis is induced upon serum stimulation, as a factor required for SV40 replication, and as an auxiliary factor for Polδ (reviewed in 9, 130, 131). PCNA is a ubiquitous, highly conserved protein. The yeast *POL30* gene encoding PCNA is essential for cell growth, and a conditional mutant shows a terminal phenotype consistent with a requirement for PCNA in chromosomal DNA replication (132, 133). In addition to its interaction with RF-C, Polδ, and Polε, human PCNA also binds to the p53-regulated inhibitor of cyclin-dependent protein kinase, called p21 or Cip1, resulting in an inhibition of SV40 DNA replication (134, 135). Presumably, one of the antiproliferative roles of p21 is inhibition of chromosomal DNA replica-tion by this mechanism.

Yeast PCNA is a homotrimer of 29 kDa subunits (88). The crystal structure of yeast PCNA has recently been determined (136). Plate 1 shows a comparison of this structure with that of the β subunit of *E. coli* DNA polymerase III holoenzyme. The β subunit is a dimer, with each monomer consisting of three structurally identical domains giving the dimer structure an overall six-fold symmetry (137). In view of the modularity of the β subunit structure, the size of the PCNA monomer (258 residues) which is approximately two thirds that of the β subunit monomer (366 residues) and its trimeric state in solution, as well as the similarity in function between the two proteins, it had been proposed that the structure of PCNA should resemble that of the β subunit (137). This prediction has been borne out as the struc-ture contains two identical domains per subunit resulting in an overall six-fold symmetry in the trimer (Plate 1). Yet, despite the virtually identical structure there is no significant sequence similarity between PCNA and the β subunit. In PCNA each of the central helices is orthogonal to the local direction of the phosphate backbone of B form double-stranded DNA which has been modeled in the center of the ring. Such an arrangement would likely restrict protein–DNA interactions to non-specific contacts with the phosphate backbone. Electrostatic potential maps show a positive electrostatic potential for the inside of the ring, favoring interactions with the phosphate backbone, whereas the rest of the molecule has an overall negative electrostatic potential (136). The PCNA sequences from organisms as divergent as *S. cerevisiae*, *S. pombe*, *D. melanogaster*, *X. laevis*, human, and *Oryza sativa* (rice) show sequence identities above 30% and can all be aligned without any gaps, except at the very carboxy terminus (132, 138–142). Because of this strong conservation, the crystal structure of yeast PCNA should serve as a reliable model for structure–function studies in other eukaryotes.

The functional conservation of eukaryotic PCNAs was first appreciated when bovine PCNA was found to stimulate yeast Polδ and vice versa (88, 143). This remarkable conservation of the domains on Polδ and PCNA required for protein–protein interaction has since been found to extend to other eukaryotes (144–146). In fact, an *S. pombe* PCNA deletion mutant is partially complemented by the human gene, indicating that interactions with other replication proteins have also been largely conserved (138).

The ring-like structure of PCNA is consistent with its biochemical properties. PCNA was originally identified as an auxiliary factor for Polδ because it stimulated the processivity of Polδ on poly(dA)·oligo(dT) (147). Interaction with or stimulation of Polδ was also observed on small oligonucleotide template primers, but not on primed circular single-stranded M13 DNA templates (148–151). However, linearization of the primed circular viral molecule in such a way that the primer was at a double-stranded end restored stimulation by PCNA on the template primer (151). This analysis showed that the PCNA trimer can load only onto linear DNA molecules by diffusion onto the double-stranded DNA end consistent with the ring-like structure of this molecule (Fig. 4). However, RF-C-independent loading at double-stranded ends is a very inefficient process in comparison with enzymatic loading by RF-C (151). Loading onto single-stranded DNA coated with a single-stranded DNA binding protein and sliding to an internal primer terminus was not observed, indicating that the primary interaction of PCNA is with double-stranded DNA (151). Loading of PCNA on circular template–primers requires RF-C. PCNA loaded onto linear DNA molecules, even when loaded by RF-C, is also likely to slip off those DNA substrates via the ends, especially if the DNA substrates are predominantly in double-stranded form promoting sliding of PCNA (152).

5.3 Replication factor C

The subunit structures of yeast and mammalian RF-C (also called activator 1) are quite similar (Table 1) (109, 112, 153–156). In addition to a large subunit, 130 kDa in human cells and 96 kDa in yeast, there are three or four small subunits in the 37–40 kDa range (157, 158). The small subunits of RF-C are all highly homologous (156, 159–163). The pairwise homology which is evident between each of the human and

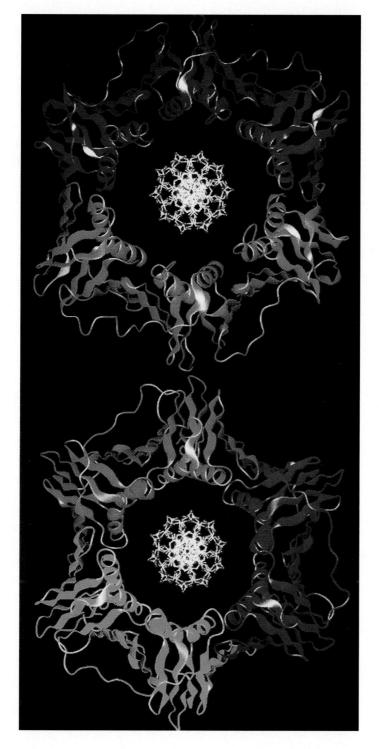

Plate 1 Ribbon representation of the polypeptide backbones of PCNA and the B subunit, with hypothetical duplex DNA. In this representation of a trimer of PCNA (left) and a dimer of B subunit (right), B strands are shown as flat ribbons and α helices as spirals. Individual monomers are distinguished by different colours (from 136, with permission).

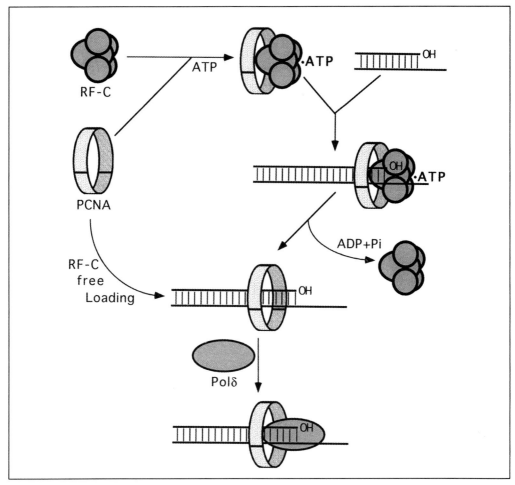

Fig. 4 Enzymatic and nonenzymatic loading of PCNA. The left side of the figure shows a nonenzymatic loading pathway of PCNA by diffusion onto the ends of double-stranded DNA. The right side shows the RF-C-mediated mechanism. In this mechanism, ATP binding is required for loading of PCNA by RF-C, whereas release of RF-C from the complex with PCNA at the primer terminus requires hydrolysis of the bound ATP. The fate of RF-C at the replication fork, either remaining on the DNA after a conformational change or dissociating from the DNA, is not known. In this figure, dissociation of RF-C is shown to indicate similar properties of the PCNA.Polδ complex, whether loaded nonenzymatically or enzymatically.

yeast small subunits indicates that these subunits were generated by gene duplication before divergence of these two organisms (Fig. 5). The small human subunit hA1-38 and a putative yeast subunit encoded by the *RFC5* (YBR0810) gene, which was identified as part of the yeast genome sequencing effort, show 38% sequence similarity, but are more distantly related to the other small subunits showing 20–25% similarity (Fig. 5). All of the yeast genes are essential for growth (133, 156, 157, 162, 163). However, in contrast to most of the other replication genes, which show cell cycle-regulated expression with maximal mRNA levels at the G1/S border of the cell cycle (reviewed in 164), expression of the *RFC2, RFC3*, and *RFC4* genes is constitutive, and the *RFC1*

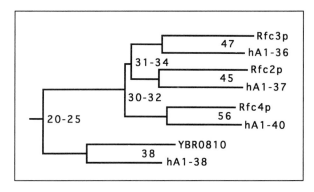

Fig. 5 Phylogenetic tree of the human and yeast small RF-C subunits. Human subunits are indicated as hA1-36 to hA1-40, whereas subunits are indicated as Rfc2p to Rfc4p and YBR0810 (Rfc 5p). The program CLUSTAL V was used to generate the alignment (180). Sequence similarities in % between pairs and groups are shown.

(*CDC44*) and *RFC5* genes lacks the upstream *cis*-acting element required for periodic expression (156, 157, 162, 163, 165). Cold-sensitive mutants in the *RFC1* (*CDC44*) gene show a terminal phenotype consistent with a defect in DNA synthesis. However, bulk DNA replication appears to be largely complete at the restrictive temperature, indicating loss of a late function in these mutants (157).

RF-C binds specifically to the template–primer junction, mediated predominantly by the large subunit (149, 155, 166). This specificity may explain the observed sequence similarity with poly(ADP-ribose) polymerase and DNA ligase which recognize similar DNA substrates (158). RF-C has a single-stranded DNA-dependent ATPase activity, which is optimal when template–primer junctions are present (109, 112, 154, 155, 166). Despite the high homology among the small subunits, particularly in the ATP-binding domain, only one of these subunits has been shown to actually bind ATP. The human 40 kDa subunit is the only subunit in RF-C to bind ATP, and ATP binding, but not its hydrolysis has also been detected in the cloned 40 kDa subunit (hA1p40) (19, 160, 166). Surprisingly, the yeast subunit most homologous to hA1p40, Rfc4p, does not bind ATP, either in the RF-C complex or as the overexpressed subunit alone (51, 162). Rather, ATP binding in yeast RF-C resides in Rfc3p, and this subunit overexpressed in *E.coli* shows single-stranded DNA-dependent ATPase activity (51, 156). Like the large subunit of RF-C, human A1p37 shows preferential binding to template–primer junctions (167). This may explain the stimulatory activity of this subunit on DNA synthesis by Polε (167). Yeast Rfc3p and Rfc4p form a tight complex. Neither Rfc2p, Rfc3p, Rfc4p, nor the Rfc3p–Rfc4p complex, seriously affects the activities of Polδ or Polε, or RF-C (156, 162, 163).

5.4 Eukaryotic DNA polymerase holoenzymes

The basic enzymatic make-up of a processive DNA polymerase holoenzyme complex at the replication fork has been remarkably conserved between eukaryotes, *E. coli*, and bacteriophage T4 (reviewed in 168). The *E. coli* and phage analogues of PCNA are the β subunit and gene 45 protein, respectively. Although no significant sequence similarity exists between these three proteins, the structures of PCNA and β subunit are virtually identical, and gene 45 protein is expected to have the same structure (Plate 1) (136). However, significant sequence similarity exists between the

RF-C subunits and subunits of the functionally analogous *E. coli* γ.δ complex and T4 gene 44/62 complex (161, 162). In these three organisms, RF-C (γ.δ complex or gene 44/62 complex) uses the energy of ATP hydrolysis to load the circular PCNA (β subunit or gene 45 protein) at the template–primer junction, which in turn complexes with Polδ or Polε (PolIII core or gene 43 protein) to form a processive replication complex.

The mechanism of Polδ holoenzyme formation has been studied in mammalian cells and in yeast and is shown in Fig. 4. Footprinting experiments have shown that RF-C recognizes both the template and the primer strand at the primer–template junction, and this footprint is extended into the double-stranded region if PCNA is also present (166). Efficient loading of PCNA requires RF-C and ATP (109, 110, 112, 149, 166, 169). Kinetic experiments employing analyses of lag phases in DNA synthesis, which measure rates of complex formation, indicate that PCNA, RF-C, and ATP form a ternary complex in the absence of DNA, and this complex is capable of loading PCNA onto the primer terminus with hydrolysis of ATP (110). The ATPase activity of RF-C is stimulated by PCNA when primer–template junctions are present. ATPγS is a potent competitive inhibitor of ATP in holoenzyme activation. Yet, ATPγS increases the binding of RF-C and PCNA to primer–template junctions and, if removed from the reaction by gel filtration, will support processive DNA synthesis by Polδ (110, 166, 169). These types of experiments are consistent with a model in which loading of PCNA by RF-C onto the primer terminus requires ATP binding, whereas release of RF-C from the PCNA clamp requires its hydrolysis. In this model release could either mean a conformational change in RF-C or its dissociation from the DNA (Fig. 4). Once PCNA has been loaded, binding of Polδ and processive DNA synthesis by the holoenzyme do not require any further ATP (110, 169).

Processive DNA synthesis by yeast Polδ holoenzyme proceeds at a rate of 110 nt/s at 30 °C, regardless of whether the single-stranded DNA has been coated with *E. coli* SSB or yeast RP-A (110). This rate compares well to the *in vivo* rate of fork movement in yeast of about 60 nt/s (170). The rate of DNA synthesis by the human Polδ holoenzyme, is much lower, about 5–10 nt/s (27). In contrast to the strong interaction between Polδ and PCNA at the primer terminus, a much weaker interaction is observed between Polε and PCNA. The yeast Polε holoenzyme replicates DNA with moderate processivity, elongating about 200–500 nt per processive cycle at 50 nt/s (110). At those pause sites Polε, but not PCNA, dissociates from the DNA. Similarly, the mammalian Polε holoenzyme has a lower stability than the Polδ complex (111, 113). These differences in processivities between Polδ and Polε holoenzyme, together with genetic data which indicate a role for Polε in DNA replication (Section 4.4), have led us to propose a model in which leading strand DNA replication is carried out by Polδ holoenzyme, whereas the elongation of primers on the lagging strand is carried out by Polε holoenzyme (Fig. 3) (110). SV40 model replication studies do not support a role for Polε in Okazaki fragment synthesis, but considering that some of the components, including the DNA helicase, are going to be quite different between SV40 and cellular systems, a different mode of cellular fork biogenesis and function can at present not be excluded.

6. Okazaki fragment maturation

Considerable progress has recently been made in our knowledge of the enzymes that mature Okazaki fragments (reviewed in 10). Early SV40 replication studies indicated that a RNase H activity, a double-stranded exonuclease activity, and DNA ligase are required for the production of covalently closed molecules (84). Recent model enzymatic studies and SV40 replication studies show that four enzymes act in sequence or in concert to process Okazaki fragments into fully ligated DNA, i.e., RNase H1, Polδ or Polε, FEN-1 nuclease, and DNA ligase 1 (86, 171–174). RNase H1 makes a single structure-specific endonucleolytic cleavage in the RNA primer, releasing the intact primer and leaving one ribonucleotide at the 5′-terminus of the RNA–DNA junction (Fig. 6) (174). The exact yeast analogue of RNase H1 has not yet been identified. A 70 kDa RNase H from yeast cuts precisely at the RNA–DNA junction, and the cutting specificities of other RNase Hs have not yet been determined (175, 176).

The 5′-3′ exonuclease studied by Bambara and co-workers is probably identical to maturation factor 1 (MF-1) from Stillman and co-workers and FEN-1 nuclease from Lieber and co-workers (172, 173, 177). Because the FEN-1 gene has been cloned this nuclease will be referred to as FEN-1 (177). The yeast analogue of FEN-1 is *RTH1*, the *RAD2* homologue (178). *RTH1* deletion mutants are viable, but temperature sensitive for growth, arresting in the S phase at the restrictive temperature (179). The enzymatic activities of FEN-1 are two-fold. In addition to an endonuclease activity specific for V(D)J recombination intermediates, FEN-1 is also a nick-specific 5′-3′ exonuclease (172, 177). From the latter specificity it follows that FEN-1 action has to be preceded by polymerase action to close the gap left by RNase H1 action. In the presence of RF-C and PCNA, gap filling by Polα is inhibited and, therefore,

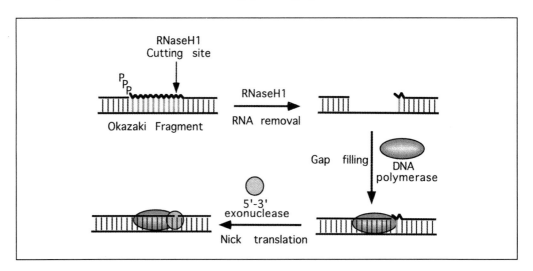

Fig. 6 Maturation of Okazaki fragments. Either DNA polymerase δ or DNA polymerase ε can carry out the gap filling and nick translation. Ligation of the nick by DNA ligase I is not shown.

either Polδ or Polε must carry out this function (86). The polymerase–FEN-1 complex carries out nick translation, until DNA ligase I gains access to seal the nick. This last event in the maturation process is specifically carried out by DNA ligase I as other DNA ligases will not substitute in this reaction (173).

7. Conclusions and perspectives

The last five years have brought a fascinating insight into the eukaryotic replication machinery in part because of the availability of the *in vitro* SV40 DNA replication system. However, at this time the limitations of this system are being revealed and point at the gaps in our knowledge of chromosomal forks. Two obvious problems which are of major interest are the identity and activities of the DNA helicase(s) which replace(s) the T antigen at the chromosomal fork, and the uncertain role of DNA polymerase ε, and, therefore, also DNA polymerase δ, in DNA replication. Obviously, these problems would be directly accessible if an *in vitro* origin-dependent chromosomal DNA replication system were available. As a number of proteins which interact with yeast origins are already known, this organism would be the obvious choice for such an *in vitro* system. In its absence, genetic manipulation in yeast is an experimental approach of high priority as it may give us further insights into these problems.

Acknowledgements

I would like to thank John Majors, Tim Lohman, and Tom Kunkel for stimulating and critical discussions during the writing of this review. The work in the author's laboratory is in part supported by grant GM32431 from the National Institutes of Health.

References

1. Challberg, M. D. and Kelly, T. J. (1989) Animal virus DNA replication. *Annu. Rev. Biochem.*, **58**, 671.
2. Stillman, B. (1989) Initiation of eukaryotic DNA replication in vitro. *Annu. Rev. Cell Biol.*, **5**, 197.
3. Burgers, P. M. J. (1989) Eukaryotic DNA polymerases alpha and delta: conserved properties and interactions, from yeast to mammalian cells. *Prog. Nucleic Acids Res. Mol. Biol.*, **37**, 235.
4. Bambara, R. A. and Jessee, C. B. (1991) Properties of DNA polymerases delta and epsilon, and their roles in eukaryotic DNA replication. *Biochim. Biophys. Acta*, **1088**, 11.
5. Campbell, J. L. and Newlon, C. S. (1991) *Chromosomal DNA replication. The molecular and cellular biology of the yeast Saccharomyces: genome dynamics, protein synthesis, and energetics*, Broach, J. R., Pringle, J. R., and Jones, E. W. (eds.). Cold Spring Harbor Laboratory Press, Cold Spring Harbor, NY.
6. Wang, T. S. F. (1991) Eukaryotic DNA polymerases. *Annu. Rev. Biochem.*, **60**, 513.
7. Hubscher, U. and Thommes, P. (1992) DNA polymerase epsilon: in search of a function. [Review]. *Trends Biochem. Sci.*, **17**, 55.

8. Morrison, A. and Sugino, A. (1993) DNA polymerase II, the epsilon polymerase of *Saccharomyces cerevisiae*. [Review]. *Prog. Nucleic Acid Res. Mol. Biol.*, **46**, 93.

9. So, A. G. and Downey, K. M. (1992) Eukaryotic DNA replication. *Crit. Rev. Biochem. Mol. Biol.*, **27**, 129.

10. Bambara, R. A. and Huang, L. (1995). Reconstitution of mammalian DNA replication. *Progr. Nucleic Acids Mol. Biol.*, **51**, 93.

11. Bramhill, D. and Kornberg, A. (1988) Duplex opening by dnaA protein at novel sequences in initiation of replication at the origin of the *E. coli* chromosome. *Cell*, **52**, 743.

12. Kornberg, A. and Baker, T. A. (1991) *DNA replication*. Freeman, New York.

13. Dean, F. B. and Hurwitz, J. (1991) Simian virus 40 large T antigen untwists DNA at the origin of DNA replication. *J. Biol. Chem.*, **266**, 5062.

14. Bell, S. P. and Stillman, B. (1992) ATP-dependent recognition of eukaryotic origins of DNA replication by a multiprotein complex. *Nature*, **357**, 128.

15. Baker, T. A., Funnell, B. E., and Kornberg, A. (1987) Helicase action of dnaB protein during replication from the *Escherichia coli* chromosomal origin in vitro. *J. Biol. Chem.*, **262**, 6877.

16. Marszalek, J. and Kaguni, J. M. (1994) DnaA protein directs the binding of DnaB protein in initiation of DNA replication in *Escherichia coli*. *J. Biol. Chem.*, **269**, 4883.

17. Mallory, J. B., Alfano, C., and McMacken, R. (1990) Host virus interactions in the initiation of bacteriophage lambda DNA replication. Recruitment of *Escherichia coli* DnaB helicase by lambda P replication protein. *J. Biol. Chem.*, **265**, 13297.

18. Liberek, K., *et al.* (1990) Physical interactions between bacteriophage and *Escherichia coli* proteins required for initiation of lambda DNA replication. *J. Biol. Chem.*, **265**, 3022.

19. Weinberg, D. H., *et al.* (1990) Reconstitution of simian virus 40 DNA replication with purified proteins. *Proc. Natl. Acad. Sci. USA*, **87**, 8692.

20. Tsurimoto, T., Melendy, T., and Stillman, B. (1990) Sequential initiation of lagging and leading strand synthesis by two different polymerase complexes at the SV40 DNA replication origin. *Nature*, **346**, 534.

21. Thorner, L. K., Lim, D. A., and Botchan, M. R. (1993) DNA-binding domain of bovine papillomavirus type 1 E1 helicase: structural and functional aspects. *J. Virol.*, **67**, 6000.

22. Seo, Y. S., Muller, F., Lusky, M., and Hurwitz, J. (1993) Bovine papilloma virus (BPV)-encoded E1 protein contains multiple activities required for BPV DNA replication. *Proc. Natl. Acad. Sci. USA*, **90**, 702.

23. Bruckner, R. C., Crute, J. J., Dodson, M. S., and Lehman, I. R. (1991) The herpes simplex virus 1 origin binding protein: a DNA helicase. *J. Biol. Chem.*, **266**, 2669.

24. Dornreiter, I., Hoss, A., Arthur, A. K., and Fanning, E. (1990) SV40 T antigen binds directly to the large subunit of purified DNA polymerase alpha. *EMBO J.*, **9**, 3329.

25. Dornreiter, I., *et al.* (1992) Interaction of DNA polymerase alpha-primase with cellular replication protein A and SV40 T antigen. *EMBO J.*, **11**, 769.

26. Collins, K. L., Russo, A. A., Tseng, B. Y., and Kelly, T. J. (1993) The role of the 70 kDa subunit of human DNA polymerase alpha in DNA replication. *EMBO J.*, **12**, 4555.

27. Tsurimoto, T. and Stillman, B. (1991) Replication factors required for SV40 DNA replication in vitro. II. Switching of DNA polymerase α and δ during initiation of leading and lagging strand synthesis. *J. Biol. Chem.*, **266**, 1961.

28. Bullock, P. A., Seo, Y. S., and Hurwitz, J. (1991) Initiation of simian virus 40 DNA synthesis in vitro. *Mol. Cell. Biol.*, **11**, 2350.

29. Sinha, N. K., Morris, C. F., and Alberts, B. M. (1980) Efficient in vitro replication of

double-stranded DNA templates by a purified T4 bacteriophage replication system. *J. Biol. Chem.*, **255**, 4290.

30. Wang, J. C. (1991) DNA topoisomerases: why so many. *J. Biol. Chem.*, **266**, 6659.
31. Yang, L., Wold, M. S., Li, J. J., Kelly, T. J., and Liu, L. F. (1987) Roles of DNA topoisomerases in simian virus 40 DNA replication in vitro. *Proc. Natl. Acad. Sci. USA*, **84**, 950.
32. Brill, S. J., DiNardo, S., Voelkel, M. K., and Sternglanz, R. (1987) Need for DNA topoisomerase activity as a swivel for DNA replication for transcription of ribosomal RNA. *Nature*, **326**, 414.
33. Matson, S. W. and Kaiser-Rogers, K. A. (1990) DNA helicases. *Annu. Rev. Biochem.*, **59**, 289.
34. Thommes, P. and Hubscher, U. (1992) Eukaryotic DNA helicases: essential enzymes for DNA transactions. [Review]. *Chromosoma*, **101**, 467.
35. Lohman, T. M. (1993) Helicase-catalyzed DNA unwinding. [Review]. *J. Biol. Chem.*, **268**, 2269.
36. Wong, I., Chao, K. L., Bujalowski, W., and Lohman, T. M. (1992) DNA-induced dimerization of the *Escherichia coli* Rep helicase. *J. Biol. Chem.*, **267**, 7596.
37. Wong, I. and Lohman, T. M. (1992) Allosteric effects of nucleotide cofactors on *Escherichia coli* Rep helicase-DNA binding. *Science*, **256**, 350.
38. Amaratunga, M. and Lohman, T. M. (1993) *Escherichia coli* rep helicase unwinds DNA by an active mechanism. *Biochemistry*, **32**, 6815.
39. Patel, S. S. and Hingorani, M. M. (1993) Oligomeric structure of bacteriophage T7 DNA primase/helicase proteins *J. Biol. Chem.*, **268**, 10668.
40. Venkatesan, M., Silver, L. L., and Nossal, N. G. (1982) Bacteriophage T4 gene 41 protein, required for the synthesis of RNA primers, is also a DNA helicase. *J. Biol. Chem.*, **257**, 12426.
41. Dong, F., Gogol, E. P., and von Hippel, P. H. (1993) Study of the bacteriophage T4 helicase by biophysical and molecular biological approaches. *Biophys. J.*, **64**, A238.
42. Cha, T. A. and Alberts, B. M. (1989) The bacteriophage T4 DNA replication fork. *J. Biol. Chem.*, **246**, 12220.
43. Reha, K. L. and Hurwitz, J. (1978) The dnaB gene product of *Escherichia coli*. I. Purification, homogeneity, and physical properties. *J. Biol. Chem.*, **253**, 4043.
44. Arai, K., Yasuda, S., and Kornberg, A. (1981) Mechanism of dnaB protein action. I. Crystallization and properties of dnaB protein, an essential replication protein in *Escherichia coli*. *J. Biol. Chem.*, **256**, 5247.
45. LeBowitz, J. H. and McMacken, R. (1986) The *Escherichia coli* dnaB replication protein is a DNA helicase. *J. Biol. Chem.*, **261**, 4738.
46. Wahle, E., Lasken, R. S., and Kornberg, A. (1989) The dnaB-dnaC replication protein complex of *Escherichia coli*. I. Formation and properties. *J. Biol. Chem.*, **264**, 2463.
47. Stahl, H., Droge, P., and Knippers, R. (1986) DNA helicase activity of SV40 large tumor antigen. *EMBO J.*, **5**, 1939.
48. Mastrangelo, I. A., *et al.* (1989) ATP-dependent assembly of double hexamers of SV40 T antigen at the viral origin of DNA replication. *Nature*, **338**, 658.
49. Gorbalenya, A. E. and Koonin, E. V. (1994) Helicases: amino acid sequence comparisons and structure–function relationships. *Curr. Opin. Struct. Biol.*, **3**, 419.
50. Ilyina, T. V., Gorbalenya, A. E., and Koonin, E. V. (1992) Organization and evolution of bacterial and bacteriophage primase–helicase systems. *J. Mol. Evol.*, **34**, 351.
51. Li, X., Yoder, B. L., and Burgers, P. M. J. (1992) A *Saccharomyces cerevisiae* DNA helicase associated with Replication Factor C. *J. Biol. Chem.*, **267**, 25321.

52. Li, X., Tan, C. K., So, A. G., and Downey, K. M. (1992) Purification and characterization of δ helicase from fetal calf thymus. *Biochemistry*, **31**, 3507.

53. Thommes, P., Ferrari, E., Jessberger, R., and Hubscher, U. (1992) Four different DNA helicases from calf thymus. *J. Biol. Chem.*, **267**, 6063.

54. Biswas, E. E., Chen, P. H., and Biswas, S. B. (1993) DNA helicase associated with DNA polymerase alpha: isolation by a modified immunoaffinity chromatography. *Biochemistry*, **32**, 13393.

55. Lehman, I. R. and Kaguni, L. S. (1989) DNA polymerase alpha. *J. Biol. Chem.*, **264**, 4265.

56. Burgers, P. M. J., *et al.* (1990) Revised nomenclature for eukaryotic DNA polymerases. *Eur. J. Biochem.*, **191**, 617.

57. Prasad, R., *et al.* (1993) Yeast open reading frame YCR14C encodes a DNA beta-polymerase-like enzyme. *Nucleic Acids Res.*, **21**, 5301.

58. Shimizu, K., *et al.* (1993) Purification and characterization of a new DNA polymerase from budding yeast *Saccharomyces cerevisiae*. A probable homolog of mammalian DNA polymerase beta. *J. Biol. Chem.*, **268**, 27148.

59. Morrison, A., *et al.* (1989) REV3, a *Saccharomyces cerevisiae* gene whose function is required for induced mutagenesis, is predicted to encode a nonessential DNA polymerase. *J. Bacteriol.*, **171**, 5659.

60. Larimer, F. W., Perry, J. R., and Hardigree, A. A. (1989) The REV1 gene of *Saccharomyces cerevisiae*: isolation, sequence, and functional analysis. *J. Bacteriol.*, **171**, 230.

61. Rajagopalan, M., *et al.* (1992) Activity of the purified mutagenesis proteins UmuC, UmuD′, and RecA in replicative bypass of an abasic DNA lesion by DNA polymerase III. *Proc. Natl. Acad. Sci. USA*, **89**, 10777.

62. Cha, T. A. and Alberts, B. M. (1986) Studies of the DNA helicase–RNA primase unit from bacteriophage T4. A trinucleotide sequence on the DNA template starts RNA primer synthesis. *J. Biol. Chem.*, **261**, 7001.

63. Hinton, D. M. and Nossal, N. G. (1987) Bacteriophage T4 DNA primase–helicase. Characterization of oligomer synthesis by T4 61 protein alone and in conjunction with T4 41 protein. *J. Biol. Chem.*, **263**, 10873.

64. Nakai, H. and Richardson, C. C. (1988) gp4 processive helicase-distributive primase. *J. Biol. Chem.*, **263**, 9818.

65. Crute, J. J. *et al.* (1989) Herpes simplex virus 1 helicase–primase: a complex of three herpes-encoded gene products. *Proc. Natl. Acad. Sci. USA*, **86**, 2186.

66. Calder, J. M. and Stow, N. D. (1990) Herpes simplex virus helicase–primase: the UL8 protein is not required for DNA-dependent ATPase and DNA helicase activities. *Nucleic Acids Res.*, **18**, 3573.

67. Conaway, R. C. and Lehman, I. R. (1982) A DNA primase activity associated with DNA polymerase alpha from *Drosophila melanogaster* embryos. *Proc. Natl. Acad. Sci. USA*, **79**, 2523.

68. Kaguni, L. S., Rossignol, J. M., Conaway, R. C., Banks, G. R., and Lehman, I. R. (1983) Association of DNA primase with the beta/gamma subunits of DNA polymerase alpha from *Drosophila melanogaster* embryos. *J. Biol. Chem.*, **258**, 9037.

69. Plevani, P., *et al.* (1985) Polypeptide structure of DNA primase from a yeast DNA polymerase–primase complex. *J. Biol. Chem.*, **260**, 7102.

70. Santocanale, C., Foiani, M., Lucchini, G., and Plevani, P. (1993) The isolated 48,000-dalton subunit of yeast DNA primase is sufficient for RNA primer synthesis. *J. Biol. Chem.*, **268**, 1343.

71. Longhese, M. P., Jovine, L., Plevani, P., and Lucchini, G. (1993) Conditional mutations

in the yeast DNA primase genes affect different aspects of DNA metabolism and inter-actions in the DNA polymerase alpha–primase complex. *Genetics*, **133**, 183.

72. Cotterill, S., Chui, G., and Lehman, I. R. (1987) DNA polymerase–primase from embryos of *Drosophila melanogaster*. The DNA polymerase subunit. *J. Biol. Chem.*, **262**, 16 100.

73. Copeland, W. C. and Wang, T. S. (1991) Catalytic subunit of human DNA polymerase alpha overproduced from baculovirus-infected insect cells. Structural and enzymologi-cal characterization. *J. Biol. Chem.*, **266**, 22 739.

74. Foiani, M., Marini, F., Gamba, D., Lucchini, G., and Plevani, P. (1994) The B subunit of the DNA polymerase alpha–primase complex in *Saccharomyces cerevisiae* executes an essential function at the initial stage of DNA replication. *Mol. Cell. Biol.*, **14**, 923.

75. Derbyshire, V., *et al.* (1988) Genetic and crystallographic studies of the 3′ –5′-exonu-cleolytic site of DNA polymerase I. *Science*, **240**, 199.

76. Bernad, A., Bianco, L., Lazaro, J. M., Martin, G., and Salas, M. (1989) A conserved 3′–5′ exonuclease active site in prokaryotic and eukaryotic DNA polymerases. *Cell*, **59**, 219.

77. Simon, M., Giot, L., and Faye, G. (1991) The 3′ to 5′ exonuclease activity located in the DNA polymerase delta subunit of *Saccharomyces cerevisiae* is required for accurate replication. *EMBO J.*, **10**, 2165.

78. Morrison, A., Bell, J. B., Kunkel, T. A., and Sugino, A. (1991) Eukaryotic DNA poly-merase amino acid sequence required for 3′–5′ exonuclease activity. *Proc. Natl. Acad. Sci. USA*, **88**, 9473.

79. Budd, M. and Campbell, J. L. (1987) Temperature-sensitive mutations in the yeast DNA polymerase I gene. *Proc. Natl. Acad. Sci. USA*, **84**, 2838.

80. Dong, Q., Copeland, W. C., and Wang, T. S. (1993) Mutational studies of human DNA polymerase alpha. Serine 867 in the second most conserved region among alpha-like DNA polymerases is involved in primer binding and mispair primer extension. *J. Biol. Chem.*, **268**, 24 175.

81. Dng, Q., Copeland, W. C., and Wang, T. S. (1993) Mutational studies of human DNA polymerase alpha. Identification of residues critical for deoxynucleotide binding and misinsertion fidelity of DNA synthesis. *J. Biol. Chem.*, **268**, 24 163.

82. Copeland, W. C. and Wang, T. S. (1993) Mutational analysis of the human DNA poly-merase alpha. The most conserved region in alpha-like DNA polymerases is involved in metal-specific catalysis. *J. Biol. Chem.*, **268**, 11 028.

83. Blasco, M. A., Lazaro, J. M., Blanco, L., and Salas, M. (1993) Phi 20 DNA polymerase active site. Residue ASP249 of conserved amino acid motif "Dx2SLYP" is critical for synthetic activities. *J. Biol. Chem.*, **268**, 24 106.

84. Ishimi, Y., Claude, A., Bullock, P., and Hurwitz, J. (1988) Complete enzymatic synthesis of DNA containing the SV40 origin of replication. *J. Biol. Chem.*, **263**, 19 723.

85. Nethanel, T. and Kaufman, G. (1990) Two DNA polymerases may be required for synthesis of the lagging DNA strand of simian virus 40. *J. Virol.*, **64**, 5912.

86. Waga, S. and Stillman, B. (1994) Anatomy of a DNA replication fork revealed by recon-stitution of SV40 DNA replication in vitro. *Nature*, **369**, 207.

87. Lee, M. Y. W. T., Tan, C.-K., Downey, K. M., and So, A. G. (1984) Further studies on calf thymus DNA polymerase δ purified to homogeneity by a new procedure. *Biochemistry*, **23**, 1906.

88. Bauer, G. A. and Burgers, P. M. J. (1988) The yeast analog of mammalian cyclin/prolif-erating-cell nuclear antigen interacts with mammalian DNA polymerase delta. *Proc. Natl. Acad. Sci. USA*, **85**, 7506.

89. Boulet, A., Simon, M., Faye, G., Bauer, G. A., and Burgers, P. M. J. (1989) Structure and function of the *Saccharomyces cerevisiae* CDC2 gene encoding the large subunit of DNA polymerase III. *EMBO J.*, **8**, 1849.

90. Chung, D. W., *et al.* (1991) Primary structure of the catalytic subunit of human DNA polymerase delta and chromosomal location of the gene. *Proc. Natl. Acad. Sci. USA*, **88**, 11 197.

91. Kesti, T., Frantti, H., and Syvaoja, J. E. (1993) Molecular cloning of the cDNA for the catalytic subunit of human DNA polymerase epsilon. *J. Biol. Chem.*, **268**, 10 238.

92. Conrad, M. N. and Newlon, C. S. (1983) *Saccharomyces cerevisiae* cdc2 mutants fail to replicate approximately one-third of their nuclear genome. *Mol. Cell. Biol.*, **3**, 1000.

93. Budd, M. E. and Campbell, J. L. (1993) DNA polymerases delta and epsilon are required for chromosomal replication in *Saccharomyces cerevisiae*. *Mol. Cell. Biol.*, **13**, 496.

94. Morrison, A., Johnson, A. L., Johnston, L. H., and Sugino, A. (1993) Pathway correcting DNA replication errors in *Saccharomyces cerevisiae*. *EMBO J.*, **12**, 1467.

95. Schaaper, R. M. (1989) *Escherichia coli* mutator mutD5 is defective in the mutHLS pathway of DNA mismatch repair. *Genetics*, **121**, 205.

96. Dresler, S. L. and Frattini, M. G. (1986) DNA replication and UV-induced DNA repair synthesis in human fibroblasts are much less sensitive than DNA polymerase alpha to inhibition by butylphenyl-deoxyguanosine triphosphate. *Nucleic Acids Res.*, **14**, 7093.

97. Brown, W. C. and Campbell, J. L. (1993) Interaction of proliferating cell nuclear antigen with yeast DNA polymerase delta. *J. Biol. Chem.*, **268**, 21 706.

98. Goulian, M., Herrmann, S. M., Sackett, J. W., and Grimm, S. L. (1990) Two forms of DNA polymerase delta from mouse cells. Purification and properties. *J. Biol. Chem.*, **265**, 16 402.

99. Chiang, C. S., Mitsis, P. G., and Lehman, I. R. (1993) DNA polymerase delta from embryos of *Drosophila melanogaster*. *Proc. Natl. Acad. Sci. USA*, **90**, 9105.

100. Aoyagi, N., Matsuoka, S., Furunobu, A., Matsukage, A., and Sakaguchi, K. (1994) *Drosophila* DNA polymerase delta. Purification and characterization. *J. Biol. Chem.*, **269**, 6045.

101. Kesti, T. and Syvaoja, J. E. (1991) Identification and tryptic cleavage of the catalytic core of HeLa and calf thymus DNA polymerase epsilon. *J. Biol. Chem.*, **266**, 6336.

102. Hamatake, R. K., *et al.* (1990) Purification and characterization of DNA polymerase II from the yeast *Saccharomyces cerevisiae*. Identification of the catalytic core and a possible holoenzyme form. *J. Biol. Chem.*, **265**, 4072.

103. Nishida, C., Reinhard, P., and Linn, S. (1988) DNA repair synthesis in human fibroblasts requires DNA polymerase delta. *J. Biol. Chem.*, **263**, 501.

104. Wang, Z., Wu, X., and Friedberg, E. C. (1993) DNA repair synthesis during base excision repair in vitro is catalyzed by DNA polymerase epsilon and is influenced by DNA polymerases alpha and delta in *Saccharomyces cerevisiae*. *Mol. Cell. Biol.*, **13**, 1051.

105. Araki, H., *et al.* (1992) DNA polymerase II, the probable homolog of mammalian DNA polymerase epsilon, replicates chromosomal DNA in the yeast *Saccharomyces cerevisiae*. *EMBO J.*, **11**, 733.

106. Araki, H., Hamatake, R. K., Johnston, L. H., and Sugino, A. (1991) DPB2, the gene encoding DNA polymerase II subunit B, is required for chromosome replication in *Saccharomyces cerevisiae*. *Proc. Natl. Acad. Sci. USA*, **88**, 4601.

107. Araki, H., *et al.* (1991) Cloning DPB3, the gene encoding the third subunit of DNA polymerase II of *Saccharomyces cerevisiae*. *Nucleic Acids Res.*, **19**, 4867.

108. Morrison, A. and Sugino, A. (1994) The 3′->5′ exonucleases of both DNA polymerases

delta and epsilon participate in correcting errors of DNA replication in *Saccharomyces cerevisiae*. *Mol. Gen. Genet.*, **242,** 289.

109. Yoder, B. L. and Burgers, P. M. J. (1991) *Saccharomyces cerevisiae* replication factor C. I. Purification and characterization of its ATPase activity. *J. Biol. Chem.*, **266,** 22 689.

110. Burgers, P. M. J. (1991) *Saccharomyces cerevisiae* replication factor C. II. Formation and activity of complexes with the proliferating cell nuclear antigen and with DNA polymerases delta and epsilon. *J. Biol. Chem.*, **266,** 22 698.

111. Lee, S. H., Kwong, A. D., Pan, Z. Q., Burgers, P. M. J., and Hurwitz, J. (1991) Synthesis of DNA by DNA polymerase ε in vitro. *J. Biol. Chem.*, **266,** 22 707.

112. Podust, V. N., Georgaki, A., Strack, B., and Hubscher, U. (1992) Calf thymus RF-C as an essential component for DNA polymerase delta and epsilon holoenzymes function. *Nucleic Acids Res.*, **20,** 4159.

113. Podust, V., Mikhailov, V., Georgaki, A., and Hubscher, U. (1992) DNA polymerase delta and epsilon holoenzymes from calf thymus. *Chromosoma*, **102,** 41.

114. Roberts, J. D., Izuta, S., Thomas, D. C., and Kunkel, T. A. (1994) Mispair-, site-, and strand-specific error rates during simian virus 40 origin-dependent replication in vitro with excess deoxythymidine triphosphate. *J. Biol. Chem.*, **269,** 1711.

115. Izuta, S., Roberts, J. D., and Kunkel, T. A. (1995) Replication error rates for G.dGTP, T.dGTP and A.dGTP mispairs and evidence for differential proofreading by leading and lagging strand DNA replication complexes in human cells. *J. Biol. Chem.*, **270,** 2595.

116. Thomas, D. C., *et al.* (1991) Fidelity of mammalian DNA replication and replicative DNA polymerases. *Biochemistry*, **30,** 11 751.

117. Kramer, W., Kramer, B., Williamson, M. S., and Fogel, S. (1980) Cloning and nucleotide sequence of DNA mismatch repair gene PMS1 from *Saccharomyces cerevisiae*: homology of PMS1 to procaryotic MutL and HexB. *J. Bacteriol.*, **171,** 5339.

118. Kumble, K. D., Iversen, P. L., and Vishwanatha, J. K. (1992) The role of primer recognition proteins in DNA replication: inhibition of cellular proliferation by antisense oligodeoxyribonucleotides. *J. Cell. Sci.*, **101,** 35.

119. Goulian, M. and Heard, C. J. (1990) The mechanism of action of an accessory protein for DNA polymerase alpha/primase. *J. Biol. Chem.*, **265,** 13 231.

120. Wold, M. S. and Kelly, T. (1988) Purification and characterization of replication protein A, a cellular protein required for in vitro replication of simian virus 40 DNA. *Proc. Natl. Acad. Sci. USA*, **85,** 2523.

121. Fairman, M. P. and Stillman, B. (1988) Cellular factors required for multiple stages of SV40 DNA replication in vitro. *EMBO J.*, **7,** 1211.

122. Brill, S. J. and Stillman, B. (1991) Replication factor-A from *Saccharomyces cerevisiae* is encoded by three essential genes coordinately expressed at S phase. *Genes Dev.*, **5,** 1589.

123. Wold, M. S., Li, J. J., and Kelly, T. J. (1987) Initiation of simian virus 40 DNA replication in vitro: large-tumor-antigen- and origin-dependent unwinding of the template. *Proc. Natl. Acad. Sci. USA*, **84,** 3643.

124. Wold, M. S., Weinberg, D. H., Virshup, D. M., Li, J. J., and Kelly, T. J. (1989) Identification of cellular proteins required for simian virus 40 DNA replication. *J. Biol. Chem.*, **264,** 2801.

125. Kenny, M. K., Lee, S. H., and Hurwitz, J. (1989) Multiple functions of human single-stranded-DNA binding protein in simian virus 40 DNA replication: single-strand stabilization and stimulation of DNA polymerases alpha and delta. *Proc. Natl. Acad. Sci. USA*, **86,** 9757.

126. Brill, S. J. and Stillman, B. (1989) Yeast replication factor-A functions in the unwinding of the SV40 origin of DNA replication. *Nature*, **342**, 92.

127. Din, S., Brill, S. J., Fairman, M. P., and Stillman, B. (1990) Cell-cycle-regulated phosphorylation of DNA replication factor A from human and yeast cells. *Genes Dev.*, **4**, 968.

128. Fotedar, R. and Roberts, J. M. (1992) Cell cycle regulated phosphorylation of RPA-32 occurs within the replication initiation complex. *EMBO J.*, **11**, 2177.

129. Kim, C., Snyder, R. O., and Wold, M. S. (1992) Binding properties of replication protein A from human and yeast cells. *Mol. Cell. Biol.*, **12**, 3050.

130. Celis, J. E., Madsen, P., Celis, A., Nielsen, H. V., and Gesser, B. (1987) Cyclin (PCNA, auxiliary protein of DNA polymerase delta) is a central component of the pathway(s) leading to DNA replication and cell division. *FEBS Lett.*, **220**, 1.

131. Tan, E. M. (1989) Antinuclear antibodies: diagnostic markers for autoimmune diseases and probes for cell biology. *Adv. Immunol.*, **33**, 167.

132. Bauer, G. A. and Burgers, P. M. J. (1990) Molecular cloning, structure and expression of the yeast proliferating cell nuclear antigen gene. *Nucleic Acids Res.*, **18**, 261.

133. Burgers, P. M., *et al.* unpublished results.

134. Waga, S., Hannon, G. J., Beach, D., and Stillman, B. (1994) The p21 inhibitor of cyclin-dependent kinases controls DNA replication by interaction with PCNA [see comments]. *Nature*, **369**, 574.

135. Flores, R. H., *et al.* (1994) Cdk-interacting protein 1 directly binds with proliferating cell nuclear antigen and inhibits DNA replication catalyzed by the DNA polymerase delta holoenzyme. *Proc. Natl. Acad. Sci. USA*, **91**, 8655.

136. Krishna, T. S., Kong, X., Gary, S., and Kuriyan, J. (1994) Crystal structure of the eukaryotic DNA polymerase processivity factor PCNA. *Cell*, **79**, 1233.

137. Kong, X. P., Onrust, R., O'Donnell, M., and Kuriyan, J. (1992) Three-dimensional structure of the beta subunit of *E. coli* DNA polymerase III holoenzyme: a sliding DNA clamp. *Cell*, **69**, 425.

138. Waseem, N. H., Labib, K., Nurse, P., and Lane, D. P. (1992) Isolation and analysis of the fission yeast gene encoding polymerase delta accessory protein PCNA. *EMBO J.*, **11**, 5111.

139. Yamaguchi, M., *et al.* (1990) *Drosophila* proliferating cell nuclear antigen (cyclin) gene: structure, expression during development, and specific binding of homeodomain proteins to its 5'-flanking region. *Mol. Cell. Biol.*, **10**, 872.

140. Leibovici, M., Gusse, M., Bravo, R., and Mechali, M. (1990) Characterization and developmental expression of Xenopus proliferating cell nuclear antigen (PCNA). *Dev. Biol.*, **141**, 183.

141. Almendral, J. M., Huebsch, D., Blundell, P. A., Macdonald, B. H., and Bravo, R. (1987) Cloning and sequence of the human nuclear protein cyclin: homology with DNA-binding proteins. *Proc. Natl. Acad. Sci. USA*, **84**, 1575.

142. Suzuka, I., Hata, S., Matsuoka, M., Kosugi, S., and Hashimoto, J. (1991) Highly conserved structure of proliferating cell nuclear antigen (DNA polymerase delta auxiliary protein) gene in plants. *Eur. J. Biochem.*, **195**, 571.

143. Burgers, P. M. J. (1988) Mammalian cyclin/PCNA (DNA polymerase delta auxiliary protein) stimulates processive DNA synthesis by yeast DNA polymerase III. *Nucleic Acids Res.*, **16**, 6297.

144. Ng, L., Prelich, G., Anderson, C. W., Stillman, B., and Fisher, P. A. (1990) *Drosophila* proliferating cell nuclear antigen. Structural and functional homology with its mammalian counterpart. *J. Biol. Chem.*, **265**, 11 948.

145. Laquel, P., Litvak, S., and Castroviejo, M. (1993) Mammalian proliferating cell nuclear antigen stimulates the processivity of two wheat embryo DNA polymerases. *Plant Physiol.*, **102,** 107.

146. Matsumoto, T., Hata, S., Suzuka, I., and Hashimoto, J. (1994) Expression of functional proliferating-cell nuclear antigen from rice (*Oryza sativa*) in *Escherichia coli*. Activity in association with human DNA polymerase delta. *Eur. J. Biochem.*, **223,** 179.

147. Tan, C. K., Castillo, C., So, A. G., and Downey, K. M. (1986) An auxiliary protein for DNA polymerase delta from fetal calf thymus. *J. Biol. Chem.*, **261,** 12310.

148. Ng, L., McConnell, M., Tan, C. K., Downey, K. M., and Fisher, P. A. (1993) Interaction of DNA polymerase delta, proliferating cell nuclear antigen, and synthetic oligo-nucleotide template-primers. Analysis by polyacrylamide gel electrophoresis-band mobility shift assay. *J. Biol. Chem.*, **268,** 13571.

149. Tsurimoto, T. and Stillman, B. (1990) Functions of replication factor C and proliferating cell nuclear antigen: functional similarity of DNA polymerase accessory proteins from human cells and bacteriophage T4. *Proc. Natl. Acad. Sci. USA*, **87,** 1023.

150. O'Day, C. L., Burgers, P. M., and Taylor, J. S. (1992) PCNA-induced DNA synthesis past cis-syn and trans-syn- I thymine dimers by calf thymus DNA polymerase delta in vitro. *Nucleic Acids Res.*, **20,** 5403.

151. Burgers, P. M. J. and Yoder, B. L. (1993) ATP-independent loading of the proliferating cell nuclear antigen requires DNA ends. *J. Biol. Chem.*, **268,** 19923.

152. Podust, L. M., Podust, V. N., Floth, C., and Hubscher, U. (1994) Assembly of DNA polymerase delta and epsilon holoenzymes depends on the geometry of the DNA template. *Nucleic Acids Res.*, **22,** 2970.

153. Tsurimoto, T. and Stillman, B. (1989) Purification of a cellular replication factor, RF-C, that is required for coordinated synthesis of leading and lagging strands during simian virus 40 DNA replication in vitro. *Mol. Cell. Biol.*, **9,** 609.

154. Lee, S. H., Kwong, A. D., Pan, Z. Q., and Hurwitz, J. (1991) Studies on the activator 1 protein complex, an accessory factor for proliferating cell nuclear antigen-dependent DNA polymerase delta. *J. Biol. Chem.*, **266,** 594.

155. Fien, K. F. and Stillman, B. (1992) Identification of replication factor C from *Saccharomyces cerevisiae*: A component of the leading strand DNA replication complex. *Mol. Cell Biol.*, **12,** 155.

156. Li, X. and Burgers, P. M. (1994) Molecular cloning and expression of the *Saccharomyces cerevisiae* RFC3 gene, an essential component of replication factor C. *Proc. Natl. Acad. Sci. USA*, **91,** 868.

157. Howell, E. A., McAlear, M. A., Rose, D., and Holm, C. (1994) CDC44: a putative nucleotide-binding protein required for cell cycle progression that has homology to subunits of replication factor C. *Mol. Cell. Biol.*, **14,** 255.

158. Bunz, F., Kobayashi, R., and Stillman, B. (1993) cDNAs encoding the large subunit of human replication factor C. *Proc. Natl. Acad. Sci. USA*, **90,** 11014.

159. Chen, M., Pan, Z.-Q., and Hurwitz, J. (1992) Studies of the cloned 37-kDa subunit of activator 1 (replication factor C) of HeLa cells. *Proc. Natl. Acad. Sci. USA*, **89,** 5211.

160. Chen, M., Pan, Z.-Q., and Hurwitz, J. (1992) Sequence and expression in *Escherichia coli* of the 40-kDa subunit of activator 1 (replication factor C) of HeLa cells. *Proc. Natl. Acad. Sci. USA*, **89,** 2516.

161. O'Donnell, M., Onrust, R., Dean, F. B., Chen, M., and Hurwitz, J. (1993) Homology in accessory proteins of replicative polymerases – *E. coli* to humans. *Nucleic Acids Res.*, **21,** 1.

162. Li, X. and Burgers, P. M. (1994) Cloning and characterization of the essential *Saccharomyces cerevisiae* RFC4 gene encoding the 37-kDa subunit of replication factor C. *J. Biol. Chem.*, **269**, 21880.

163. Noskov, V., *et al.* (1994) The RFC2 gene encoding a subunit of replication factor C of *Saccharomyces cerevisiae*. *Nucleic Acids Res.*, **22**, 1527.

164. Johnston, L. H. and Lowndes, N. F. (1992) Cell cycle control of DNA synthesis in budding yeast. [Review]. *Nucleic Acids Res.*, **20**, 2403.

165. Mannhaupt, G., Stucka, R., Ehnle, S., Vetter, I., and Feldmann, H. (1994) Analysis of a 70 kb region from the right arm of yeast chromosome II. Unpublished, GenBank Accession No. X78993.

166. Tsurimoto, T. and Stillman, B. (1991) Replication factors required for SV40 DNA replication in vitro. I. DNA structure-specific recognition of a primer–template junction by eukaryotic DNA polymerases and their accessory proteins. *J. Biol. Chem.*, **266**, 1950.

167. Pan, Z. Q., Chen, M., and Hurwitz, J. (1993) The subunits of activator 1 (replication factor C) carry out multiple functions essential for proliferating-cell nuclear antigen-dependent DNA synthesis. *Proc. Natl. Acad. Sci. USA*, **90**, 6.

168. Stillman, B. (1994) Smart machines at the DNA replication fork. [Review]. *Cell*, **78**, 725.

169. Lee, S. H. and Hurwitz, J. (1990) Mechanism of elongation of primed DNA by DNA polymerase delta, proliferating cell nuclear antigen, and activator 1. *Proc. Natl. Acad. Sci. USA*, **87**, 5672.

170. Rivin, C. J. and Fangman, W. L. (1980) Replication fork rate and origin activation during the S phase of *Saccharomyces cerevisiae*. *J. Cell. Biol.*, **85**, 108.

171. Turchi, J. J. and Bambara, R. A. (1993) Completion of mammalian lagging strand DNA replication using purified proteins. *J. Biol. Chem.*, **268**, 15136.

172. Murante, R. S., Huang, L., Turchi, J. J., and Bambara, R. A. (1994) The calf 5'- to 3'-exonuclease is also an endonuclease with both activities dependent on primers annealed upstream of the point of cleavage. *J. Biol. Chem.*, **269**, 1191.

173. Waga, S., Bauer, G., and Stillman, B. (1994) Reconstitution of complete SV40 DNA replication with purified replication factors. *J. Biol. Chem.*, **269**, 10923.

174. Huang, L., Kim, Y., Turchi, J. J., and Bambara, R. A. (1994) Structure-specific cleavage of the RNA primer from Okazaki fragments by calf thymus RNase H1. *J. Biol. Chem.*, **269**, 25922.

175. Karwan, R. and Wintersberger, U. (1986) Yeast ribonuclease H(70) cleaves RNA–DNA junctions. *FEBS Lett.*, **206**, 189.

176. Karwan, R. and Wintersberger, U. (1988) In addition to RNase H(70) two other proteins of *Saccharomyces cerevisiae* exhibit ribonuclease H activity. *J. Biol. Chem.*, **263**, 14970.

177. Harrington, J. J. and Lieber, M. R. (1994) The characterization of a mammalian DNA structure-specific endonuclease. *EMBO J.*, **13**, 1235.

178. Jacquier, A., Legrain, P., and Dujon, B. (1992) Sequence of a 10.7 kb segment of yeast chromosome XI identifies the APN1 and the BAF1 loci and reveals one tRNA gene and several new open reading frames including homologs to RAD2 and kinases. *Yeast*, **8**, 121.

179. Prakash, S., Sung, P., and Prakash, L. (1993) DNA repair genes and proteins of *Saccharomyces cerevisiae*. [Review]. *Annu. Rev. Genet.*, **27**, 33.

180. Higgins, D. G. and Sharp, P. M. (1989) Fast and sensitive multiple sequence alignments on a microcomputer. *Comput. Appl. Biosci.*, **5**, 151.

181. Navas, T. A., Zhou, Z., and Elledge, S. J. (1995) DNA polymerase epsilon links the DNA replication machinery to the S phase checkpoint. *Cell*, **80**, 29.

2 | SV40 DNA replication

DANIEL HERENDEEN and THOMAS J. KELLY

1. Introduction

Much of what is currently known about the mechanisms of DNA replication in eukaryotic cells has come from studying SV40 and related viruses. SV40 is a small DNA-containing virus that is indigenous to Asian macaques. It produces a lytic infection in permissive primate cells in culture and it is capable of transforming nonpermissive rodent cells to a tumorigenic state. The virus has served as a powerful model system for understanding many aspects of the molecular biology of eukaryotic cells, including transcription, DNA replication, and the regulation of growth (1–4).

SV40 offers a number of advantages for studying eukaryotic DNA replication. The viral genome is relatively simple in structure, consisting of a 5 kb circular duplex DNA molecule with a single origin of DNA replication. SV40 DNA replication takes place in the nucleus of the host cell where the viral genome is complexed with histones to form a nucleoprotein structure (minichromosome) indistinguishable from cellular chromatin. Replication of the SV40 genome requires the participation of only one viral protein, the SV40 large T antigen, so that most of the enzymatic activities necessary for DNA synthesis are provided by cellular proteins. As a result there are many similarities between viral and cellular DNA replication. Recent work suggests that this similarity may even extend to some of the regulatory mechanisms that control DNA replication (5, 6).

The development of a cell-free system for SV40 DNA replication made it possible to study the biochemistry of eukaryotic DNA replication. In its original form the system consisted of a soluble extract derived from cells infected with SV40 (7). Such an extract is capable of mediating the complete replication of DNA molecules containing the SV40 origin of DNA replication. In the years since its development, dissection of the SV40 cell-free replication system by many laboratories has resulted in the identification of cellular replication proteins and the characterization of their roles in the replication pathway. It has been possible to reconstitute SV40 DNA replication with purified proteins and to probe the detailed mechanisms of initiation and DNA chain elongation. Recent studies indicate that the cellular proteins involved in SV40 DNA replication constitute a universal set that is required for DNA replication in all eukaryotes from yeast to man.

2. Organization and expression of the SV40 genome

The SV40 genome is a covalently closed circular duplex DNA molecule that contains five genes. As shown in Fig. 1 the genome is functionally organized as two divergent transcriptional units, referred to as the early and late regions (1, 8, 9). Both the early and the late promoter are localized in a 300 bp regulatory region which also includes the viral origin of DNA replication. The early promoter closely resembles cellular promoters and is competent to direct transcription in the absence of any viral gene product. Thus, early mRNAs, encoding the viral proteins, small t antigen and large T antigen, are synthesized and processed immediately after the viral genome reaches the nucleus of an infected host cell. The small t antigen is not essential for viral multiplication in permissive cells and will not be discussed further here. The large T antigen is a regulatory protein that orchestrates all subsequent events in the virus life cycle.

Because the large T antigen contains a nuclear localization signal (the first such signal to be recognized), it efficiently accumulates in the nucleus following its synthesis (10, 11). Four distinct biological activities of T antigen have been observed.

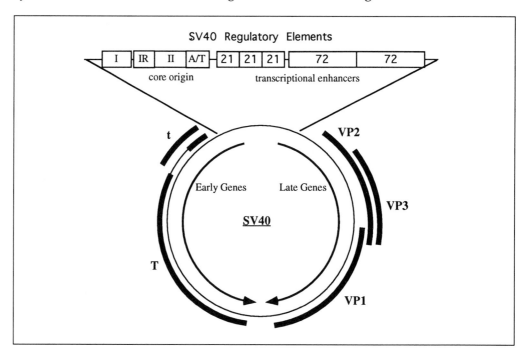

Fig. 1 Diagram of the circular SV40 genome. Early and late mRNAs are depicted as arrows pointing in the 5′ to 3′ direction. The regions encoding the early and late gene products are depicted as thick arcs and labeled accordingly. The intron of the large T antigen is shown as a thin arc connecting the T-encoding regions. The SV40 regulatory element controlling transcription and replication is magnified above. The principle features of the transcriptional enhancers are noted: the 21 bp GC-rich repeats contain SP1 binding motifs and the 72 bp repeats contain binding sites for several proteins. T antigen binding site I and the components of the tripartite core origin (inverted repeat, IR; site II; A/T-rich element) are also shown.

1. *Activation of cellular genes.* T antigen activates the expression of a number of cellular genes, including those that encode enzymes involved in DNA replication directly (12, 13). This phenomenon is probably important for efficient virus multiplication, particularly when the virus infects resting (G0) cells which do not normally express DNA polymerases and other replication proteins. The mechanism responsible for transactivation of cellular gene expression by T antigen is not completely understood. However, recent work suggests that the ability of T antigen to bind to members of the retinoblastoma (RB) protein family may play an important role (14, 15). The RB proteins bind and inhibit the activity of E2F, a transcription factor that appears to activate the expression of genes involved in DNA synthesis (16). By sequestering RB, T antigen prevents inhibition of E2F activity and promotes expression of such genes. T antigen also binds and inhibits the activity of the cellular p53 protein (17–19). The p53 protein is a transcription factor that appears to function as an inhibitor of cell cycle progression, particularly following DNA damage (20, 21). The function of the interaction between T antigen and p53 in virus multiplication is not clear.

2. *Repression of early transcription.* T antigen represses the expression of its own gene by binding to a specific sequence element (site I) that overlaps the start sites for early transcription in the regulatory region (22). The role of this autoregulatory activity is presumably to optimize the levels of T antigen during the virus multiplication cycle. The protein acts catalytically in DNA replication and transcription, so it is not needed in large amounts like the virion capsid proteins which are synthesized during the late phase of the infection.

3. *Initiation of SV40 DNA replication.* As described in greater detail below, T antigen binds to a specific sequence element in the viral origin and brings about the initiation of DNA replication. The protein plays several key roles in this process. It catalyzes the initial opening of the duplex at the viral origin and then functions as a helicase to unwind the DNA at the replication forks. It also mediates the assembly of a functional replication complex via specific protein–protein interactions with cellular DNA polymerase α–primase and other cellular replication proteins.

4. *Activation of late transcription.* The late promoter, unlike its early counterpart, is relatively inactive in the absence of viral gene products, so that the genes encoding the structural proteins of the virion are not expressed during the early phase of the infectious cycle. Abundant evidence now indicates that T antigen plays an essential role in the activation of the late promoter (23, 24). The biochemical mechanism of activation is not yet clear, but appears to be indirect. T antigen probably functions to increase the activity of one or more cellular transcription factors (e.g. the factor TEF-1) which are required for efficient utilization of the late promoter (25–27). Thus, by turning off its own expression and activating the expression of the late genes, T antigen functions as part of a sophisticated genetic switch that ensures the orderly progression of the virus multiplication cycle.

During the late phase of SV40 infection viral DNA replication continues, and copious quantities of the three capsid proteins, VP1, VP2, and VP3, are also synthesized. Progeny virions are spontaneously assembled in the nucleus from capsid proteins and newly synthesized viral genomes (8). In a typical infection of monkey cells in culture as many as 10^5 virions per cell are produced (28).

3. The SV40 origin of DNA replication

The SV40 origin of replication is located within the noncoding regulatory region of the viral genome (see Fig. 1). The functional organization of the origin is complex, consisting of a 64 bp core, which is absolutely required for initiation of DNA synthesis, and several ancillary elements, which facilitate, but are not essential for initiation. The core sequence itself is tripartite (Fig. 2). At its center is a 27 bp recognition sequence for SV40 T antigen referred to as site II. Site II contains four GAGGC pentamers which are bound tightly by T antigen (29, 30). Since the pentamers are organized as a perfect inverted repeat, site II actually consists of two equivalent half-sites related by a dyad axis. T antigen assembles as a hexameric complex over each half site, forming a bi-lobed structure visible by electron microscopy (31). Immediately adjacent to site II on the late side of the core origin is a 17 bp tract containing exclusively A and T residues. The A/T tract is characterized by significant curvature in solution as evidenced by abnormal mobility during gel electrophoresis (32). On the early side of site II is a 15 bp imperfect palindromic sequence element. All three elements of the core origin are required for initiation of SV40 DNA replication. The A/T tract and the early palindrome are not essential for T antigen binding to the origin, but they are essential for the initial opening of the duplex which is mediated by the bound T antigen (32, 33). Base substitution mutations at almost all positions within the tripartite core origin are deleterious to replication efficiency (29, 32–34).

In addition to the core origin, the viral regulatory region contains auxiliary sequence elements that increase the efficiency of initiation of DNA replication. One of these elements is T antigen-binding site I which is located on the early side of the core origin. This is the same site that functions as an operator for repression of early gene expression as described above. Binding of T antigen to site I stimulates

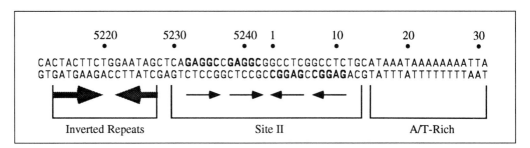

Fig. 2 The SV40 tripartite core origin sequence. GAGGC pentamer repeats of site II are in bold and indicated with thin arrows. The inverted repeats (imperfect palindrome) are indicated with thicker arrows. The 17 bp A/T-rich sequence is also designated. The top strand is oriented 5′ to 3′, with the late transcription unit to the right (as in Fig. 1). The accepted base pair numbering of the SV40 genome is above.

initiation of DNA replication several fold both *in vitro* and *in vivo* (35–37). The mechanism of this stimulation is not clear, but the effect may be due in part to augmentation of DNA unwinding by T antigen bound at the core origin.

Two auxiliary sequence elements involved in transcription of the viral genome, the Sp1-binding sites (within the 21 bp repeats) and the SV40 enhancers (see Fig. 1), also have stimulatory effects on viral DNA replication. As in the case of site I the SV40 transcriptional elements are not essential for DNA replication, but both have large effects on replication activity and represent major determinants of the efficiency of viral DNA replication *in vivo* (38). Transfection experiments have demonstrated that insertion of the recognition sites for several known transcription factors, including some that do not bind to the SV40 DNA normally (e.g., NF1), adjacent to the core origin can significantly increase the replication activity of the latter (39, 40). However, the magnitude of the stimulatory effect of a given transcription factor on DNA replication does not correlate well with the ability of the same factor to enhance transcription (38). Thus, it seems unlikely that the stimulation of DNA replication is secondary to activation of transcription itself. Rather, it seems most likely that transcriptional activator proteins interact with 'basal' replication factors (e.g., T antigen and/or cellular replication proteins) to facilitate the assembly or the activity of the initiation complex. The possible mechanism involved in the stimulatory effect of transcription factors will be discussed further below.

Origins of replication have been identified in yeast and in other viruses. Although such origins are less well characterized, they all appear to share some general features with the SV40 origin (38, 41, 42; see Chapters 3 and 4). In particular, they contain the recognition sites for specific proteins that determine the specificity of initiation. Some of these initiator proteins, like SV40 T antigen, may harbor intrinsic enzymatic activities (e.g., helicase) important for catalyzing the early steps in initiation of DNA replication. Others may lack such activities themselves, but function to recruit the appropriate enzymes to the origin. Viral and cellular origins are often A/T-rich and contain easily unwound regions that are probably crucial for the initial opening of the duplex (43). Finally, all origins that have been carefully studied to date contain binding sites for one or more cellular transcriptional factors. In many cases these binding sites have been demonstrated to contribute significantly to the overall efficiency of DNA replication. We will return to a discussion of the possible biochemical roles of such factors in a later section of this chapter.

4. The SV40 T antigen

4.1 Domain organization

Analysis of the structure and function of the SV40 T antigen indicates that the protein is organized into several functional domains (see Fig. 3) (4–6, 44). The limits of these domains have been defined by analysis of T antigen mutants, by characterization of segments of the protein that are resistant to protease attack, and by characterization of the action of several T-specific monoclonal antibodies. The

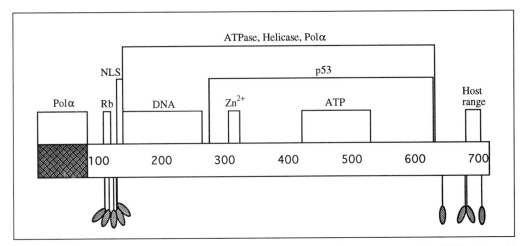

Fig. 3 Domain organization of SV40 large T antigen. Numbering is from the N-terminus of the 708 amino acid protein. Functional ATPase and helicase domains, as well as binding domains for Polα, Rb, p53, ATP, and DNA are indicated. Additionally, the nuclear localization signal (NLS; aa126–132), zinc finger (Zn²⁺; aa302–320), and a region involved in the host range (aa673–699) of SV40 are depicted (10, 11, 202, 203). 'Lollipop' structures below flag the sites for phosphorylation at serines and threonines (Ser106, Ser112, Ser120, Ser123, Thr124, Ser639, Ser677, Ser679, and Thr701). Shaded section indicates the N-terminal region (aa1–82) shared with small t antigen, encoded by the first exon of the large T antigen mRNA.

DNA-binding domain of T antigen resides between residues 131 and 259 (3, 45, 46). Peptides containing only this domain can be generated in bacteria and have been shown to bind to both site I and the core origin. Other regions of the protein are not absolutely required for DNA binding, but can significantly affect the interactions of T antigen with the origin (47). These interactions are also modulated by phosphorylation of certain residues near the N- and C-termini of T antigen (see Section 8.1).

The limits of the helicase/ATPase domain are less well defined, but deletion studies indicate that it is contained within the segment bounded by residues 131–627 (48, 49). Affinity labeling techniques have placed the single ATP-binding site in T antigen between residues 418 and 528 (50). This segment contains a classical consensus purine nucleotide binding motif.

T antigen interacts specifically with several cellular proteins, some of which were described earlier. The region of the protein that binds to p53 lies within the ATPase domain (residues 272–625) (3), while the region that interacts with RB is in the N-terminal domain adjacent to the DNA-binding domain (51). T antigen also binds the cellular DNA polymerase α–primase, and this interaction probably plays a central role in initiation of DNA replication (see Section 7.1). Two independent regions of the protein have been reported to bind DNA polymerase α–primase (Polα), one in the N-terminal domain and one in the ATPase domain (52, 53). Polα and p53 have been shown to compete for binding of T antigen (54). Finally, T antigen binds to RPA, another cellular protein required for initiation of DNA synthesis, but the residues essential for this interaction have not yet been localized (55).

4.2   Assembly of T antigen at the origin of replication

As mentioned above, T antigen binds to the recognition pentamer sequence GAGGC, which is present in four copies within the core origin. In the presence of ATP, T antigen oligomerizes into two hexamers, one bound to each half-site of the origin (31, 56). ATP promotes the oligomerization of T antigen in the absence of DNA as well, but the hexamers formed under these conditions are not competent to bind to the origin (56). This observation, together with electron microscopic and other structural studies of T antigen bound to the origin (31, 57), has led to the suggestion that T antigen hexamers form complexes in which the protein surrounds the DNA. The oligomerization of T antigen to form hexamers is a highly cooperative process. Several lines of evidence indicate that there are also cooperative interactions between the two hexamers bound at the origin and that these hexamer–hexamer interactions are critical for local unwinding of the duplex during initiation of DNA replication (6, 58, 59).

The assembly of the T antigen double hexamer at the origin results in significant alterations in the local DNA structure. Experiments with chemical probes indicate that the early palindrome is partially melted and that the structure of the A/T-rich segment differs substantially from B-form DNA (60, 61). The net effect of these distortions in DNA structure is an untwisting of the DNA by 1–2 helical turns (61–63). The hydrolysis of ATP is not required for either the assembly of the double hexamer or the perturbation of the local DNA structure, since both events occur in the presence of nonhydrolyzable analogues of ATP. The role of ATP is presumably to stabilize a particular conformation of the T antigen molecule that is competent for oligomerization and DNA binding.

4.3   DNA helicase activity

The discovery of the helicase activity of T antigen was a significant advance in our understanding of the role of the protein in SV40 DNA replication. Early work had indicated that T antigen is present at replication forks and is probably required for fork movement, since anti-T antigen monoclonal antibodies block DNA chain elongation *in vitro* (64). However, the key observation, made by Stahl *et al.*, was that T antigen is capable of displacing oligonucleotides that are hydrogen bonded to circular single-stranded DNA (ssDNA) molecules (65). Subsequent work demonstrated that T antigen translocates in the 3′ to 5′ direction along a single strand, melting duplex regions that it encounters (66, 67). The unidirectional movement of the T antigen helicase activity is absolutely dependent upon the hydrolysis of ATP. When a ssDNA-binding protein is present to prevent reannealing of the unwound DNA strands, T antigen is capable of mediating the complete unwinding of double-stranded DNA regions that are several thousand base pairs in length (62, 66, 67). With ATP as a cofactor T antigen can also melt partially duplex segments consisting of an RNA oligonucleotide hydrogen bonded to a longer DNA strand (68). With other cofactors, such as GTP, UTP, or CTP, T antigen can even unwind partially double-stranded RNA molecules (68). The role of this RNA helicase activity, if any, in virus multiplication is unknown.

It is important to emphasize the fact that T antigen differs from previously described prokaryotic and eukaryotic DNA helicases in that it is able to initiate DNA unwinding from internal sites on completely duplex DNA molecules. This occurs most efficiently when T antigen assembles on the viral origin of DNA replication. However, under conditions of low ionic strength and high protein concentration, T antigen can initiate unwinding of duplex molecules of random sequence, albeit at rather low efficiency (66, 69). Such nonspecific unwinding events probably occur when two hexamers of T antigen fortuitously assemble at immediately adjacent sites. The ability of T antigen double hexamer to open a duplex makes possible the critical first step leading to the establishment of the replication forks. As we shall see later, it is this step that is regulated to control the overall efficiency of DNA replication.

5. Cellular replication proteins

As indicated above, the establishment of a cell-free SV40 DNA replication assay opened a way to characterize the cellular replication apparatus (7). Initial experiments demonstrated that cell extracts from permissive cells could catalyze DNA replication which was dependent upon both the SV40 origin and T antigen. As in *in vivo* replication of SV40 DNA, *in vitro* replication initiated at a discrete site near the origin and proceeded in a bidirectional and semiconservative manner (7, 37, 70, 71). The utility of the SV40 *in vitro* system for identifying the participating cellular replication factors was immediately apparent, and exhaustive biochemical characterization ensued. Crude cell lysates were fractionated into multiple components which could be reassembled to reconstitute DNA replication activity (71–77). Eventually 10 highly purified host replication proteins were identified which, together with T antigen, could carry out the complete process of SV40 replication (Table 1).

Each of the replication proteins shown in Table 1 participates in one or more of four enzymatic processes that occur during the SV40 DNA replication reaction:

1. *DNA unwinding.* Semiconservative DNA replication requires the separation of the two strands of the parental duplex. This process occurs first during the initiation reaction at the SV40 origin and then continuously during replication fork movement. DNA unwinding requires the functions of T antigen, replication protein A (RPA), and topoisomerase I. T antigen is the ATP-driven engine that catalyzes the denaturation of duplex DNA. RPA binds to the unwound single strands to prevent renaturation of the DNA. Topoisomerase I relieves the superhelical tension created by DNA unwinding during fork progression.

2. *Priming.* The synthesis of all DNA chains during SV40 DNA replication is primed by short RNA oligomers. The priming reaction requires DNA polymerase α–primase (Polα), RPA, and T antigen. Two of the subunits of Polα catalyze the synthesis of the RNA primers on the template DNA. A third subunit extends the

Table 1 Proteins required for SV40 DNA replication

Protein	Subunit composition	Replication activity
SV40 T antigen	90 kDa	Origin binding/unwinding, helicase
Replication protein A (RPA)	70, 32, 14 kDa	Single-stranded DNA binding
DNA polymerase alpha (Polα)	180, 68, 58, 48 kDa	DNA polymerase, primase
DNA polymerase delta (Polδ)	124, 48 kDa	DNA polymerase, 3' to 5' exonuclease
Proliferating cell nuclear antigen (PCNA)	37 kDa	Increases Polδ processivity, serves as sliding clamp
Replication factor C (RF-C)	140, 41, 40, 38, 37 kDa	Increases Polδ processivity, loads PCNA on DNA
Topoisomerase I	100 kDa	Relieves torsional stress in DNA
Topoisomerase II	140 kDa	Unlinks daughter duplexes, relieves torsional stress in DNA
RNase H1	68–90 kDa	Endonuclease specific for hybrid RNA/DNA primers
MF1/FEN-1	44 kDa	5' to 3' exonuclease
DNA ligase I	85–125 kDa	Joins Okazaki fragments

primer into a short DNA chain. RPA and T antigen interact with Polα in a highly specific manner, and these interactions appear to be required for initiation of SV40 DNA replication. Similar interactions may also play an important role during primer synthesis on the lagging DNA strand.

3. *Elongation of DNA chains.* The elongation of newly synthesized primer RNA/DNA molecules appears to be catalyzed by DNA polymerase δ (Polδ), which has a proofreading 3′ → 5′ exonuclease activity. RF-C and PCNA are required to assemble Polδ into a highly processive machine capable of incorporating thousands of nucleotides without dissociating from the template. RPA also serves to enhance the rate of Polδ assembly at primer termini. Under certain *in vitro* reaction conditions the weakly processive Polα can catalyze the synthesis of long DNA chains in the absence of Polδ, RF-C, and PCNA, but it is doubtful that the enzyme serves this role during SV40 DNA replication *in vivo*

4. *Maturation of replication products.* Several enzymatic activities are required to remove RNA primers and complete the synthesis of covalently closed progeny DNA molecules. Current evidence indicates that RNase H1 and FEN-1 (MF1) act to remove the RNA primers. The polymerase responsible for filling the resulting gaps has not been identified with certainty. Either DNA Polδ or DNA Polα presumably fulfils this role *in vitro*. It is possible that a third DNA polymerase (Polε) carries out this function *in vivo*. Following gap filling, the abutting DNA strands are joined by DNA ligase I, and catenated daughter molecules are untangled by topoisomerase II.

The fact that the cellular proteins listed in Table 1 have been conserved through evolution from yeast to man strongly supports the notion that they are involved in replication of eukaryotic genomes (see Chapter 1). Indeed, genetic experiments in yeast have demonstrated that many of these proteins function during S phase (DNA synthesis) of the cell cycle and are indispensable for cell viability (78–90).

SV40 DNA replication is, therefore, an excellent model for the study of cellular DNA replication. Even replicative events that are specific for the viral origin and the viral T antigen have provided useful information about the probable functions of their cellular counterparts.

6. Mechanism of SV40 DNA replication

The four sequential stages in the replication of SV40 DNA defined by *in vitro* studies are summarized in Figs 4 and 5. Each of these stages will be described in some detail below. It should be kept in mind that separation of the reaction into stages is somewhat arbitrary since several of the enzymatic processes can occur simultaneously on the same replication intermediate.

6.1 Unwinding of the origin

As described previously, origin unwinding is preceded by the binding of two T antigen hexamers to site II of the core origin (Fig. 4a). The binding of T antigen is ATP dependent and causes perturbations in the origin-proximal DNA structure (60, 61). Following assembly of the double hexamer, local unwinding of the origin can be initiated in the presence of RPA, the cellular ssDNA-binding protein (SSB) (91–93). The unwinding reaction requires ATP hydrolysis, unlike the binding step which can occur in the presence of nonhydrolyzable ATP analogues. Conceptually, the unwinding of the origin can be divided into two sequential steps: (i) the initial opening of the duplex in the immediate vicinity of the origin, and (ii) the extension of the initial opening by the T antigen helicase activity. The initial duplex opening appears to be dependent upon specific interactions between the two T antigen hexamers (58, 59). These critical interactions presumably drive a major rearrangement of the hexamers and the bound DNA strands. Once origin opening occurs, each of the two T antigen hexamers functions as classical DNA helicase, translocating in the 3' to 5' direction along one of the two DNA strands, melting the duplex DNA in front of it. Since the two hexamers translocate on opposite DNA strands, they must move in opposite directions from the origin. In essence, this series of events establishes the bidirectional motion of the replication forks (Figs 4b–d). The T antigen hexamer is a highly processive helicase capable of unwinding thousands of nucleotide pairs without dissociating. However, efficient unwinding of the circular SV40 genome requires the activity of topoisomerase I to relieve the accumulation of superhelical tension (94).

On templates containing a single fork, it is clear that a single T antigen hexamer is sufficient to catalyze DNA unwinding (95, 96). However, recent electron microscopic evidence suggests that during bidirectional unwinding, the two T antigen hexamers may actually remain associated (97). Under these circumstances, the double hexamer, functioning as a unit, would reel in the duplex DNA from both sides and expel the unwound DNA at its center. In this kind of model the two forks would be in close proximity, perhaps allowing better coordination of the replicative process. It remains to be determined whether the interactions between T antigen hexamers are stable enough to maintain the double hexamer throughout replication.

It is likely that under most conditions, origin unwinding is the rate-limiting step for SV40 DNA replication (62, 98). It is at this stage that the entire replicative process is regulated by protein phosphorylation and dephosphorylation (see below). It is suspected that the same step may be a major point of control during cellular DNA replication as well.

The second protein required for unwinding of SV40 DNA is RPA. RPA is a three-subunit protein complex that binds single-stranded DNA. The ssDNA-binding capacity is imparted by the largest (70 kDa) subunit (72, 99, 100). The principal role of RPA in origin unwinding is to sequester the unwound single strands, preventing their reassociation (Fig. 4b). This function can be performed by the isolated 70 kDa subunit of RPA in the absence of the other two subunits. A number of heterologous SSB proteins, including those encoded by *E. coli* or phage T4, can substitute for RPA in the unwinding reaction. This finding suggests that the unwinding reaction *per se* does not require specific protein–protein interactions between T antigen and RPA (62, 83, 92, 99, 100). In contrast, only mammalian RPA will function during the initiation of SV40 DNA synthesis on the unwound DNA (92, 100, 101). Thus, it is likely that RPA plays a more specific role in subsequent steps in the replication process, perhaps providing sites of interaction for the polymerase machinery.

6.2 Initiation of DNA synthesis by DNA polymerase α–primase

The unwound DNA strands coated with RPA are the templates for initiation of DNA synthesis by DNA polymerase α–primase (Polα). Polα plays a central role in the initiation of DNA replication because it is the only activity in eukaryotic cells capable of starting new DNA chains (102, 103; Chapter 1). This four-subunit enzyme contains two distinct catalytic activities, primase and DNA polymerase. Under certain experimental conditions it is possible to dissociate the two activities and separate them by chromatographic techniques. The primase activity is associated with a stable subcomplex consisting of 48 and 58 kDa subunits (p48 and p58) (104, 105). Further analysis of the functions of these subunits indicates that p48 contains the primase catalytic subunit (106–109). The p58 subunit stabilizes the primase activity of the p48 subunit, and probably serves to link the primase to the DNA polymerase subcomplex (106, 108, 109). The DNA polymerase activity of Polα resides in the largest (180 kDa) subunit which is tightly associated with the fourth subunit of 68 kDa. The latter subunit, also called the B subunit, has been shown to increase the rate of association of p180 with the primase subcomplex complex (110). The B subunit may also aid in the localization of Polα to sites for primer synthesis, as it is capable of associating with SV40 T antigen (111).

Interestingly, mammalian Polα contains no proofreading exonuclease (112, 113). This lack of proofreading capability may have only minor consequences on the overall fidelity of SV40 DNA replication, since most of the DNA synthesis is probably

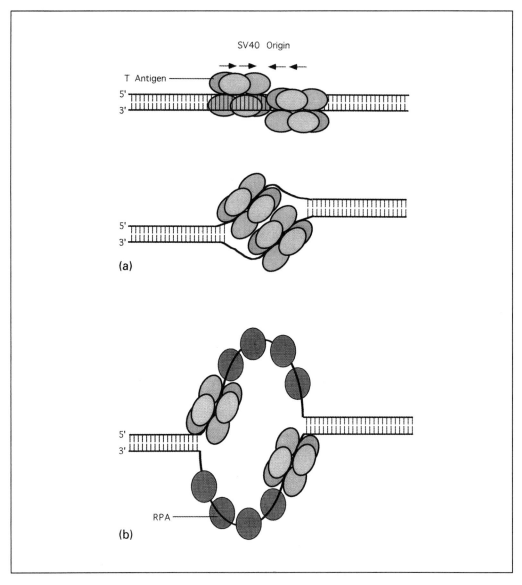

Fig. 4 Schematic representation of the mechanism of SV40 DNA replication. (a) Two T antigen hexamers bind the origin and act cooperatively to locally distort the DNA structure (below). Arrows indicate GAGGC repeats of T antigen-binding site II. (b) Formation of a proto-replication bubble as RPA sequesters the single-stranded DNA and the T antigen hexamer-helicases unwind the DNA in opposite directions. (c) DNA polymerase α initiates RNA primer synthesis (top strand) with its primase component (pentagon). The Polα large catalytic subunit (rod-shaped) polymerizes DNA onto the end of the RNA primer (bottom strand). The B subunit of Polα (triangle) has been drawn to indicate the proposed physical interaction with T antigen during initiation of DNA synthesis. (d) The dissociation of Polα from the newly synthesized RNA/DNA primer allows the assembly of RF-C, PCNA, and Polδ onto the DNA (top strand). Polδ then carries out processive leading strand synthesis and Polα serves to initiate lagging strand synthesis (bottom strand). RNA primers are depicted as crooked lines and DNA is depicted as thicker smooth lines. Cartoons of all proteins are annotated. Nucleotide requirements of T antigen, RF-C, primase, and the DNA polymerases are not encompassed in the drawing.

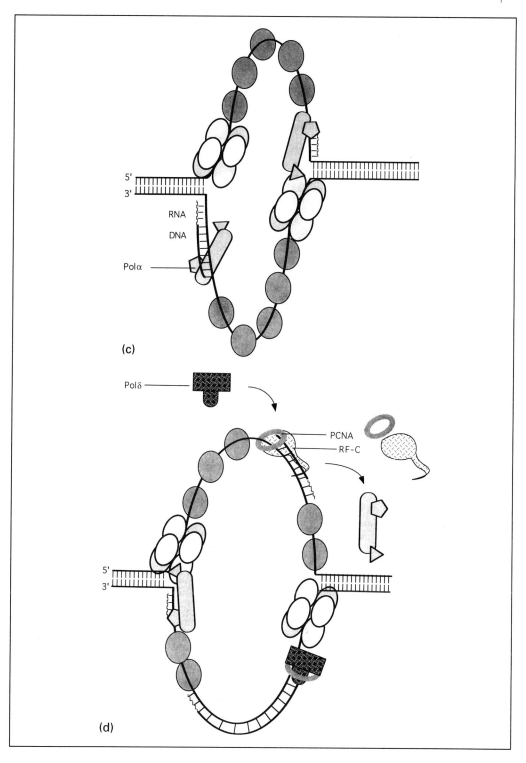

5'
3'

RNA

DNA

Polα

(c)

Polδ

PCNA
RF-C

5'
3'

(d)

carried out by a processive polymerase with proofreading capabilities. In addition, the DNA synthesized by Polα may be largely removed during the process of DNA maturation (discussed below). Alternatively, there may be proofreading activities that work in conjunction with Polα that have not been identified. Weak exonuclease activities associated with the large Polα subunits of *D. melanogaster* and *S. cerevisiae* have been detected, although their roles in proofreading remain obscure (114, 115).

The primase activity of Polα synthesizes RNA oligomers of 8–10 nucleotides (nt), which serve to prime DNA polymerization (Fig. 4c) (116–118). The DNA polymerase activity of Polα is weakly processive, capable of incorporating only 15–25 nt before dissociating from the primer terminus (119, 120). Thus, the initial product synthesized by Polα consists of the RNA primer joined to a short DNA segment ('initiator DNA') which is subsequently extended by the elongation machinery (see below). The synthesis of RNA primers by Polα appears to occur at preferred sites in the template. Such sites generally contain a T residue in the second and third positions (117, 118), consistent with the early observation that poly(dT) is a preferred homopolymer template for Polα–primase (121). The sites where DNA synthesis is initiated during SV40 DNA replication have been mapped to regions immediately flanking the viral origin. The most proximal of these initiation sites are about 30 and 60 nt away from the core origin boundaries (on the early and late sides, respectively) (117). The first 'initiator DNAs' synthesized by Polα serve to prime leading strand synthesis by the highly processive DNA elongation apparatus. Subsequently, Polα serves to initiate DNA synthesis on the lagging strands as the replication forks advance (Fig. 4d) (119). The initiation of DNA synthesis on the lagging strands occurs repeatedly in a discontinuous fashion, and the newly synthesized DNA chains are generally referred to as Okazaki fragments after their discoverer.

As outlined above, the initiation reaction is an ordered sequence of events involving seven different polypeptide chains (T antigen, three RPA subunits, and four DNA polymerase α–primase subunits). There are a number of reasons to believe that the reaction sequence is directed by highly specific protein–protein interactions among the participants. First, the initiation of SV40 DNA replication displays strong species specificity both *in vivo* and *in vitro* (71, 83, 122). The role of human RPA in the reaction cannot be fulfilled by prokaryotic SSBs or yeast RPA, and, even murine DNA polymerase α–primase cannot replace the human polymerase. Second, a variety of direct binding assays have detected the physical association of T antigen with DNA polymerase α–primase both in crude extracts and with purified proteins (52–54, 111, 123). Complexes between T antigen and RPA and between DNA polymerase α–primase and RPA have also been documented (55, 124). Finally, recent experiments with model substrates strongly suggest that the observed physical interactions among the initiation proteins may be functionally significant. For example, it has been demonstrated that T antigen can dramatically stimulate the activity of human DNA polymerase α–primase on both primed and unprimed single-stranded DNA templates (124, 125) The stimulatory effect requires physical contact between the two proteins, suggesting that the complex of T antigen and DNA polymerase α–primase may function as a mobile priming engine (primosome) similar to those observed in prokaryotic systems (126, 127). The SV40 priming engine functions most efficiently in the context of DNA templates that are coated with homologous (primate) RPA. When the template is coated with other SSBs, including yeast RPA, the synthesis of primers by Polα is repressed, even in the presence of T antigen (124).

The physical interactions between T antigen and Polα probably serve to recruit the latter to the replication fork, thereby increasing the rate at which new DNA chains are initiated. This consideration is important, not only at the origin during the initiation phase of the replication reaction, but also at the replication forks during lagging strand synthesis (see below). The interactions among the three initiation proteins probably also play a major role in coordinating priming with DNA unwinding. Complex formation between T antigen and Polα ensures that primer synthesis occurs preferentially at fork-proximal locations, rather than randomly along the unwound DNA. Finally, the physical interactions between T antigen and Polα probably account for the experimental observations that Polα stimulates T antigen binding to the origin and slows the rate of T antigen-mediated DNA unwinding (128, 129).

6.3 Elongation of DNA strands

The first initiator DNAs synthesized by Polα are elongated by the highly processive DNA polymerase δ holoenzyme, becoming the leading DNA strands of the two opposite moving replication forks (Fig. 4d). The Polδ core enzyme is a two-subunit complex with 5′→3′ DNA polymerase and 3′→5′ exonuclease activities (103, 130). Both of these activities reside within the 124 kDa subunit (131, 132). The function of the 48 kDa Polδ subunit is unknown. The assembly of processive Polδ complexes at primer termini requires the activity of two Polδ accessory proteins, replication factor C (RF-C) and proliferating cell nuclear antigen (PCNA) (133–135). RF-C is a multi-subunit DNA-dependent ATPase that recognizes primer–template junctions and assembles PCNA trimers onto the DNA in a reaction that is dependent upon ATP hydrolysis (135). The assembled PCNA binds Polδ and is thought to function as a 'sliding clamp', preventing the dissociation of Polδ from the DNA during polymerization. The effect of RF-C and PCNA on Polδ activity is dramatic, increasing processivity from ~5 to ~2000 nt (136). E. coli and bacteriophage T4 encode functional homologues of Polδ and its processivity factors, RF-C and PCNA, which have been subjected to extensive study (137). In each of these prokaryotic systems the RF-C homologue serves to load the PCNA homologue onto the DNA. The PCNA homologue, once loaded, is capable of tracking along the DNA like a bead on a string even in the absence of its cognate DNA polymerase (138–140). The crystal structure of E. coli DNA polymerase III β dimer (the PCNA homologue) reveals a doughnut-shaped protein assembly with a central hole large enough to accommodate a duplex DNA molecule (141). It is easy to see how such a structure could slide along DNA for long distances without dissociating. It was recently demonstrated that the three-dimensional structure of a trimer of PCNA is very similar to that of the E. coli β dimer, even though the two proteins have essentially no amino acid sequence homology (142).

Unlike the strict requirement for primate Polα in initiation of SV40 DNA replication, there does not seem to be a strong species specificity to the elongation reaction. For example, processive polymerases derived from prokaryotic sources (*E. coli* and T4) can mediate the elongation of DNA chains from primer termini generated by Polα during SV40 DNA replication (75, 143). On the other hand, there is a high degree of specificity to the interactions between the individual components that are required to generate a given processive polymerase complex (75, 143). Thus, the roles of RF-C and PCNA in increasing the processivity of Polδ cannot be fulfilled by their prokaryotic counterparts. There may also be species specific interactions between the processive elongation machinery and the SSB coating the template. In a recent study of human, *E. coli*, and T4 processivity factors, the prokaryotic factors functioned preferentially with their cognate SSB proteins (140). The human factors, RF-C and PCNA, did not exhibit a strong preference for a particular SSB, but were the only processivity factors that functioned efficiently on templates coated with human RPA. These data strongly suggest that the overall efficiency of the elongation reaction depends to a significant extent on specific protein–protein interactions among the participating factors.

The switch from Polα to Polδ holoenzyme that occurs after synthesis of initiator DNA is probably a simple consequence of competition between the polymerases for the primer terminus (75, 144). Because of its low processivity Polα normally vacates the primer terminus after synthesis of a relatively short DNA chain. Although in principle Polα could rebind the primer terminus and synthesize another short chain, an alternative possibility is the binding of RF-C, followed by the assembly of the PCNA–Polδ complex (see Fig. 4d). It is evident that the more processive Polδ–PCNA complex will eventually win this competition for the primer terminus because, once assembled on the DNA, the complex does not dissociate until replication of the strand is complete. *In vitro* studies have provided a direct demonstration that RF-C and PCNA have the ability to compete with Polα for the primer terminus even in the absence of Polδ (75, 144). While it is possible that RF-C and PCNA actively participate in the removal of Polα from the DNA, this is not essential since the passive dissociation of Polα alone would favor efficient switching of polymerases on the newly synthesized DNA. It has been shown that when Polδ is not present, Polα can actually carry out the complete synthesis of both the leading and lagging DNA strands. In this somewhat artificial situation (termed the 'mono-polymerase' system) Polα gradually (and very inefficiently) extends each primer terminus by repeated rounds of association, DNA synthesis, and dissociation (74, 75).

A similar polymerase-switching mechanism is probably employed during lagging strand DNA synthesis. While it is clear that Polα initiates the synthesis of Okazaki fragments on the lagging stand template, several lines of evidence indicate that the nascent chains are actually completed by a more processive DNA polymerase (119, 120, 145). First, addition of antibodies against PCNA to an *in vitro* DNA replication reaction results in shortening of Okazaki fragments (119). This observation strongly suggests that elongation of initiator DNA on the lagging strand requires a

PCNA-dependent DNA polymerase such as Polδ or, perhaps, Polε (see below). Second, studies on the fidelity of SV40 replication indicate that both leading and lagging strands are duplicated with similar accuracy, suggesting that most of the DNA synthesis on both strands is mediated by a polymerase(s) with proofreading capabilities (146). If the lagging strand were completely synthesized by Polα, which lacks a proofreading exonuclease, the replication error frequency would be expected to be at least 30-fold greater than observed (147). Finally, reconstitution experiments using purified proteins have directly implicated Polδ in lagging strand synthesis. When Polδ or its processivity factors, RF-C and PCNA, is omitted from the replication reaction, the average size of Okazaki fragments is significantly reduced and closed circular DNA products are not produced (73, 75, 145, 148). This and the other lines of evidence discussed above indicate that Polα is not solely responsible for replication of the lagging strand, and that Polδ can participate in this process *in vitro*.

Although biochemical analysis of the SV40 replication system identified Polδ as the major polymerase involved in DNA chain elongation, recent studies have opened the possibility that a second DNA polymerase, Polε, may also participate in this process. Polε appears to be present in all eukaryotes from yeast to man and is essential for viability (103). Like Polδ, Polε has intrinsic proofreading exonuclease and is capable of forming a processive complex with PCNA. Moreover, the enzyme can substitute for Polδ (albeit rather inefficiently) during *in vitro* DNA synthesis (149). The role of Polε *in vivo* is still something of a mystery. It has been implicated in DNA repair, but it is not yet clear that it plays a replicative role as well, although there are a few suggestions that this may be the case (150). In yeast the terminal phenotype of mutants defective in the Polε gene (*POL2* mutants) is similar to that of mutants defective in the completion of DNA synthesis (151). In addition, Polε from calf thymus copurifies with an exonuclease (FEN-1/MFI) involved in Okazaki fragment maturation (see below) (152). On the basis of this latter observation it has been suggested that Polε might function in lagging strand synthesis *in vivo*, and Polδ might function exclusively in leading strand synthesis. However, further work will clearly be required to determine whether or not this is the case. A recent study of misincorporation frequencies during SV40 DNA replication *in vivo* revealed the existence of characteristic mutational hotspots on the leading and lagging strands (153). Additional exploration of this phenomenon could lead to a better understanding of the role of the three DNA polymerases (α, δ, and ε) if each enzyme exhibits a specific misincorporation 'signature'.

The speed of the SV40 replication fork seems to be limited by the rate of T antigen helicase movement (about 200 bp/min) (119, 129, 154). Polδ is capable of synthesizing DNA at over three times this rate on singly primed templates (129). The slow movement of T antigen relative to Polδ virtually ensures that unwinding of the duplex does not outpace leading strand synthesis. The rate of fork movement during cellular DNA replication has been estimated to be at least 5–10 times greater than that observed during SV40 DNA replication (155). Thus, cellular helicases must function more efficiently than T antigen in unwinding the duplex.

6.4 Maturation of DNA products

The final steps in SV40 DNA replication of SV40 include the removal of the RNA primers, the filling of the resulting gap, and the joining together of nascent DNA fragments. Completion of these steps is required for the formation of covalently closed circular progeny DNA molecules identical to the parental DNA (74, 76). Reconstitution experiments indicate that RNA primer removal is carried out by two enzymes, RNase H1 and FEN-1(MFI), a 5′ → 3′ exonuclease. The gap-filling function can be fulfilled by either Polα or Polδ, although in the presence of RFC and PCNA, Polδ generally wins the competition for primer termini (see above) (145). The nicks remaining after primer removal and gap filling are sealed by the action of DNA ligase I (76). Interestingly, another eukaryotic ligase, DNA ligase III, cannot fulfil this function, indicating that recognition of maturation sites may involve direct interaction between DNA ligase I and the replication machinery (76).

The molecular processes leading to the removal of primer RNA and the production covalently joined DNA fragments have been extensively studied using purified mammalian enzymes and model DNA structures designed to mimic nascent Okazaki fragments on the lagging strand (Fig. 5). These studies demonstrated that RNase H1 has an endonucleolytic activity that specifically cleaves the RNA primer one nucleotide upstream of the RNA/DNA junction (156). The single ribonucleotide remaining at the 5′ end of the nascent DNA is efficiently removed by the 5′–3′ exonuclease activity of FEN-1 (157). FEN-1 exonuclease activity appears to be closely coordinated with the gap-filling process. Thus, the extension of an upstream primer by DNA polymerase dramatically enhances the removal of the downstream primer by FEN-1 (152, 158). With model substrates any one of the three polymerases, Polα, Polδ, or Polε, can carry out the gap-filling operation to generate a nick that can be sealed by DNA ligase I (145, 152, 158). Interestingly, PCNA also has a significant stimulatory effect on the activity of the FEN-1 exonuclease (158). This observation suggests the possibility that PCNA might serve as a molecular tag for maturation sites. For example, following gap filling by Polδ–PCNA, the Polδ might dissociate leaving PCNA behind. PCNA could then serve to recruit the RNA primer removal machinery and a gap-filling DNA polymerase (i.e., Polδ or Polε). In the *E. coli* replication mechanism, it is known that the β dimer (PCNA homologue) remains bound to the DNA following completion of Okazaki fragment synthesis, although it is not yet clear whether the abandoned dimer plays a role in subsequent replicative events (159).

As mentioned above, the advancement of the replication forks requires the action of a topoisomerase to prevent the build-up of superhelical tension in the replication intermediates as the parental duplex is unwound. DNA topoisomerase I can fulfil this role by introducing transient nicks that allow free rotation about a phosphodiester bond. Because of this activity, topoisomerase I is often described as a 'swivelase'. Although the parental DNA strands are efficiently unwound by this mechanism, not all of the topological links between the two parental strands are removed during the replication process. Thus, the immediate products of SV40

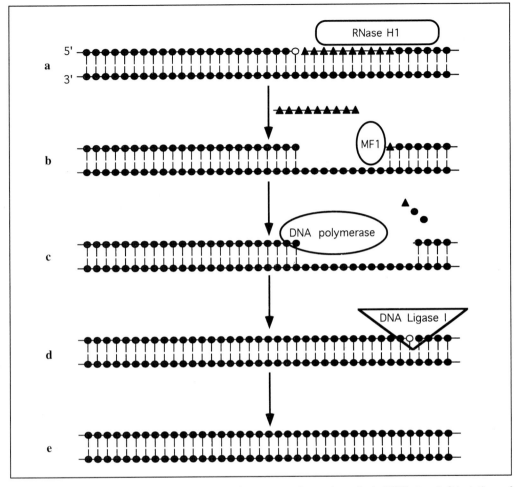

Fig. 5 Maturation of Okazaki fragments. Top strand represents the newly replicated DNA strand. Orientations of the strands are indicated at the left and the nucleotide residues are shown as circles (dNMP) or triangles (rNMP and 5' rNTP). Nicks are denoted by the 3' dNMP being unshaded (white circle in a and d). (a) The RNA primer-containing strand is the substrate for cleavage by RNase H1 endonuclease, resulting (b) in the removal of the RNA moiety except for the most 3' ribonucleotide. (c) MF1 exonuclease removes the remaining rNMP and adjacent dNMP residues. (d) DNA polymerase extends the upstream DNA chain to fill the gap and create a nick. (e) DNA ligase I covalently seals the nick.

DNA replication are two covalently closed circular DNA molecules that remain multiply intertwined. These catenated dimers are resolved into individual circular monomers by the action of DNA topoisomerase II (94). Topoisomerase II introduces transient double-strand breaks that allow the passage of a second DNA duplex (160, 161). The enzyme appears to play a similar role in cellular DNA replication, allowing the intertwined DNA of two newly synthesized daughter chromosomes to separate during mitosis (86, 162).

7. Protein–protein interactions of the initiator proteins

It should be evident at this point that the replication of the SV40 genome is a highly complex process, involving the coordinated actions of a large number of proteins. A complete understanding of the replication mechanism will require detailed analysis of the physical and functional interactions among these proteins. Such interactions ensure that the various steps that lead from initiation of DNA synthesis at the origin to completion of daughter molecules occur in an efficient and orderly fashion. We have already discussed the functional significance of interactions among the proteins involved in initiation, T antigen, RPA and Polα (discussed in Section 6.2). Here we discuss recent studies aimed at defining the critical physical interactions among the replication proteins. This area of research is in its infancy, but is likely to provide important insight into the regulation of DNA replication in the future.

7.1 T antigen binds DNA polymerase α

The physical interaction between T antigen and Polα was first detected by co-immunoprecipitation experiments with crude cell extracts (52). Subsequently, several groups, making use of highly purified proteins, demonstrated that the interaction is direct (53, 54, 111, 123). There is evidence that two different Polα subunits, p180 and p68, have affinity for T antigen. The T antigen-binding domain of p180 was mapped to residues 195–313. Truncated proteins containing this domain inhibit SV40 DNA replication *in vitro*, possibly by competing with intact Polα for T antigen binding (163). A region of T antigen that interacts with p180 was mapped to residues 1–83 by direct binding assays (53). However, studies of anti-T antigen monoclonal antibodies that inhibit the binding of p180 suggest the existence of another interacting region that corresponds roughly to the helicase/ATPase domain (123).

Recent studies have demonstrated a potentially important interaction between T antigen and the 68 kDa B subunit of DNA polymerase α (111). The biochemical function of the B subunit has been a mystery in spite of extensive studies on the biochemical properties of the Polα holoenzyme. Bacterially expressed B subunit is capable of binding to SV40 T antigen, either in solution or when T antigen is bound to the viral origin. Analysis of mutant forms of the B subunit indicated that the N-terminal 240 amino acids are sufficient to mediate complex formation. Importantly, the B subunit greatly enhances the formation of a complex containing T antigen and recombinant p180, suggesting that it serves as a tether to promote the stable interaction of the two proteins. These physical interactions are functionally significant, since the ability of T antigen to stimulate the activity of the catalytic subunit on model substrates is highly dependent upon the presence of the B subunit. It seems likely that the interactions mediated by the B subunit play an important role in SV40 DNA replication by promoting initiation of DNA synthesis at the origin and facilitating the subsequent priming and synthesis of DNA chains on the lagging strand template. The B subunit may play a similar role in cellular DNA replication. However, further work will clearly be required to test these ideas.

7.2 RPA binds T antigen and DNA polymerase α

Physical interactions between RPA and the other initiation proteins were discovered using purified proteins in sensitive ELISA-based binding experiments (55). This work demonstrated that purified RPA can form complexes with both DNA polymerase α and T antigen. In the former case the major interaction was localized to the 70 kDa subunit of RPA and one or both primase subunits (p48 and p58) of DNA Polα. The functional consequences of the RPA–Polα interaction are not yet clear. although it has been demonstrated that RPA can specifically stimulate DNA synthesis by purified Polα in certain assays. The interaction between RPA and T antigen is not well characterized. The isolated 70 kDa subunit of RPA does not bind T antigen, so it is possible that binding is mediated by one of the smaller RPA subunits whose biochemical functions are obscure at this point.

7.3 The species-specificity quandary

The identification of specific protein–protein interactions between T antigen and DNA polymerase α corroborates conclusions drawn from analysis of the species specificity of SV40 replication. These studies demonstrated that extracts derived from mouse cells do not support efficient SV40 DNA replication *in vitro* unless they are supplemented with primate (human) Polα (71, 122). The converse is true for replication of the closely related murine polyomavirus. Human cell extracts do not normally support the replication of polyomavirus DNA, but will do so if supplemented with murine Polα (164, 165). Given the data on the physical interactions of the initiation proteins described above, one possible explanation for these biological observations might be that mouse Polα interacts with polyomavirus T antigen, but does not interact with SV40 T antigen. Surprisingly, this is not the case. SV40 T antigen binds to mouse Polα and even stimulates polymerase activity in primer extension assays. In spite of this fact, mouse Polα fails to support the initiation of SV40 replication in the presence of T antigen and RPA (122). These results suggest that the species specificity of initiation is not simply determined by which proteins can interact, but is strongly dependent upon the details of the interactions. For example, the precise spatial relationship of Polα and T antigen may be critical for efficient initiation. Alternatively, the species specificity of the initiation reaction may be determined by physical interactions among the initiation proteins that have not yet been identified. Interestingly, recent work suggests that the species specificity of polyomavirus DNA replication is determined by interactions between the primase subunits of murine Polα and polyomavirus T antigen (165). Similar interactions in the SV40 system may have simply escaped detection, particularly since many of the key experiments that probed SV40 T antigen–primase interactions were performed with calf thymus Polα which does not function well in SV40 DNA replication (53).

8. Regulation of SV40 DNA replication

8.1 Regulation by protein phosphorylation

The major proteins involved in initiation of SV40 DNA replication, T antigen, RPA and DNA polymerase α–primase, are all phosphoproteins. T antigen contains at least nine serine and threonine residues that are subject to phosphorylation by various protein kinases (166). The phosphorylated residues reside in two clusters, one near the N-terminus and the other near the C-terminus (Fig. 3). A large body of evidence now indicates that the activity of T antigen in DNA replication is tightly regulated both positively and negatively by phosphorylation of a subset of these residues (5, 6). The phosphorylation states of RPA and DNA polymerase α–primase vary in a reproducible way during the cell cycle, suggesting that their activities may also be controlled in part by protein phosphorylation (167, 168). Since the basic pathway of SV40 DNA replication is understood in some detail, the viral system offers an opportunity to explore the biochemical mechanisms responsible for regulation of DNA replication by phosphorylation and possibly other modifications. It seems certain that insights obtained in this system will be applicable to the more complex events that occur at cellular origins of replication. Indeed, there is some experimental evidence suggesting viral and cellular regulatory mechanisms may overlap to some extent. Physiological studies with synchronized cell populations indicate that SV40 DNA replication cannot occur until the host cell has passed through the boundary between the G1 and S phases of the cell cycle, suggesting that the SV40 replication machinery is responsive to some of the normal mechanisms that control cellular entry into S phase (169). Moreover, extracts prepared from primate cells in the G1 phase are much less efficient in supporting SV40 DNA synthesis *in vitro* than are extracts from S phase cells (170). Significantly, the replication of SV40 DNA in G1 extracts is markedly stimulated by incubation with either protein phosphatase 2A or cyclin-dependent protein kinase (p34^{cdc2}), consistent with a critical role for protein phosphorylation in activating viral DNA synthesis at S phase (171, 172).

Several lines of evidence indicate that the replication activity of T antigen is absolutely dependent upon phosphorylation at threonine 124 by a cyclin-dependent kinase. Mutation of Thr 124 to alanine (T124A) renders T antigen incompetent for SV40 DNA replication *in vivo* or *in vitro* (173). Wild-type T antigen produced in unphosphorylated form in bacteria also fails to support DNA replication *in vitro*, but replication activity can be restored by incubation with a purified kinase of the p34^{cdc2} family (174). Phosphorylation of Thr124 affects the interaction of T antigen with the origin. Although unphosphorylated T antigen will form double hexamers at the origin, the cooperativity of the binding reaction and the stability of the resulting complexes are reduced (174, 175). However, the major effect of phosphorylation is not on double hexamer formation, but upon the ability of bound T antigen to promote DNA unwinding at the origin. The T124A mutant or wild-type T antigen produced in *E. coli* are profoundly defective in unwinding plasmid DNA molecules containing the viral origin of replication (175, 176). The defect is not a consequence of a reduction in the helicase activity of T antigen which is unaffected by phosphorylation. Rather, it appears that the bound T antigen hexamers are unable to mediate initial opening of the duplex. This presumably requires a major conformational change that is dependent upon phosphorylation.

Phosphorylation of Ser120 and Ser123 (and perhaps Ser679) has the opposite effect on T antigen activity from phosphorylation of Thr124. The presence of phosphoryl groups at these sites strongly inhibits initiation of DNA replication *in vitro* (171, 177, 178). Again, the major impact of phosphorylation is on the initial duplex opening by bound T antigen hexamers. A protein kinase capable of inhibiting DNA replication by specific phosphorylation of Ser120 and Ser123 has been purified from mammalian cells and identified as casein kinase I (179). The inhibition can be reversed by incubation with certain oligomeric forms of cellular protein phosphatase 2A (PP2A) which preferentially remove phosphates from Ser120 and Ser123 (59, 171, 180). PP2A is normally a heterotrimeric protein consisting of a 34 kDa catalytic subunit (C) complexed to a 65 kDa A subunit and one of several possible B subunits (181). The inhibitory phosphates are specifically removed by PP2A containing a 72 kDa B subunit (182). Interestingly, PP2A containing a different B subunit (55 kDa) can specifically dephosphorylate Thr124, which inactivates T antigen (182). Thus, alterations in the subunit composition of PP2A could potentially have quite different effects on SV40 DNA replication. It is worth noting here that the SV40 small t antigen has been shown to bind tightly to PP2A and to modulate phosphatase activity, although the biological significance of this observation is not yet clear (183, 184).

Figure 6 summarizes the available information on T antigen phosphorylation in the form of a simple model. In this model, the intermediate labeled the 'competent complex' plays the central role in the regulation of initiation. This intermediate consists of a double hexamer of T antigen phosphorylated at Thr124, but not at Ser120 or Ser123. The experimental evidence indicates that this intermediate is uniquely competent to open the duplex. The concentration of the key intermediate can be increased or decreased by specific protein phosphorylation/dephosphorylation to control the rate of the initiation reaction. Recent work suggests that duplex opening by the competent complex requires specific interactions between the two hexameric units of T antigen bound to the origin (175, 176). It has been hypothesized that the presence of phosphoryl groups at Thr124 allows favorable interactions between the hexamers. These interactions provide the driving force for a conformational change in the double hexamer complex that results in local melting of the DNA. The open duplex then serves as a substrate for the intrinsic helicase activity of T antigen which catalyzes more extensive unwinding of the origin region. In this model, phosphorylation of the inhibitory sites at Ser120 and Ser123 perturbs the interactions between the hexamers preventing the required conformational change.

Mutations in Ser120, Ser123, and Thr124 all reduce the efficiency of viral DNA replication in infected cells, consistent with a critical role for T antigen phosphorylation *in vivo* (173). However, the biological role of phosphorylation of T antigen in

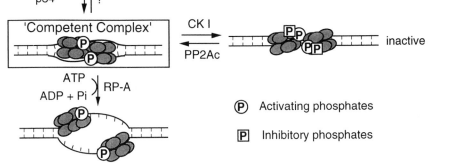

Fig. 6 Model of how site-specific phosphorylation regulates cooperative interaction between T antigen hexamers at the SV40 origin. The key for activating phosphates (at Thr124) and inhibitory phosphates (at Ser120 and Ser123) is shown. The assembly of T antigen hexamers into a 'competent complex' requires phosphorylation by p34^{cdc2} kinase. Inhibitory phosphates are regulated by the activities of casein kinase I (CK I) and protein phosphatase 2A (PP2A). T antigen and DNA are depicted as in Fig. 4.

regulating viral DNA replication during the viral multiplication is not yet clear. The most likely possibility is that phosphorylation events function to coordinate viral DNA replication with the cell cycle. For example, phosphorylation by the appropriate kinases may provide a means for T antigen to sense that the cell is in a state that is favorable for viral DNA replication. Further work will be required to determine whether this or a number of other possibilities is correct.

As indicated above, the cellular proteins that are required for initiation of SV40 DNA replication may also be subject to regulation by protein phosphorylation. In both yeast and mammalian cells the 32 kDa subunit of RPA (RPA32) is phosphorylated in a cell cycle-dependent manner (167). The protein is unphosphorylated in G1, but becomes highly phosphorylated beginning in S phase. RPA32 is also phosphorylated during SV40 DNA replication *in vitro* (185). Although RPA32 is a substrate for cyclin-dependent protein kinases *in vitro* (186), recent work suggests that phosphorylation during SV40 DNA replication is due to DNA-PK, a novel protein kinase which is activated by DNA as a cofactor (187). The data suggest that phosphorylation of the 32 kDa subunit occurs when RPA binds to single-stranded regions in replication intermediates. Thus, phosphorylation of RPA appears to be a consequence of DNA replication and probably plays no role in triggering the onset of either viral or cellular DNA synthesis. One possibility is that phosphorylation of the 32 kDa subunit may alter the biochemical properties of RPA so that it functions more efficiently in DNA chain elongation. It has also been suggested that phosphorylation of RPA might be part of a checkpoint mechanism that signals the presence of replication intermediates to the cell cycle machinery (187).

8.2 Role of transcription factors

It is a striking observation that many viral and cellular origins of DNA replication contain binding sites for transcription factors which contribute significantly to origin function (38). Indeed, in the case of SV40 and related viruses, the presence of such sites is critical for efficient replication in infected cells. As mentioned above, most of the available data suggest that transcription factors do not stimulate DNA replication by promoting transcription in the vicinity of the origin. It seems more likely that the stimulatory effects are due to interactions between transcription factors and replication factors bound at the origin.

In vitro studies suggest that the presence of transcriptional elements may be particularly important in the context of chromatin. The presence of binding sites for Sp1 or the SV40 enhancers has little effect on replication efficiency when the template is naked DNA (36, 37, 188). However, when the template is assembled into a minichromosome, initiation of DNA replication is repressed, presumably because nucleosomes interfere with the assembly of the initiation complex. This general repressive effect of chromatin can be largely reversed by the binding of a transcription factor (e.g., NFI) in the vicinity of the core origin (40). The mechanism of this effect is not yet completely clear. One possibility is that occupancy of a site near the origin by a transcription factor simply interferes with local nucleosome assembly, thereby preventing repression. However, the observation that the activation domain, as well as the DNA binding domain, of the transcription factor is required for stimulation of DNA replication suggests that protein–protein interactions are required (189). Thus, the best model at present is that there is a competition between nucleosome assembly and formation of a competent initiation complex, and that stabilizing contacts between the transcription factor and replication proteins perturb the competition in favor of the latter. It also seems quite possible at this point that interactions between a transcription factor and initiation proteins could facilitate steps in the initiation reaction that occur after formation of the initiation complex, such as duplex opening. The important protein–protein interactions have not been identified, but recent work has demonstrated that the activation domains of several transcription factors (e.g., p53, VP16) bind tightly to the large subunit of RPA (190–192). Consistent with a functional role for this interaction, it has been demonstrated that mutations in the VP16 activation domain which reduce the binding of RPA also reduce the stimulatory effect of the factor on polyomavirus DNA replication (191).

In addition to the SV40 transcriptional elements, the presence of T antigen-binding site I on the early side of the core origin significantly increases the efficiency of viral DNA replication both *in vivo* and *in vitro* (35, 36, 193–195). Binding of T antigen to site I stimulates replication *in vitro* whether the template is naked DNA or chromatin (36, 196) (Cheng and Kelly, unpublished data). The effect appears to be due to the ability of T antigen bound at site I to augment unwinding of the duplex by T antigen bound at the core origin. T antigen also binds weakly within the Sp1 binding sites on the late side of the core origin and mediates a small increase in DNA replication efficiency (188). However, the magnitude of the stimulation by T antigen is far less than that due to Sp1.

9. Concluding remarks

Over the past decade considerable progress has been made towards the understanding of eukaryotic DNA replication. The study of SV40 as a model replication system has been invaluable in the identification and characterization of many cellular replication proteins. In this chapter we have tried to emphasize both the advances and many of the remaining questions concerning the SV40 replication mechanism. The role of Polε, the regulation of T-mediated origin unwinding, and the unknown function of many of the protein subunits are but a few of the mysteries to be resolved in the coming years. The SV40 system is also proving to be useful in emerging fields of study that were not discussed here. The discovery of a replication-enhanced chromatin assembly factor was made via SV40 *in vitro* replication experiments (197). Recently p53-induced proteins, involved in DNA damage-responsive checkpoint regulation of DNA replication, (p21$^{CIP1/WAF1}$ and GADD45) have been shown to bind PCNA (198–201). Sequestration of PCNA by p21$^{CIP1/WAF1}$ was shown to result in the inhibition of SV40 DNA replication *in vitro* (198, 200, 201). Certainly, one of the ultimate future conquests will be to understand cellular replication initiation events. Breakthroughs in this area may soon be obtained from studies on lower eukaryotes, where origin DNA sequences and origin-recognition complexes (ORC) have been identified (see Chapter 3).

References

1. Livingston, D. M. and Bikel, I. (1985) Replication of papovaviruses. In *Virology*. Fields, B. N. (ed.), Raven Press, New York, p. 393.
2. Challberg, M. D. and Kelly, T. J. (1989) Animal virus DNA replication. *Annu. Rev. Biochem.*, **58**, 671.
3. Mole, S. E., Gannon, J. V., Ford, M. J., and Lane, D. P. (1987) Structure and function of SV40 large-T antigen. *Phil. Trans. R. Soc. Lond.*, **317**, 455.
4. Fanning, E. and Knippers, R. (1992) Structure and function of simian virus 40 large tumor antigen. *Annu. Rev. Biochem.*, **61**, 55.
5. Prives, C. (1990) The replication functions of SV40 T antigen are regulated by phosphorylation. *Cell*, **61**, 735.
6. Fanning, E. (1994) Control of SV40 DNA replication by protein phosphorylation: a model for cellular DNA replication? *Trends Cell Biol.*, **4**, 250.
7. Li, J. J. and Kelly, T. J. (1984) Simian virus 40 DNA replication in vitro. *Proc. Natl, Acad. Sci. USA*, **81**, 6973.
8. Acheson, N. H. (1980) Lytic cycle of SV40 and polyoma virus. In *DNA tumor viruses*. Tooze, J. (ed.), Cold Spring Harbor Laboratory Press, Cold Spring Harbor, NY, p. 125.
9. Griffin, B. E. (1980) Structure and genomic organization of SV40 and polyoma virus. In *DNA tumor viruses*. Tooze, J. (ed.), Cold Spring Harbor Laboratory Press, Cold Spring Harbor, NY, p. 61.

10. Kalderon, D., Richardson, W. D., Markham, A. F., and Smith, A. E. (1984) Sequence requirements for nuclear location of simian virus 40 large-T antigen. *Nature*, **311**, 33.

11. Lanford, R. E. and Butel, J. S. (1984) Construction and characterization of an SV40 mutant defective in nuclear transport of T antigen. *Cell*, **37**, 801.

12. Postel, E. H. and Levine, A. J. (1976) The requirement of simian virus 40 gene A product for the stimulation of cellular thymidine kinase activity after viral infection. *Virology*, **73**, 206.

13. Schutzbank, T., Robinson, R., Oren, M., and Levine, A. J. (1982) SV40 large tumor antigen can regulate some cellular transcripts in a positive fashion. *Cell*, **30**, 481.

14. DeCaprio, J. A., Ludlow, J. W., Figge, J., Shew, J. Y., Huang, C. M., Lee, W. H., Marsilio, E., Paucha, E., and Livingston, D. M. (1988) SV40 large tumor antigen forms a specific complex with the product of the retinoblastoma susceptibility gene. *Cell*, **54**, 275.

15. Ludlow, J. W., Shon, J., Pipas, J. M., Livingston, D. M., and DeCaprio, J. A. (1990) The retinoblastoma susceptibility gene product undergoes cell cycle-dependent dephosphorylation and binding to and release from SV40 large T. *Cell*, **60**, 387.

16. Nevins, J. R. (1992) E2F: a link between the Rb tumor suppressor protein and viral oncoproteins. *Science*, **258**, 424.

17. Bargonetti, J., Reynisdottir, I., Friedman, P. N., and Prives, C. (1992) Site-specific binding of wild-type p53 to cellular DNA is inhibited by SV40 T antigen and mutant p53. *Genes Dev.*, **6**, 1886.

18. Lane, D. P. and Crawford, L. V. (1979) T antigen is bound to a host protein in SV40-transformed cells. *Nature*, **278**, 261.

19. Linzer, D. I. and Levine, A. J. (1979) Characterization of a 54k dalton cellular SV40 tumor antigen present in SV40-transformed cells and uninfected embryonal carcinoma cells. *Cell*, **17**, 43.

20. Levine, A. J. (1993) The tumor suppressor genes. *Annu. Rev. Biochem.*, **62**, 623.

21. Hartwell, L. H. and Kastan, M. B. (1994) Cell cycle control and cancer genes. *Science*, **266**, 1821.

22. Rio, D. C. and Tjian, R. (1983) SV40 T antigen binding site mutations that affect autoregulation. *Cell*, **32**, 1227.

23. Brady, J. and Khoury, G. (1985) trans Activation of the simian virus 40 late transcription unit by T-antigen. *Mol. Cell. Biol.*, **5**, 1391.

24. Keller, J. M. and Alwine, J. C. (1984) Activation of the SV40 late promoter: direct effects of T antigen in the absence of viral DNA replication. *Cell*, **36**, 381.

25. Casaz, P., Sundseth, R., and Hansen, U. (1991) trans Activation of the simian virus 40 late promoter by large T antigen requires binding sites for the cellular transcription factor TEF-1. *J. Virol.*, **65**, 6535.

26. Gruda, M. C. and Alwine, J. C. (1991) Simian virus 40 (SV40) T-antigen transcriptional activation mediated through the Oct/SPH region of the SV40 late promoter. *J. Virol.*, **65**, 3553.

27. Scieller, P., Omilli, F., Borde, J., and May, E. (1991) Characterization of SV40 enhancer motifs involved in positive and negative regulation of the constitutive late promoter activity; effect of T-antigen. *Virology*, **181**, 783.

28. Black, P. H., Crawford, E. M., and Crawford, L. V. (1964) The purification of simian virus 40. *Virology*, **24**, 381.

29. Deb, S., Tsui, S., Koff, A., DeLucia, A. L., Parsons, R., and Tegtmeyer, P. (1987) The T-antigen-binding domain of the simian virus 40 core origin of replication. *J. Virol.*, **61**, 2143.

30. Delucia, A., Lewton, B., Tjian, R., and Tegtmeyer, P. (1983) Topography of simian virus 40 A protein–DNA complexes: arrangement of pentanucleotide interaction sites at the origin of replication. *J. Virol.*, **46**, 143.

31. Mastrangelo, I. A., Hough, P. V., Wall, J. S., Dodson, M., Dean, F. B., and Hurwitz, J. (1989) ATP-dependent assembly of double hexamers of SV40 T antigen at the viral origin of DNA replication. *Nature*, **338**, 658.

32. Deb, S., DeLucia, A. L., Koff, A., Tsui, S., and Tegtmeyer, P. (1986) The adenine-thymine domain of the simian virus 40 core origin directs DNA bending and coordinately regulates DNA replication. *Mol. Cell. Biol.*, **6**, 4578.

33. Dean, F. B., Borowiec, J. A., Ishimi, Y., Deb, S., Tegtmeyer, P., and Hurwitz, J. (1987) Simian virus 40 large tumor antigen requires three core replication origin domains for DNA unwinding and replication in vitro. *Proc. Natl. Acad. Sci. USA*, **84**, 8267.

34. Deb, S., DeLucia, A. L., Baur, C. P., Koff, A., and Tegtmeyer, P. (1986) Domain structure of the simian virus 40 core origin of replication. *Mol. Cell. Biol.*, **6**, 1663.

35. DeLucia, A. L., Deb, S., Partin, K., and Tegtmeyer, P. (1986) Functional interactions of the simian virus 40 core origin of replication with flanking regulatory sequences. *J. Virol.*, **57**, 138.

36. Li, J. J., Peden, K. W. C., Dixon, R. A. F., and Kelly, T. J. (1986) Functional organization of the simian virus 40 origin of DNA replication. *Mol. Cell. Biol.*, **6**, 1117.

37. Stillman, B. W., Gerard, R. D., Guggenheimer, R. A., and Gluzman, Y. (1985) T antigen and template requirements for SV40 DNA replication in vitro. *EMBO J.*, **4**, 2933.

38. DePamphilis, M. L. (1993) Eukaryotic DNA replication: anatomy of an origin. *Annu. Rev. Biochem.*, **62**, 29.

39. Hoang, A. T., Wang, W., and Gralla, J. D. (1992) The replication activation potential of selected RNA polymerase II promoter elements at the simian virus 40 origin. *Mol. Cell. Biol.*, **12**, 3087.

40. Cheng, L. and Kelly, T. J. (1989) Transcriptional activator nuclear factor I stimulates the replication of SV40 minichromosomes in vivo and in vitro. *Cell*, **59**, 541.

41. Vassilev, L. T. and DePamphilis, M. L. (1992) Guide to identification of origins of DNA replication in eukaryotic cell chromosomes. *Crit. Rev. Biochem. Mol. Biol.*, **27**, 445.

42. Campbell, J. L. and Newlon, C. S. (1991) Chromosomal DNA Replication. In *The molecular and cellular biology of the yeast saccharomyces*. Broach, J. R., Pringle, J. R., and Jones, E. W. (eds.), Cold Spring Harbor Laboratory Press, Cold Spring Harbor, NY, vol. 1, p. 41.

43. Lin, S. and Kowalski, D. (1994) DNA helical instability facilitates initiation at the SV40 replication origin. *J. Mol. Biol.*, **235**, 496.

44. Pipas, J. M. (1992) Common and unique features of T antigens encoded by the polyomavirus group. *J. Virol.*, **66**, 3979.

45. Strauss, M., Argani, P., Mohr, I. J., and Gluzman, Y. (1987) Studies on the origin-specific DNA-binding domain of simian virus 40 large T antigen. *J. Virol.*, **61**, 3326.

46. Arthur, A. K., Höss, A., and Fanning, E. (1988) Expression of simian virus 40 T antigen in *Escherichia coli*: localization of T-antigen origin DNA-binding domain to within 129 amino acids. *J. Virol.*, **62**, 1999.

47. Hoss, A., Moarefi, I. F., Fanning, E., and Arthur, A. K. (1990) The finger domain of simian virus 40 large T antigen controls DNA-binding specificity. *J. Virol.*, **64**, 6291.

48. Clark, R., Peden, K., Pipas, J. M., Nathans, D., and Tjian, R. (1983) Biochemical activities of T-antigen proteins encoded by simian virus 40 A gene deletion mutants. *Mol. Cell. Biol.*, **3**, 220.

49. Wun, K. K. and Simmons, D. T. (1990) Mapping of helicase and helicase substrate-binding domains on simian virus 40 large T antigen. *J. Virol.*, **64,** 2014.

50. Bradley, M. K., Smith, T. F., Lathrop, R. H., Livingston, D. M., and Webster, T. A. (1987) Consensus topography in the ATP binding site of the simian virus 40 and polyoma large tumor antigens. *Proc. Natl. Acad. Sci. USA*, **84,** 4026.

51. DeCaprio, J. A., Ludlow, J. W., Lynch, D., Furukawa, Y., Griffin, J., Piwnica, W. H., Huang, C. M., and Livingston, D. M. (1989) The product of the retinoblastoma susceptibility gene has properties of a cell cycle regulatory element. *Cell*, **58,** 1085.

52. Smale, S. T. and Tjian, R. (1986) T-antigen-DNA polymerase alpha complex implicated in simian virus 40 DNA replication. *Mol. Cell. Biol.*, **6,** 4077.

53. Dornreiter, I., Hoss, A., Arthur, A. K., and Fanning, E. (1990) SV40 T antigen binds directly to the large subunit of purified DNA polymerase alpha. *EMBO J.*, **9,** 3329.

54. Gannon, J. V. and Lane, D. P. (1987) p53 and DNA polymerase alpha compete for binding to SV40 T antigen. *Nature*, **329,** 456.

55. Dornreiter, I., Erdile, L. F., Gilbert, I. U., Von, W. D., Kelly, T. J., and Fanning, E. (1992) Interaction of DNA polymerase alpha–primase with cellular replication protein A and SV40 T antigen *EMBO J.*, **11,** 769.

56. Dean, F. B., Borowiec, J. A., Eki, T., and Hurwitz, J. (1992) The simian virus 40 T antigen double hexamer assembles around the DNA at the replication origin. *J. Biol. Chem.*, **267,** 14 129.

57. Borowiec, J. A., Dean, F. B., Bullock, P. A., and Hurwitz, J. (1990) Binding and unwinding – how T antigen engages the SV40 origin of DNA replication. *Cell*, **60,** 181.

58. Parsons, R. E., Stenger, J. E., Ray, S., Welker, R., Anderson, M. E., and Tegtmeyer, P. (1991) Cooperative assembly of simian virus 40 T-antigen hexamers on functional halves of the replication origin. *J. Virol.*, **65,** 2798.

59. Virshup, D. M., Russo, A. A., and Kelly, T. J. (1992) Mechanism of activation of simian virus 40 DNA replication by protein phosphatase 2A. *Mol. Cell. Biol.*, **12,** 4883.

60. Parsons, R., Anderson, M. E., and Tegtmeyer, P. (1990) Three domains in the simian virus 40 core origin orchestrate the binding, melting, and DNA helicase activities of T antigen. *J. Virol.*, **64,** 509.

61. Borowiec, J. A. and Hurwitz, J. (1988) Localized melting and structural changes in the SV40 origin of replication induced by T-antigen. *EMBO J.*, **7,** 3149.

62. Wold, M. S., Li, J. J., and Kelly, T. J. (1987) Initiation of simian virus 40 DNA replication in vitro: Large-tumor-antigen and origin-dependent unwinding of the template. *Proc. Natl. Acad. Sci. USA*, **84,** 3643.

63. Dean, F. B., Bullock, P., Murakami, Y., Wobbe, C. R., Weissbach, L., and Hurwitz, J. (1987) Simian virus 40 (SV40) DNA replication: SV40 large T antigen unwinds DNA containing the SV40 origin of replication. *Proc. Natl. Acad. Sci. USA*, **84,** 16.

64. Stahl, H., Droge, P., Zentgraf, H., and Knippers, R. (1985) A large-tumor-antigen-specific monoclonal antibody inhibits DNA replication of simian virus 40 minichromosomes in an in vitro elongation system. *J. Virol.*, **54,** 473.

65. Stahl, H., Droge, P., and Knippers, R. (1986) DNA helicase activity of SV40 large tumor antigen. *EMBO J.*, **5,** 1939.

66. Goetz, G. S., Dean, F. B., Hurwitz, J., and Matson, S. W. (1988) The unwinding of duplex regions in DNA by the simian virus 40 large tumor antigen-associated DNA helicase activity. *J. Biol. Chem.*, **263,** 383.

67. Wiekowski, M., Schwartz, M. W., and Stahl, H. (1988) Simian virus 40 large T antigen DNA helicase: Characterization of the ATPase-dependent DNA unwinding activity and its substrate requirements. *J. Biol. Chem.*, **263,** 436.

68. Scheffner, M., Knippers, R., and Stahl, H. (1989) RNA unwinding activity of SV40 large T antigen. *Cell*, **57**, 955.

69. Scheffner, M., Wessel, R., and Stahl, H. (1989) Sequence independent duplex DNA opening reaction catalysed by SV40 large tumor antigen. *Nucleic Acids Res.*, **17**, 93.

70. Li, J. J. and Kelly, T. J. (1985) Simian virus 40 DNA replication in vitro: Specificity of initiation and evidence for bidirectional replication. *Mol. Cell. Biol.*, **5**, 1238.

71. Murakami, Y., Wobbe, C. R., Weissbach, L., Dean, F. B., and Hurwitz, J. (1986) Role of DNA polymerase alpha and DNA primase in simian virus 40 DNA replication in vitro. *Proc. Natl. Acad. Sci. USA*, **83**, 2869.

72. Wold, M. S., Weinberg, D. H., Virshup, D. M., Li, J. J., and Kelly, T. J. (1989) Identification of cellular proteins required for simian virus 40 DNA replication. *J. Biol. Chem.*, **264**, 2801.

73. Weinberg, D. H., Collins, K. L., Simancek, P., Russo, A., Wold, M. S., Virshup, D. M., and Kelly, T. J. (1990) Reconstitution of simian virus 40 DNA replication with purified proteins. *Proc. Natl. Acad. Sci. USA*, **87**, 8692.

74. Ishimi, Y., Claude, A., Bullock, P., and Hurwitz, J. (1988) Complete enzymatic synthesis of DNA containing the SV40 origin of replication. *J. Biol. Chem.*, **263**, 19723.

75. Eki, T., Matsumoto, T., Murakami, Y., and Hurwitz, J. (1992) The replication of DNA containing the simian virus 40 origin by the monopolymerase and dipolymerase systems. *J. Biol. Chem.*, **267**, 7284.

76. Waga, S., Bauer, G., and Stillman, B. (1994) Reconstitution of complete SV40 DNA replication with purified replication factors. *J. Biol. Chem.*, **269**, 10923.

77. Tsurimoto, T., Fairman, M. P., and Stillman, B. (1989) Simian virus 40 DNA replication in vitro: identification of multiple stages of initiation. *Mol. Cell. Biol.*, **9**, 3839.

78. Pizzagalli, A., Valsasnini, P., Plevani, P., and Lucchini, G. (1988) DNA polymerase I gene of *Saccharomyces cerevisiae*: nucleotide sequence, mapping of a temperature-sensitive mutation, and protein homology with other DNA polymerases. *Proc. Natl. Acad. Sci. USA*, **85**, 3772.

79. Lucchini, G., Francesconi, S., Foiani, M., Badaracco, G., and Plevani, P. (1987) Yeast DNA polymerase–DNA primase complex; cloning of PRI 1, a single essential gene related to DNA primase activity. *EMBO J.*, **6**, 737.

80. Foiani, M., Santocanale, C., Plevani, P., and Lucchini, G. (1989) A single essential gene, PRI2, encodes the large subunit of DNA primase in *Saccharomyces cerevisiae*. *Mol. Cell. Biol.*, **9**, 3081.

81. Foiani, M., Marini, G., Gamba, D., Lucchini, G., and Plevani, P. (1994) The B subunit of the DNA polymerase alpha–primase complex in *Saccharomyces cerevisiae* executes an essential function at the initial stage of DNA replication. *Mol. Cell. Biol.*, **14**, 923.

82. Boulet, A., Simon, M., Faye, G., Bauer, G. A., and Burgers, P. M. (1989) Structure and function of the *Saccharomyces cerevisiae* CDC2 gene encoding the large subunit of DNA polymerase III. *EMBO J.*, **8**, 1849.

83. Brill, S. J. and Stillman, B. (1991) Replication factor-A from *Saccharomyces cerevisiae* is encoded by three essential genes coordinately expressed at S phase. *Genes Dev.*, **5**, 1589.

84. Heyer, W.-D., Rao, M. R. F., Erdile, L. F., Kelly, T. J., and Kolodner, R. D. (1990) An essential *Saccharomyces cerevisiae* single-stranded DNA binding protein is homologous to the large subunit of human RP-A. *EMBO J.*, **9**, 2321.

85. Johnston, L. H., Barker, D. G., and Nurse, P. (1986) Cloning and characterization of the *Schizosaccharomyces pombe* DNA ligase gene CDC17. *Gene*, **41**, 321.

86. DiNardo, S., Voelkel, K., and Sternglanz, R. (1984) DNA topoisomerase II mutant of *Saccharomyces cerevisiae*: topoisomerase II is required for segregation of daughter molecules at the termination of DNA replication. *Proc. Natl. Acad. Sci. USA*, **81**, 2616.

87. Bauer, G. A. and Burgers, P. M. (1990) Molecular cloning, structure and expression of the yeast proliferating cell nuclear antigen gene. *Nucleic Acids Res.*, **18**, 261.

88. Li, X. and Burgers, P. M. (1994) Cloning and characterization of the essential *Saccharomyces cerevisiae* RFC4 gene encoding the 37-kDa subunit of replication factor C. *J. Biol. Chem.*, **269**, 21 880.

89. Li, X. and Burgers, P. M. (1994) Molecular cloning and expression of the *Saccharomyces cerevisiae* RFC3 gene, an essential component of replication factor C. *Proc. Natl. Acad. Sci. USA*, **91**, 868.

90. Howell, E. A., McAlear, M. A., Rose, D., and Holm, C. (1994) CDC44: a putative nucleotide-binding protein required for cell cycle progression that has homology to subunits of replication factor C. *Mol. Cell. Biol.*, **14**, 255.

91. Wold, M. S. and Kelly, T. J. (1988) Purification and characterization of replication protein A, a cellular protein required for in vitro replication of simian virus 40 DNA. *Proc. Natl. Acad. Sci. USA*, **85**, 2523.

92. Kenny, M. K., Lee, S. H., and Hurwitz, J. (1989) Multiple functions of human single-stranded-DNA binding protein in simian virus 40 DNA replication: single-strand stabilization and stimulation of DNA polymerases alpha and delta. *Proc. Natl. Acad. Sci. USA*, **86**, 9757.

93. Fairman, M. P. and Stillman, B. (1988) Cellular factors required for multiple stages of SV40 DNA replication in vitro. *EMBO J.*, **7**, 1211.

94. Yang, L., Wold, M. S., Li, J. J., Kelly, T. J., and Liu, L. F. (1987) Roles of DNA topoisomerases in SV40 DNA replication in vitro. *Proc. Natl. Acad. Sci. USA*, **84**, 950.

95. Wessel, R., Schweizer, J., and Stahl, H. (1992) Simian virus 40 T-antigen DNA helicase is a hexamer which forms a binary complex during bidirectional unwinding from the viral origin of DNA replication. *J. Virol.*, **66**, 804.

96. SenGupta, D. J. and Borowiec, J. A. (1992) Strand-specific recognition of a synthetic DNA replication fork by the SV40 large tumor antigen. *Science*, **256**, 1656.

97. Wessel, R., Ramsperger, U., Stahl, H., and Knippers, R. (1992) The interaction of SV40 large T antigen with unspecific double-stranded DNA: an electron microscopic study. *Virology*, **189**, 293.

98. Wobbe, C. R., Weissbach, L., Borowiec, J. A., Dean, F. B., Murakami, Y., Bullock, P., and Hurwitz, J. (1987) Replication of SV40-origin-containing DNA with purified proteins. *Proc. Natl. Acad. Sci. USA*, **84**, 1834.

99. Kenny, M. K., Schlegel, U., Furneaux, H., and Hurwitz, J. (1990) The role of human single-stranded DNA binding protein and its individual subunits in simian virus 40 DNA replication. *J. Biol. Chem.*, **265**, 7693.

100. Erdile, L. F., Heyer, W.-D., Kolodner, R., and Kelly, T. J. (1991) Characterization of cDNA encoding the 70-kDa subunit of human replication protein A (RP-A), a single-stranded DNA binding protein involved in DNA replication and recombination. *J. Biol. Chem.*, **266**, 12090

101. Virshup, D. M. and Kelly, T. J. (1989) Purification of replication protein C, a cellular protein involved in the initial stages of SV40 DNA replication in vitro. *Proc. Natl. Acad. Sci. USA*, **86**, 3584.

102. Lehman, I. R. and Kaguni, L. S. (1989) DNA polymerase alpha. *J. Biol. Chem.*, **264**, 4265.

103. Wang, T. S. (1991) Eukaryotic DNA polymerases. *Annu. Rev. Biochem.*, **60**, 513.

104. Kaguni, L. S., Rossignol, J. M., Conaway, R. C., Banks, G. R., and Lehman, I. R. (1983) Association of DNA primase with the beta/gamma subunits of DNA polymerase alpha from *Drosophila melanogaster* embryos. *J. Biol. Chem.*, **258**, 9037.

105. Tseng, B. Y. and Ahlem, C. N. (1982) DNA primase activity from human lymphocytes. Synthesis of oligoribonucleotides that prime DNA synthesis. *J. Biol. Chem.*, **257**, 7280.

106. Copeland, W. C. and Wang, T. S. (1993) Enzymatic characterization of the individual mammalian primase subunits reveals a biphasic mechanism for initiation of DNA replication. *J. Biol. Chem.*, **268**, 26179.

107. Bakkenist, C. J. and Cotterill, S. (1994) The 50-kDa primase subunit of *Drosophila melanogaster* DNA polymerase alpha. Molecular characterization of the gene and functional analysis of the overexpressed protein. *J. Biol. Chem.*, **269**, 26759.

108. Santocanale, C., Foiani, M., Lucchini, G., and Plevani, P. (1993) The isolated 48,000-dalton subunit of yeast DNA primase is sufficient for RNA primer synthesis. *J. Biol. Chem.*, **268**, 1343.

109. Stadlbauer, R., Brueckner, A., Rehfuess, C., Eckerskorn, C., Lottspeich, F., Forster, V., Tseng, B. Y., and Nasheuer, H. P. (1994) DNA replication in vitro by recombinant DNA-polymerase-alpha-primase. *Eur. J. Biochem.*, **222**, 781.

110. Brooke, R. G. and Dumas, L. B. (1991) Reconstitution of the *Saccharomyces cerevisiae* DNA primase–DNA polymerase protein complex in vitro. *J. Biol. Chem.*, **266**, 10093.

111. Collins, K. L., Russo, A. A., Tseng, B. Y., and Kelly, T. J. (1993) The role of the 70 kDa subunit of human DNA polymerase alpha in DNA replication. *EMBO J.*, **12**, 4555.

112. Reyland, M. E. and Loeb, L. A. (1987) On the fidelity of DNA replication. Isolation of high fidelity DNA polymerase–primase complexes by immunoaffinity chromatography. *J. Biol. Chem.*, **262**, 10824.

113. Copeland, W. C. and Wang, T. S. (1991) Catalytic subunit of human DNA polymerase alpha overproduced from baculovirus-infected insect cells. Structural and enzymological characterization. *J. Biol. Chem.*, **266**, 22739.

114. Brooke, R. G., Singhal, R., Hinkle, D. C., and Dumas, L. B. (1991) Purification and characterization of the 180- and 86-kilodalton subunits of the *Saccharomyces cerevisiae* DNA primase–DNA polymerase protein complex. *J. Biol. Chem.*, **266**, 3005.

115. Cotterill, S. M., Reyland, M. E., Loeb, L. A., and Lehman, I. R. (1987) A cryptic proofreading 3′–5′ exonuclease associated with the polymerase subunit of the DNA polymerase–primase from *Drosophila melanogaster*. *Proc. Natl. Acad. Sci. USA*, **84**, 5635.

116. Kuchta, R. D., Reid, B., and Chang, L. M. (1990) DNA primase. Processivity and the primase to polymerase alpha activity switch. *J. Biol. Chem.*, **265**, 16158.

117. Bullock, P. A., Tevosian, S., Jones, C., and Denis, D. (1994) Mapping initiation sites for simian virus 40 DNA synthesis events in vitro. *Mol. Cell. Biol.*, **14**, 5043.

118. Hay, R. T., Hendrickson, E. A., and DePamphilis, M. L. (1984) Sequence specificity for the initiation of RNA-primed simian virus 40 DNA synthesis in vivo. *J. Mol. Biol.*, **175**, 131.

119. Bullock, P. A., Seo, Y. S., and Hurwitz, J. (1991) Initiation of simian virus 40 DNA synthesis in vitrol *Mol. Cell. Biol.*, **11**, 2350.

120. Nethanel, T. and Kaufmann, G. (1990) Two DNA polymerases may be required for synthesis of the lagging DNA strand of simian virus 40. *J. Virol.*, **64**, 5912.

121. Fisher, P. A., Chen, J. T., and Korn, D. (1981) Enzymological characterization of KB cell DNA polymerase-alpha. Regulation of template binding by nucleic acid base composition. *J. Biol. Chem.*, **256**, 133.

122. Schneider, C., Weisshart, K., Guarino, L. A., Dornreiter, I., and Fanning, E. (1994) Species-specific functional interactions of DNA polymerase alpha–primase with SV40 T antigen require SV40 origin DNA. *Mol. Cell. Biol.*, **14,** 3176.

123. Gannon, J. V. and Lane, D. P. (1990) Interactions between SV40 T antigen and DNA Polymerase α. *New Biologist*, **2,** 84.

124. Melendy, T. and Stillman, B. (1993) An interaction between replication protein A and SV40 T antigen appears essential for primosome assembly during SV40 DNA replication. *J. Biol. Chem.*, **268,** 3389.

125. Collins, K. L. and Kelly, T. J. (1991) Effects of T antigen and replication protein A on the initiation of DNA synthesis by DNA polymerase alpha-primase. *Mol. Cell. Biol.*, **11,** 2108.

126. Barry, J. and Alberts, B. (1994) Purification and characterization of bacteriophage T4 gene 59 protein. *J. Biol. Chem.*, **269,** 33 049.

127. Allen, G. J. and Kornberg, A. (1993) Assembly of the primosome of DNA replication in *Escherichia coli. J. Biol. Chem.*, **268,** 19 204.

128. Murakami, Y. and Hurwitz, J. (1993) DNA polymerase alpha stimulates the ATP-dependent binding of simian virus tumor T antigen to the SV40 origin of replication. *J. Biol. Chem.*, **268,** 11 018.

129. Murakami, Y. and Hurwitz, J. (1993) Functional interactions between SV40 T antigen and other replication proteins at the replication fork. *J. Biol. Chem.*, **268,** 11 008.

130. Lee, M. Y., Tan, C. K., So, A. G., and Downey, K. M. (1980) Purification of deoxyribonucleic acid polymerase delta from calf thymus: partial characterization of physical properties. *Biochemistry*, **19,** 2096.

131. Lee, M. Y., Jiang, Y. Q., Zhang, S. J., and Toomey, N. L. (1991) Characterization of human DNA polymerase delta and its immunochemical relationships with DNA polymerase alpha and epsilon. *J. Biol. Chem.*, **266,** 2423.

132. Yang, C. L., Chang, L. S., Zhang, P., Hao, H., Zhu, L., Toomey, N. L., and Lee, M. Y. (1992) Molecular cloning of the cDNA for the catalytic subunit of human DNA polymerase delta. *Nucleic Acids Res.*, **20,** 735.

133. Prelich, G., Tan, C. K., Kostura, M., Mathews, M. B., So, A. G., Downey, K. M., and Stillman, B. (1987) Functional identity of proliferating cell nuclear antigen and a DNA polymerase-delta auxiliary protein. *Nature*, **326,** 517.

134. Tan, C. K., Castillo, C., So, A. G., and Downey, K. M. (1986) An auxiliary protein for DNA polymerase-delta from fetal calf thymus. *J. Biol. Chem.*, **261,** 12 310.

135. Tsurimoto, T. and Stillman, B. (1991) Replication factors required for SV40 DNA replication in vitro. I. DNA structure-specific recognition of a primer–template junction by eukaryotic DNA polymerases and their accessory proteins. *J. Biol. Chem.*, **266,** 1950.

136. Syvaoja, J., Suomensaari, S., Nishida, C., Goldsmith, J. S., Chui, G. S., Jain, S., and Linn, S. (1990) DNA polymerases alpha, delta, and epsilon: three distinct enzymes from HeLa cells. *Proc. Natl. Acad. Sci. USA*, **87,** 6664.

137. Kuriyan, J. and O'Donnell, M. (1993) Sliding clamps of DNA polymerases. *J. Mol. Biol.*, **234,** 915.

138. Stukenberg, P. T., Studwell, V. P., and O'Donnell, M. (1991) Mechanism of the sliding beta-clamp of DNA polymerase III holoenzyme. *J. Biol. Chem.*, **266,** 11 328.

139. Herendeen, D. R., Kassavetis, G. A., and Geiduschek, E. P. (1992) A transcriptional enhancer whose function imposes a requirement that proteins track along DNA. *Science*, **256,** 1298.

140. Tinker, R. L., Kassavetis, G. A., and Geiduschek, E. P. (1994) Detecting the ability of

viral, bacterial and eukaryotic replication proteins to track along DNA. *EMBO J.*, **13,** 5330.

141. Kong, X. P., Onrust, R., O'Donnell, M., and Kuriyan, J. (1992) Three-dimensional structure of the beta subunit of *E. coli* DNA polymerase III holoenzyme: a sliding DNA clamp. *Cell*, **69,** 425.

142. Krishna, T. S., Kong, X. P., Gary, S., Burgers, P. M., and Kuriyan, J. (1994) Crystal structure of the eukaryotic DNA polymerase processivity factor PCNA. *Cell*, **79,** 1233.

143. Tsurimoto, T., Melendy, T., and Stillman, B. (1990) Sequential initiation of lagging and leading strand synthesis by two different polymerase complexes at the SV40 DNA replication origin. *Nature*, **346,** 534.

144. Tsurimoto, T. and Stillman, B. (1991) Replication factors required for SV40 DNA replication in vitro. II. Switching of DNA polymerase alpha and delta during initiation of leading and lagging strand synthesis. *J. Biol. Chem.*, **266,** 1961.

145. Waga, S. and Stillman, B. (1994) Anatomy of a DNA replication form revealed by reconstitution of SV40 DNA replication in vitro. *Nature*, **369,** 207.

146. Roberts, J. D., Thomas, D. C., and Kunkel, T. A. (1991) Exonucleolytic proofreading of leading and lagging strand DNA replication errors. *Proc. Natl. Acad. Sci. USA*, **88,** 3465.

147. Roberts, J. D. and Kunkel, T. A. (1988) Fidelity of a human cell DNA replication complex. *Proc. Natl. Acad. Sci. USA*, **85,** 7064.

148. Tsurimoto, T. and Stillman, B. (1989) Purification of a cellular replication factor, RF-C, that is required for coordinated synthesis of leading and lagging strands during simian virus 40 DNA replication in vitro. *Mol. Cell. Biol.*, **9,** 609.

149. Lee, S. H., Pan, Z. Q., Kwong, A. D., Burgers, P. M., and Hurwitz, J. (1991) Synthesis of DNA by DNA polymerase epsilon in vitro. *J. Biol. Chem.*, **266,** 22707.

150. Nishida, C., Reinhard, P., and Linn, S. (1988) DNA repair synthesis in human fibroblasts requires DNA polymerase delta. *J. Biol. Chem.*, **263,** 501.

151. Araki, H., Ropp, P. A., Johnson, A. L., Johnston, L. H., Morrison, A., and Sugino, A. (1992) DNA polymerase II, the probable homolog of mammalian DNA polymerase epsilon, replicates chromosomal DNA in the yeast *Saccharomyces cerevisiae*. *EMBO J.*, **11,** 733.

152. Siegal, G., Turchi, J. J., Myers, T. W., and Bambara, R. A. (1992) A 5′ to 3′ exonuclease functionally interacts with calf DNA polymerase epsilon. *Proc. Natl. Acad. Sci. USA*, **89,** 9377.

153. Roberts, J. D., Izuta, S., Thomas, D. C., and Kunkel, T. A. (1994) Mispair-, site-, and strand-specific error rates during simian virus 40 origin-dependent replication in vitro with excess deoxythymidine triphosphate. *J. Biol. Chem.*, **269,** 1711.

154. Tapper, D. P., Anderson, S., and DePamphilis, M. L. (1979) Maturation of replicating simian virus 40 DNA molecules in isolated nuclei by continued bidirectional replication to the normal termination region. *Biochim. Biophys. Acta*, **565,** 84.

155. Edenberg, H. J. and Huberman, J. A. (1975) Eukaryotic chromosome replication. *Annu. Rev. Genet.*, **9,** 245.

156. Huang, L., Kim, Y., Turchi, J. J., and Bambara, R. A. (1994) Structure-specific cleavage of the RNA primer from Okazaki fragments by calf thymus RNase HI. *J. Biol. Chem.*, **269,** 25922.

157. Turchi, J. J., Huang, L., Murante, R. S., Kim, Y., and Bambara, R. A. (1994) Enzymatic completion of mammalian lagging-strand DNA replication. *Proc. Natl. Acad. Sci. USA*, **91,** 9803.

158. Turchi, J. J. and Bambara, R. A. (1993) Completion of mammalian lagging strand DNA replication using purified proteins *J. Biol. Chem.*, **268,** 15136.

159. Stukenberg, P. T., Turner, J., and O'Donnell, M. (1994) An explanation for lagging strand replication: polymerase hopping among DNA sliding clamps. *Cell*, **78**, 877.

160. Watt, P. M. and Hickson, I. D. (1994) Structure and function of type II DNA topoisomerases. *Biochem. J.*, **303**, 681.

161. Anderson, H. J. and Roberge, M. (1992) DNA topoisomerase II: a review of its involvement in chromosome structure, DNA replication, transcription and mitosis. *Cell. Biol. Int. Rep.*, **16**, 717.

162. Ishida, R., Sato, M., Narita, T., Utsumi, K. R., Nishimoto, T., Morita, T., Nagata, H., and Andoh, T. (1994) Inhibition of DNA topoisomerase II by ICRF-193 induces polyploidization by uncoupling chromosome dynamics from other cell cycle events. *J. Cell. Biol.*, **126**, 1341.

163. Dornreiter, I., Copeland, W. C., and Wang, T. (1993) Initiation of simian virus 40 DNA replication requires the interaction of a specific domain of human DNA polymerase a with large T antigen. *Mol. Cell. Biol.*, **13**, 809.

164. Murakami, Y., Eki, T., Yamada, M., Prives, C., and Hurwitz, J. (1986) Species-specific in vitro synthesis of DNA containing the polyoma virus origin of replication. *Proc. Natl. Acad. Sci. USA*, **83**, 6347.

165. Eki, T., Enomoto, T., Masutani, C., Miyajima, A., Takada, R., Murakami, Y., Ohno, T., Hanaoka, F., and Ui, M. (1991) Mouse DNA primase plays the principal role in determination of permissiveness for polyomavirus DNA replication. *J. Virol.*, **65**, 4874.

166. Scheidtmann, K. H., Echle, B., and Walter, G. (1982) Simian virus 40 large T antigen is phosphorylated at multiple sites clustered in two separate regions. *J. Virol.*, **44**, 116.

167. Din, S., Brill, S. J., Fairman, M. P., and Stillman, B. (1990) Cell-cycle-regulated phosphorylation of DNA replication factor A from human and yeast cells. *Journal Genes Dev.*, **4**, 968.

168. Nasheuer, H. P., Moore, A., Wahl, A. F., and Wang, T. S. (1991) Cell cycle-dependent phosphorylation of human DNA polymerase alpha. *J. Biol. Chem.*, **266**, 7893.

169. Pages, J., Manteuil, S., Stehelin, D., Fiszman, M., Marx, M., and Girard, M. (1973) Relationship between replication of simian virus 40 DNA and specific events of the host cell cycle. *J. Virol.*, **12**, 99.

170. Roberts, J. M. and D'Urso, G. (1988) An origin unwinding activity regulates initiation of DNA replication during the mammalian cell cycle. *Science*, **241**, 1486.

171. Virshup, D. M., Kauffman, M. G., and Kelly, T. J. (1989) Activation of SV40 DNA replication in vitro by cellular protein phosphatase 2A. *EMBO J.*, **8**, 3891.

172. D'Urso, G., Marraccino, R. L., Marshak, D. R., and Roberts, J. M. (1990) Cell cycle control of DNA replication by a homologue from human cells of the p34^{cdc2} protein kinase. *Science*, **250**, 786.

173. Schneider, J. and Fanning, E. (1988) Mutations in the phosphorylation sites of simian virus 40 (SV40) T antigen alter its origin DNA-binding specificity for sites I or II and affect SV40 DNA replication activity. *J. Virol.*, **62**, 1598.

174. McVey, D., Brizuela, L., Mohr, I., Marshak, D. R., Gluzman, Y., and Beach, D. (1989) Phosphorylation of large tumour antigen by cdc2 stimulates SV40 DNA replication. *Nature*, **341**, 503.

175. McVey, D., Ray, S., Gluzman, Y., Berger, L., Wildeman, A. G., Marshak, D. R., and Tegtmeyer, P. (1993) cdc2 phosphorylation of threonine 124 activates the origin-unwinding functions of simian virus 40 T antigen. *J. Virol.*, **67**, 5206.

176. Moarefi, I. F., Small, D., Gilbert, I., Hopfner, M., Randall, S. K., Schneider, C., Russo, A. A., Ramsperger, U., Arthur, A. K., Stahl, H., Kelly, T. J., and Fanning, E. (1993)

Mutation of the cyclin-dependent kinase phosphorylation site in simian virus 40 (SV40) large T antigen specifically blocks SV40 origin DNA unwinding. *J. Virol.*, **67**, 4992.

177. Grässer, F. A., Mann, K., and Walter, G. (1987) Removal of serine phosphates from simian virus 40 large T antigen increases its ability to stimulate DNA replication in vitro but has no effect on ATPase and DNA binding. *J. Virol.*, **61**, 3373.

178. Mohr, I. J., Stillman, B., and Gluzman, Y. (1987) Regulation of SV40 DNA replication by phosphorylation of T antigen. *EMBO J.*, **6**, 153.

179. Cegielska, A. and Virshup, D. M. (1993) Control of simian virus 40 DNA replication by the HeLa cell nuclear kinase casein kinase I. *Mol. Cell. Biol.*, **13**, 1202.

180. Cegielska, A., Moarefi, I., Fanning, E., and Virshup, D. M. (1994) T antigen kinase inhibits simian virus 40 DNA replication by phosphorylation of intact T antigen on serines 120 and 123. *J. Virol.*, **68**, 269.

181. Mumby, M. C. and Walter, G. (1993) Protein serine/threonine phosphatases: structure, regulation, and functions in cell growth. *Physiol. Rev.*, **73**, 673.

182. Cegielska, A., Shaffer, S., Derua, R., Goris, J., and Virshup, D. M. (1994) Different oligomeric forms of protein phosphatase 2A activate and inhibit simian virus 40 DNA replication. *Mol. Cell. Biol.*, **14**, 4616.

183. Pallas, D. C., Shahrik, L. K., Martin, B. L., Jaspers, S., Miller, T. B., Brautigan, D. L., and Roberts, T. M. (1990) Polyoma small and middle T antigens and SV40 small t antigen form stable complexes with protein phosphatase 2A. *Cell*, **60**, 167.

184. Sontag, E., Fedorov, S., Kamibayashi, C., Robbins, D., Cobb, M., and Mumby, M. (1993) The interaction of SV40 small tumor antigen with protein phosphatase 2A stimulates the map kinase pathway and induces cell proliferation. *Cell*, **75**, 887.

185. Fotedar, R. and Roberts, J. M. (1992) Cell cycle regulated phosphorylation of RPA-32 occurs within the replication initiation complex. *EMBO J.*, **11**, 2177.

186. Dutta, A. and Stillman, B. (1992) cdc2 family kinases phosphorylate a human cell DNA replication factor, RPA, and activate DNA replication. *EMBO J.*, **11**, 2189.

187. Brush, G. S., Anderson, C. W., and Kelly, T. J. (1994) The DNA-activated protein kinase is required for the phosphorylation of replication protein A during simian virus 40 DNA replication. *Proc. Natl. Acad. Sci. USA*, **91**, 12 520.

188. Guo, Z. S. and DePamphilis, M. L. (1992) Specific transcription factors stimulate simian virus 40 and polyomavirus origins of DNA replication. *Mol. Cell. Biol.*, **12**, 2514.

189. Cheng, L. Z., Workman, J. L., Kingston, R. E., and Kelly, T. J. (1992) Regulation of DNA replication in vitro by the transcriptional activation domain of GAL4-VP16. *Proc. Natl. Acad. Sci. USA*, **89**, 589.

190. Li, R. and Botchan, M. R. (1993) The acidic transcriptional activation domains of VP16 and p53 bind the cellular replication protein A and stimulate in vitro BPV-1 DNA replication. *Cell*, **73**, 1207.

191. He, Z., Brinton, B. T., Greenblatt, J., Hassell, J. A., and Ingles, C. J. (1993) The trans-activator proteins VP16 and GAL4 bind replication factor A. *Cell*, **73**, 1223.

192. Dutta, A., Ruppert, J. M., Aster, J. C., and Winchester, E. (1993) Inhibition of DNA replication factor RPA by p53 [see comments]. *Nature*, **365**, 79.

193. Lee-Chen, G.-J. and Woodworth-Gutai, M. (1986) Simian virus 40 DNA replication: functional organization of regulatory elements. *Mol. Cell. Biol.*, **6**, 3086.

194. Hertz, G. Z. and Mertz, J. E. (1986) Bidirectional promoter elements of simian virus 40 are required for efficient replication of the viral DNA. *Mol. Cell. Biol.*, **6**, 3513.

195. Guo, Z. S., Gutierrez, C., Heine, U., Sogo, J. M., and Depamphilis, M. L. (1989) Origin

auxiliary sequences can facilitate initiation of simian virus 40 DNA replication in vitro as they do in vivo. *Mol. Cell. Biol.*, **9**, 3593.

196. Guo, Z. S., Heine, U., and DePamphilis, M. L. (1991) T-antigen binding to site I facilitates initiation of SV40 DNA replication but does not affect bidirectionality. *Nucleic Acids Res.*, **19**, 7081.

197. Smith, S. and Stillman, B. (1989) Purification and characterization of CAF-I, a human cell factor required for chromatin assembly during DNA replication in vitro. *Cell*, **58**, 15.

198. Li, R., Waga, S., Hannon, G. J., Beach, D., and Stillman, B. (1994) Differential effects by the p21 CDK inhibitor on PCNA-dependent DNA replication and repair. *Nature*, **371**, 534.

199. Smith, M. L., Chen, I. T., Zhan, Q., Bae, I., Chen, C. Y., Gilmer, T. M., Kastan, M. B., O'Connor, P. M., and Fornace, A. J. (1994) Interaction of the p53-regulated protein Gadd45 with proliferating cell nuclear antigen [see comments]. *Science*, **266**, 1376.

200. Waga, S., Hannon, G. J., Beach, D., and Stillman, B. (1994) The p21 inhibitor of cyclin-dependent kinases controls DNA replication by interaction with PCNA [see comments]. *Nature*, **369**, 574.

201. Flores, R. H., Kelman, Z., Dean, F. B., Pan, Z. Q., Harper, J. W., Elledge, S. J., O'Donnell, M., and Hurwitz, J. (1994) Cdk-interacting protein 1 directly binds with proliferating cell nuclear antigen and inhibits DNA replication catalyzed by the DNA polymerase delta holoenzyme. *Proc. Natl. Acad. Sci. USA*, **91**, 8655.

202. Loeber, G., Stenger, J. E., Ray, S., Parsons, R. E., Anderson, M. E., and Tegtmeyer, P. (1991) The zinc finger region of simian virus 40 large T antigen is needed for hexamer assembly and origin melting. *J. Virol.*, **65**, 3167.

203. Pipas, J. M. (1985) Mutations near the carboxyl terminus of the simian virus 40 large tumor antigen alter viral host range. *J. Virol.*, **54**, 569.

3 | The initiation of DNA replication in the yeast *Saccharomyces cerevisiae*

YORK MARAHRENS and BRUCE STILLMAN

1. Introduction

The discovery that DNA consists of two complementary antiparallel strands associated by hydrogen bonds immediately led to the realization that the genetic material duplicates by using the preexisting strands as templates (1). DNA polymerase was discovered shortly thereafter (2) and marked the onset of increasingly detailed enzymological studies of the template-driven reaction whose mechanism is now understood in great detail (Chapter 1). Insight into how DNA replication is initiated was gained much more slowly. Most progress has been made by studying the replication of the genomes of prokaryotes, bacteriophages and some eukaryote viruses, and the basic mechanism has been deduced for some of these organisms. How eukaryote cell chromosomes initiate DNA replication remains an unsolved problem.

Because the initiation of DNA replication is one of the key control points for chromosome replication, it is important that this process be understood. The initiation event must be regulated to ensure that every chromosome is replicated once and only once per cell division. Furthermore, the initiation of DNA replication must be inhibited when cells are programmed not to divide. Another important control imposed on the start of DNA synthesis is the link of cell cycle and checkpoint controls so that replication does not occur when the cell is not ready for division (see Chapter 7).

Two important concepts in DNA replication that arose from studies of bacteria and their viruses (called bacteriophages or phages) are the replicator and initiator. It was noted in the late 1950s and early 1960s that if genes of the bacterial chromosome became incorporated in extrachromosomal DNA circles called episomes or into the genomes of bacteriophages, the incorporated genes replicated together with the episome or bacteriophage genome (3, 4). This indicated that not all regions of bacterial chromosomes were capable of inducing autonomous DNA replication. François Jacob and Sydney Brenner hypothesized that a limited number of specific *cis*-acting elements might control DNA replication much like the operator elements

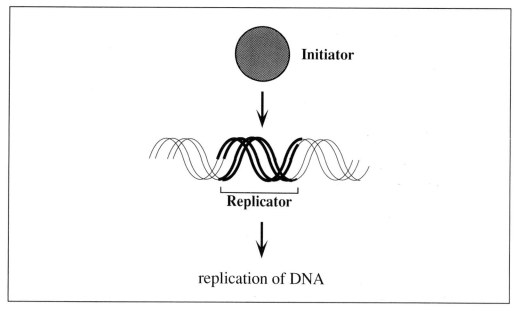

Fig. 1 The replicon model. Proposed by Jacob and Brenner (3), this model posits that a site-specific DNA-binding protein called the initiator interacts with a specific DNA sequence called the replicator. The initiator, when bound to the replicator, acts as a 'landing pad' for other proteins that initiate DNA replication.

that were thought to control bacterial transcription. They adapted to the initiation of DNA replication, concepts from the then leading model for bacterial transcriptional initiation, the operon model (5). According to their model for DNA replication, which they called the replicon model, replication initiates when a positive *trans*-acting factor they called an initiator recognizes a *cis*-acting element in the DNA called a replicator (Fig. 1). The DNA replicated as a result of the initiator acting at a specific replicator was called a replicon (6). Unfortunately the replicator terminology has been lost from most of the current literature and has been replaced by the term 'origin of DNA replication'. But the origin of DNA replication is most commonly used to mean the location where DNA replication actually begins, and this site does not necessarily coincide with the replicator. In the interest of clarity, therefore, the site at which the initiation of DNA synthesis occurs shall be referred to as the origin of DNA replication (ori) to distinguish this from the replicator element that controls the initiation event (7). In practical terms, the replicator is defined genetically where the ori is defined by physical measurements that locate the actual site for initiation of DNA synthesis.

It has been demonstrated that the replicon model holds true for the replication of DNA in bacterial chromosomes and plasmids and also for the replication of some bacteriophage and eukaryote virus genomes. Important for the validation of the replicon model was the mapping of initiation sites (oris). Oris were localized in some viruses by partial denaturation mapping (8, 9) and in the *E. coli* chromosome by gene dosage studies (10, 11). It was later demonstrated that DNA fragments encompassing the initiation sites also contain the replicator that governs ori function, since these fragments induce autonomous replication when cloned onto a plasmid (12). The isolation of bacterial, phage, and viral replicator sequences allowed these important sequences to be defined precisely. Genetic studies *in vivo* and investigation of phage and plasmid DNA replication *in vitro* led to the identification and characterization of initiator proteins that recognized and bound to the replicators, as well as many other additional proteins that are required for the initiation of DNA replication. These studies ultimately led to a detailed understanding of how replicators and their associated proteins function. Although the replication of nearly all prokaryotic, bacteriophage, and viral genomes is controlled by replicators and initiators, they work through a surprising diversity of molecular mechanisms.

How DNA replication initiates in eukaryotes remains an enigma. The replicon model has been difficult to demonstrate in most eukaryotes because at chromosomal regions where initiation events are known to occur, the precise sites where oris form vary from one cell to the next (see Chapter 4). Moreover, genetic experiments in most eukaryotes (particularly metazoan species) have failed to identify specific DNA sequences that control the initiation of DNA replication. The situation in some metazoans is more complicated because the frequency along a chromosome where initiation occurs changes during development. Furthermore, in the early embryos of frogs, any DNA injected into the rapidly dividing embryo will replicate in synchrony with the cell chromosomes, leading to the suggestion that replicators might not be not necessary in these quickly dividing cells (see Chapter 6).

Over the past 10–15 years, however, progress towards understanding the initiation of DNA replication in eukaryotes has accelerated. This is primarily due to insight gleaned from studies of *S. cerevisiae* which have led to the validation of the replicon model in this eukaryote (for a recent review that puts this into perspective, see 13). Much of the molecular machinery for DNA replication has been identified in *S. cerevisiae*, setting the stage for an understanding of the initiation process.

2. Plasmid replication in yeast

2.1 Identification of ARSs

The discovery that certain genomic DNA fragments allow the plasmids to replicate extrachromosomally in *S. cerevisiae* (14, 15) was a watershed for the eukaryotic DNA field. Such fragments were called autonomously replicating sequences (ARSs) because it was not known whether they represented true chromosomal replicators. As we will see, some of these ARS sequences were found to be identical to chromosomal replicators. ARS elements were subsequently found in other yeasts, including *S. pombe* (see Chapter 8), but attempts to obtain specific ARS elements in multicellular eukaryotes have failed (see Chapter 4).

One attractive hypothesis to explain why ARSs are readily obtained in yeasts but have been difficult to find in mammalian cells postulates that the deciding difference is due to the nuclear membrane remaining intact throughout the entire cell cycle in yeast cells, while in cells from other species, the nuclear membrane disappears entirely at the onset of mitosis. According to this hypothesis, plasmid replication occurs in other eukaryotes, but in the absence of a nuclear membrane to keep them corralled, the plasmids are rapidly lost each time a cell divides. Studies with yeast point to at least two additional features that may be important for stable transmission of plasmids in metazoans. Autonomously replicating plasmids in yeast are quite unstable unless a centromere is added (16–18). Unfortunately, centromeres of multicellular organisms are too large to be included on simple pasmids. Second, increasing plasmid size in yeast results in an additional increase in plasmid stability (19). Nevertheless, even the largest plasmids are lost at considerably higher rates than native chromosomes (18).

2.2 Physical mapping of oris shows that ARSs determine sites of initiation in plasmids

A second breakthrough for the DNA replication field was the simultaneous development of two techniques to map oris (20, 21). Although these techniques were developed to map oris in *S. cerevisiae*, they have proven to be valuable tools for mapping oris in other organisms, including mammals (see Chapter 4; 22–24). Both techniques use physical methods that examine one locus at a time and determine whether oris arise in proliferating cells. If an ori forms at a locus, the DNA strands are separated and the synthesis of new DNA commences, using the old DNA strands as a template. Small 'eye-shaped structures' called replication bubbles form at the very onset of DNA replication and mark the site of the ori, which then enlarges as replication continues. In contrast, if replication does not initiate at the locus, this region of the chromosome will be replicated by a single replication fork derived from elsewhere in the DNA. Both techniques examine specific sites and determine whether they contain DNA replication bubbles or single replication forks. In practice, DNA is isolated from cultures of proliferating cells, purified and cleaved with a restriction endonuclease. A very small proportion of the restriction fragments contain replication intermediates that can be separated from the linear DNA fragments by two-dimensional (2D) agarose gel electrophoresis.

One ori mapping method, called the neutral/neutral 2D gel method, identifies either DNA replication bubbles or replication forks (21). DNA fragments containing intact replication bubbles or forks are separated from each other and nonreplicating linear DNA by their unique electrophoretic properties due to their distinctive shapes. Genomic digests are first subjected to low voltage electrophoresis. This separates restriction fragments primarily by their differences in mass. Forked (Y) and bubble-containing restriction fragments have greater masses than the linear molecules of the same sequence and therefore migrate more slowly in the gel. The DNA fragments are then further resolved by high voltage electrophoresis in a direction perpendicular to the first. At the higher voltage, both mass and shape play a major role in the migration of DNA, causing molecules that previously comigrated to separate from each other. Linear molecules migrate the fastest, Y-shaped molecules

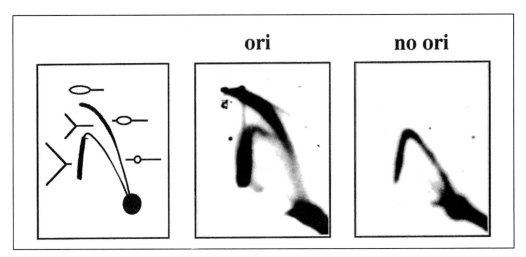

Fig. 2 Physical mapping of an origin of DNA replication by neutral/neutral, two-dimensional agarose gel electrophoresis. DNA from a growing culture of cells is isolated and digested with a specific restriction endonuclease. The resulting fragments are separated by agarose gel electrophoresis in two dimensions that optimize separation of DNAs with differing masses and shapes. The DNA is then blotted to a filter paper and probed with a specific DNA probe. In the diagram shown in the left panel, if the fragment has an origin of DNA replication located within it, initiation of DNA replication will result in bubble structures that migrate as a 'bubble arc'. If there is no origin, the DNA replication intermediates will migrate as a 'fork' or 'Y arc'. If the ori is not centred on the fragment, this will result in a bubble to Y transition when the replication fork reaches one end of the fragment before the other. The panels in the middle and the right show examples of this 2D gel technique with and without an ori, respectively. The technique, developed by Brewer and Fangman (21), has a number of variations that can detect different DNA replication intermediates. The data in this figure were adapted from an analysis of *ARS1* (39).

migrate at an intermediate rate, and bubble-containing molecules migrate the slowest. (The contribution of molecular shape to migration in agarose gels of DNA fragments having the same mass can also be seen following electrophoresis of linear (form III), nicked (form II), and supercoiled (form I), circular plasmids.) After electrophoresis, the migration of a specific fragment is examined by DNA hybridization to a specific radioactive probe. A replication origin located within the boundaries of a fragment will generate a distinctive pattern called a bubble arc (Fig. 2). All other DNA fragments are passively replicated from an ori that is located outside the fragment, so their replication intermediates are Y-shaped molecules which produce Y-arcs that can easily be discriminated (21).

The second ori mapping method, called the neutral/alkaline 2D gel method, identifies the newly synthesized strands derived from replication bubbles that have been denatured (20). Genomic digests are first subjected to agarose gel electrophoresis as in the neutral/neutral method. The DNA is then subjected to agarose gel electrophoresis in a direction perpendicular to the first, but this time under denaturing conditions and then hybridized to a radioactive DNA probe. Probes specific to the left end, the right end, and the central region of the DNA fragment of interest are employed in separate hybridizations. If the DNA fragment is passively replicated from one end to the other, then the fastest migrating (shortest) replication

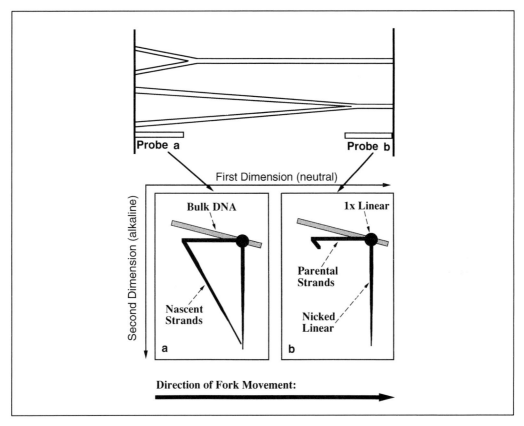

Fig. 3 Physical mapping of DNA replication intermediates by neutral/alkaline two-dimensional agarose gel electrophoresis. In this technique, DNA from a growing population of cells is isolated and digested with a restriction endonuclease. The fragments are first separated by agarose gel electrophoresis under neutral pH conditions and then in a second dimension under alkaline pH conditions. The DNA is then blotted to a filter paper and probed with a specific DNA probe. The figure shows a diagram of a DNA fragment that does not contain an ori and therefore replicates from one end to the other. The bulk DNA is shown by the gray stripes and the specific DNA fragment detected by the probe is in black. Two different probes, a and b, detect different length nascent DNA strands and therefore indicate the direction of fork movement. This figure is based on the technique of Huberman *et al.* (20) and is taken from Liang *et al.* (24).

intermediates will be detected by a probe that recognizes one of the ends of the fragment (Fig. 3). If the DNA fragment is replicated from an internal ori, then the fastest migrating replication intermediates will be detected by a probe specific for the central region of the restriction fragment.

Both 2D gel techniques have been used to demonstrate that oris colocalize with ARSs in plasmids (20, 21). The neutral/neutral technique, which has a greater resolution than the neutral/alkaline technique, has been used to show that ARSs and oris colocalize to within 400 bp of each other (21). It is still not known whether the actual initiation event occurs at the ARS itself, at specific site(s) near the ARS, or anywhere within a region near the ARS. Recent studies on the initiation of replication of the SV40 genome have mapped the start sites for DNA synthesis to a broad region that surrounds the SV40 replicator (25).

2.3 General structure of ARS elements

With the knowledge that ARSs determine the location of an ori in a plasmid, attention was initially focused on the structure of these autonomous replicating sequences. The expectation was that the chromosomal replicators from yeast would be highly conserved in sequence. This is because the replicators from prokaryotes, phages, and viruses tend to contain multiple, highly conserved DNA sequences. For example, the sequences of chromosomal replicators from a number of bacteria, including the *E. coli* oriC, are highly conserved, even among bacteria that diverged up to 100 million years ago (26). Furthermore, prokaryotic and viral replicators commonly contain two or more copies of a conserved sequence that functions as the recognition sequence for the initiator protein (26). Yeast replicators, however, proved to be different.

Comparison of the sequences of several ARSs showed that they were all very A/T rich and that only a single conserved, but quite degenerate, 11 base pair sequence 5'-(A/T)TTTA(T/C)(A/G)TTT(A/T)-3' referred to as ARS consensus sequence (ACS), is common to all ARSs (27, 28). Absent the repeated initiator binding sites that were found in other replicators, suggesting that if ARSs represented chromosomal replicators, then eukaryotic replicators differed significantly in structure from their prokaryotic and viral counterparts. This lack of sequence conservation in ARSs led to the idea that initiation of DNA replication in eukaryotes was not very sequence specific, but further analysis of ARSs showed this not to be the case (see below).

Deletion analyses of several ARSs have revealed that in every ARS a match to the ACS is required for ARS function (29–36). In no case, however, was the ACS sufficient for ARS function. Indeed, only a subset of matches to the ACS identified in the yeast genome were associated with ARS activity on plasmids. This is because all ARSs also contain, in addition to an ACS, a second essential component within a 100–200 bp region of DNA located directly 3' to the T-rich strand of the ACS (29–31). The small region that contains a match to the ACS (but also requires additional sequence, see below) has been referred to as domain A, while the broad region adjacent to domain A was called domain B (32). This designation was described first for *ARS1* but has subsequently been found in all ARSs. Deletion analyses of domain B, however, failed to define specific DNA sequences within the B region.

Although deletion analyses failed to define additional sequences important for ARS function, they did reveal other aspects about the function of ARSs. ARS sequences are very sensitive to their spacing with respect to one another (37–39). Moreover, ARSs are very sensitive to the nature of the DNA sequences that surround them. For example, one segment of DNA was shown to exhibit ARS activity when cloned into a plasmid in one orientation, but not when cloned into the same plasmid in the opposite orientation (33). Many sequences, including vector sequences of bacterial origin, could substitute for portions of the domain B (31) and yet other flanking sequences reduced or even eliminated ARS function (40).

A conserved feature of ARS elements and the replicators from bacteria and viruses is the presence of DNA sequences that can be easily unwound. The *E. coli* oriC replicator has three 13-mer sequences that are essential for initiation of DNA replication and these repeats are readily unwound when placed in a highly super-coiled plasmid (41). These 13-mers also unwind when the *E. coli* initiator protein, dnaA, binds to oriC, a process that requires some negative supercoiling in the DNA (42). Easily unwound DNA sequences are also a feature of yeast ARS elements (43, 44). Indeed, other easily unwound DNA sequences can be placed adjacent to the A element that contains the ACS and ARS activity can be recovered. These studies have led to the concept that a conserved feature of yeast replicators is a DNA unwinding element (DUE) that can facilitate initiation of DNA replication. As dis-cussed below, however, ease of unwinding is not the only property of the B domain in ARSs.

3. Chromosomal replicators: proximal and distal elements control the initiation of DNA replication in a chromosome

3.1 A subset of ARS elements are chromosomal replicators

Use of the 2D gel technology has demonstrated convincingly that some ARS sequences function as replicators in their normal location on yeast chromosomes and therefore determine the location of chromosomal oris (39, 44–50). Surprisingly, oris were not seen at many of the chromosomal ARS loci tested (29, 51, 52). It is poss-ible that some ARS sequences represent chromosomal replicators that are activated infrequently in the chromosome and consequently replication bubbles cannot be detected by the relatively insensitive 2D gel assay. Another possibility is that some of the ARSs only function as chromosomal replicators under certain conditions. For example, in *Drosophila*, a specialized replicator called the amplification control element (ACE) is only active in certain cell types and only at a specific time during development (22, 53). The number of oris used to replicate *Drosophila* chromosomes also varies tremendously throughout development (54, 55) and this may also be due to changes in replicator utilization. No case of this type, however, has been dis-covered in yeast. It has even been demonstrated that the same yeast replicators are used during the mitotic and meiotic DNA S phases (56).

When functional ARS sequences are deleted from the chromosome and inserted elsewhere, ori activity is abolished at the native locus (39, 44, 46–50) and is usually established at the new position (38, 39, 57–60) demonstrating that ARS sequences are essential and sufficient for chromosomal replicator function. Interestingly, no effect on DNA replication or chromosome stability has ever been observed following the elimination of a single replicator. More surprisingly, systematic deletion of all known replicators over more than 200 kb of *S. cerevisiae* chromosome III produced only small effects on chromosome stability (57). This suggests that there are far more replicators than is necessary to replicate the genome in a single S phase. Perhaps replicators perform other functions in the chromosome and an example of this has recently been uncovered (see below) (61, 62).

3.2 Context influences replicator efficiency and timing

Relocated replicators are often subject to position effects indicating that more distally located elements also control replicator function. Controls on the timing during S phase when a replicator functions is a good example. If the normally early activated *ARS1* replicator (Fig. 4) is moved to a position near the late activated *ARS501* replicator, *ARS1* becomes late activated and less efficient (Fig. 4) (59, 60). Distal elements therefore control time and efficiency of replicator activation. Interestingly, oriC of *E. coli* is also subject to position effects (26).

One source of position effects is interference between replicators located on the same molecule. In a yeast strain where *ARS1* and *ARS501* reside within a few kilobases of one another, each replicator suppresses the other (Fig. 4) (60). In another strain, where two copies of *ARS1* are located within a few kilobases of one another, one copy dominates over the other (39). Since the sequence of the two copies is identical, flanking sequences must dictate which copy is used.

It is not known why replicators sometimes interfere with one another. One possibility is that a mechanism in the cell that prevents overreplication of DNA is responsible for replicator suppression. The prevention of overreplication may be necessary if two replicators on the same molecule differ in their times of activation because one replicator could then cause the region to be duplicated before the second replicator is activated. A second possibility is that a replicator may establish a chromatin structure that spreads to, and inactivates, other replicator(s). A third possibility relies on the notion that replicators must associate with specific sites in the nucleus in order to be activated. These sites may be spaced too far apart for two closely spaced replicators to associate simultaneously and both be activated.

Other distal elements that control replicator function in yeast have not been identified. Such elements have, however, been localized to yeast telomeres (59, 63) and to an internal location at or near the *KEX2 ARS* (38). When a replicator that was normally activated early in S phase of the cell cycle was moved to a position near a telomere, it was still active but now replicated late in S phase (59, 60). Interestingly, telomeres can also regulate the expression of nearby genes (64), suggesting that the same chromatin elements may be exerting an influence on both promoters and replicators.

Two examples are also known in mammals where the same locus controls both the gene expression and replicator timing in a region. The locus control region (LCR) associated with the human β-globin locus is required for gene expression over a 50 kb region (65). In erythroid cells where the LCR is active, gene expression

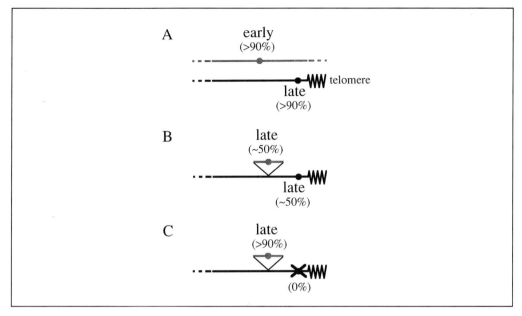

Fig. 4 Effects of chromosomal context on a yeast replicator. (a) Each replicator in *S. cerevisiae* has a character-istic time of activation in S phase. The *ARS1* replicator of choromosome IV (shown in gray) is activated early in S phase, while the *ARS501* replicator near the telomere of chromosome V (shown in black), is activated late in S phase. (b) Chromosomal context is a determinant of both the time and efficiency of replicator activation. The early activated *ARS1* replicator, transposed to a position near the ARS501 replicator, becomes late activated and suffers a loss of replicator efficiency. Interestingly, the *ARS501* replicator also suffers a loss of efficiency when the *ARS1* replicator is inserted nearby. (c) The closely spaced *ARS1* and *ARS501* replicators interfere with one another. If one of the two replicators is removed, the other becomes more active. Late activation of both replicators appears to be due to their proximity to a telomere (not shown).

occurs and a replicator in this region is active early in S phase (66). In nonerythroid cells where the LCR is inactive and in cells where the LCR is missing due to a deletion, all genes in the region are turned off and the replicator is activated late in S phase (66). The second example is the X inactivation center (*XIC*), a *cis*-acting locus which ensures that X-linked genes are expressed in equal amounts in males and females through the transcriptional inactivation of an entire X chromosome in females (67, 68). When the *XIC* locus is active, almost all the genes on that X chromosome are turned off and nearly the entire chromosome is replicated late in S phase (69, 70). These observations point to a complex regulation of the initiation of DNA replica-tion that is mediated through replicator and other chromosomal elements.

4. Detailed structure of chromosomal replicators

4.1 Chromosomal replicators are modular

The first detailed structure and functional analysis other than deletion mutagenesis was performed on the A element that contained the ARS consensus sequence (28).

Point mutations were first introduced into the conserved DNA sequences and ARS activity was measured by a plasmid stability assay. Next, specific mutations in the ACS were introduced into chromosomal replicators and their activity was measured by 2D agarose gel electrophoresis (39, 44, 49, 71). These analyses confirmed previous DNA sequences analyses that the ACS was essential (27). Because the ACS was the only apparent conserved feature of yeast replicators, it was thought that no other specific DNA sequences were required.

The detailed structure of the entire yeast replicator was subsequently determined, but first, a way was needed to circumvent the position effects that had limited the resolution of deletion analyses. Problems with position effects had been previously encountered during the characterization of some transcription promoters. One effective solution was the systematic mutagenesis of a promoter with short substitution mutations, most commonly called linker-scan mutagenesis (72). This approach also proved to work very effectively for replicators. Extensive linker-scanning mutagenesis was first performed on plasmid-borne ARSs. This assay measured the degree to which mutant plasmids are retained in a population of yeast cells after growth in conditions that did not select for the plasmid. Several generations of non-selective growth enhanced the effect of mutations that cause only small reductions in replicator efficiency in each generation. Then selected mutations that altered plasmid ARS function were introduced into the chromosome using homologous recombination to overcome the second obstacle, the technical difficulty of performing detailed mutational analyses in the chromosome.

Mutational analyses performed on *ARS1*, *ARS307*, and *ARS121*, showed that ARSs are modular (35, 73–75). In *ARS1*, which has been studied in the most detail, linker substitutions defined four short sequence elements called A, B1, B2, and B3 (Fig. 5) (73) while *ARS307* (Fig. 5) and *ARS121* have been shown to contain at least three components each (35, 74, 75). Element A of *ARS1* corresponds to the previously defined domain that is essential for ARS function (32). Domain B of *ARS1* was found to contain three important elements, B1, B2, and B3. Collectively these three sequences were also essential (73), but mutations in any one B element only had a partial effect on the efficiency of ARS function (Figure 5B). A synthetic DNA sequence that contained these four elements spaced exactly as in their native ARS sequence, but surrounded by unrelated sequence worked almost as well as the native ARS sequence indicating that A, B1, B2, and B3 are all that is necessary to confer efficient autonomous replication on the plasmid (73).

Introduction of specific mutations that had been identified in the plasmid assay into replicators in the yeast chromosome showed that the same sequences also function in the chromosome. It has been shown at a number of chromosomal replicators that the A element is essential for replicator function (39, 44, 46, 49). Deletion analysis showed that a chromosomal replicator also requires the essential B domain and that certain heterologous sequences could substitute (44). A more detailed analysis at the chromosomal *ARS1* and *ARS307* loci revealed that, as for the plasmid borne replicators, elements B1, B2, and B3 each participated in replicator function, but that only collectively were they essential (39, 75). The ARS assay was therefore validated

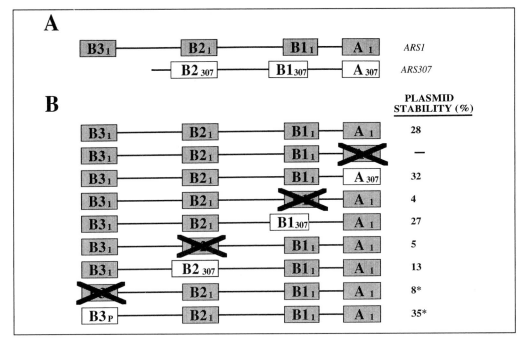

Fig. 5 Structure of yeast replicators. (a) Modular structures of the replicators *ARS1* and *ARS307* as defined by linker substitution mutations. Element B3 is a replication enhancer that has been identified at a number of replicators, but not at *ARS307*. (b) The *ARS1* replicator contains one essential sequence element (A) and three important elements (B1, B2, and B3) which are interchangeable with the corresponding elements of other replicators. In contrast, heterologous B elements are not interchangeable (not shown). This figure is modified from Rao *et al.* (74). A single asterisk indicates that the plasmid stability was obtained in a separate experiment in which the plasmid stability of the wild-type ARS1 plasmid was 36% (73). Element B3 is the binding site for the protein Abf1p. Element B3p is an Abf1p-binding site derived from an *S. cerevisiae* promoter and not from another replicator (73).

as a valuable experimental system for studying the structure and function of chromosomal replicators.

4.2 The functional elements are conserved among chromosomal replicators

Despite the initial lack of apparent DNA sequence conservation in the B domain, the four elements defined at *ARS1* are now known to be conserved functionally among yeast replicators (Fig. 5). As mentioned earlier, the core (ACS) of the A element is so highly conserved that it was identified by sequence inspection. These elements are equivalent to one another and can be exchanged between replicators without loss of function (35, 74). Surprisingly, reiterating an A element or a sequence homologous to the A element generates an ARS (34, 76). In contrast, multiple copies of any single B element fail to stimulate ARS activity when linked to a single A element, indicating that each of the B elements is functionally distinct from the others (73). It therefore appears that a single A element has an intrinsic, but low basal ori-forming activity but that this activity cannot be detected unless either multiple copies of the A element are present or the A element is associated with auxiliary or enhancing sequences.

Until recently, the status of the B1 and B2 elements as conserved ARS elements was controversial. Unlike the A element, these elements could not be identified at other replicators by sequence comparison. In contrast to element B3 (see below), elements B1 and B2 could not be identified at other replicators by their protein-binding properties. These elements were therefore defined at each replicator by their function. However, it was first shown that the *ARS1* region containing B1 and B2 could be substituted by the corresponding region from another replicator (*ARS121*) (35). Linker substitution mutagenesis at another chromosomal replicator (*ARS307*) defined two elements at positions that corresponded to the positions of the B1 and B2 elements in *ARS1* (74, 75). Interchanging elements between *ARS1* and *ARS307* showed that the elements at the B1 position are functionally equivalent to each other and that the elements at the B2 position are also functionally conserved (Fig. 5). B1 and B2 elements were, however, functionally distinct from one another (74).

The B3 element is also less conserved than the A element and was not identified by sequence inspection but by mutational analyses (36, 73). A protein called Abf1p that bound to the B3 sequence (see below) was purified before it was clear that the binding site was important for replicator function (77, 78). At least one chromosomal replicator contains two copies of this element, both of which appear to be functional (36). B3 elements can be interchanged without loss of function (35, 73) and can function at a distance and orientation independent manner (36). For these reasons, the B3 element is considered a replication enhancer analogous to transcriptional enhancers.

An unusual chromosomal replicator is located at a locus (called *HMR E*) which also functions as a transcriptional silencer in the regulation of the mating type of the yeast cell (79, 80) (see below). Elements A and B3 have been identified at this replicator, as well as an additional element called element E (81). Element E resembles a B element in that it is important, but not essential, for replicator function (49). Although element E has not been identified at other chromosomal replicators, it has been identified at two other mating type silencers which function as ARS elements in a plasmid, but neither of which functions as a chromosomal replicator (82).

5. Abf1p and Rap1p function in the initiation of DNA replication, transcriptional activation, and silencing

5.1 Identification of Abf1p and Rap1p

The characterization of important replicator sequences facilitated the identification of key replicator binding proteins, leading to a new era of unprecedented progress in understanding eukaryotic replication. Although the study of these initiation proteins is still in its infancy, it has already yielded a number of surprises and done much to reshape our view of the relationship between replication in initiation and other cellular processes.

The first proteins that bind directly to eukaryotic replicators were discovered as activities in whole cell extracts that generated gel mobility shifts with short DNA fragments that contained either the B3 or E elements. The activity that gel-shifted B3 element was purified as a single polypeptide of 87 kDa is called ARS-binding factor 1 (Abf1p) (77, 78). Yeast strains containing mutations in the *ABF1* gene lose plasmids at higher frequencies than the wild-type controls (83). The activity that gel shifted the *HMR* E sequence has also been purified as a 120 kDa single subunit protein called repressor/activator protein 1 (Rap1p) (84, 85). Cloning and sequencing of the genes encoding Abf1p and Rap1p showed the two proteins to be partially related (85, 86). Interestingly, the Abf1p-binding site (B3) at *ARS1* can be substituted by a Rap1p-binding site (E) without any loss of ARS function (73), suggesting that Abf1p and Rap1p may perform similar roles at yeast replicators (Fig. 6a).

5.2 Abf1p and Rap1p are transcriptional regulators

Rap1p also binds telomeres that are responsible for the replication and maintenance of the ends of the chromosomes (Fig. 6c) (87, 88). In contrast to conventional DNA replicators, which initiate semiconservative DNA replication, telomere replication involves a specialized polymerase called telomerase, that adds short $[C(1-3)A]_n$ repeats *de novo*, without using a DNA template (89). The repeats of yeast telomeres are typically 300–450 bp long and are thought to contain one Rap1p-binding site (for example, CACCCACACACC) every 18 bp (90). Mutations in the *RAP1* gene or overproduction of a portion of Rap1p changes telomere length (91–94) indicating that Rap1p is also involved in either the replication or maintenance of telomeres.

Abf1p- and Rap1p-binding sites have also been found to be important components of the silencers that transcriptionally repress broad regions of the chromosome (Fig. 6b) (77, 79, 82, 95). The silencers flank the mating type genes at two loci (*HML* and *HMR*) that the yeast cell does not want transcribed. Silencers are responsible for the establishment and maintenance of a chromatin structure that prevents access of *trans*-acting factors to the DNA. Each silencer is associated with ARS activity on plasmids (96, 97); however, only one of the four ARSs, called *HMR* E ARS, has detectable replicator activity in the chromosome (49, 52). Rap1p also plays an important role in a related form of transcriptional silencing that spreads into the chromosome from yeast telomeres (92–94, 98). Like silencing at *HML* and *HMR*, an inactive chromatin structure is established at the yeast telomeres and spreads several kilobases into the chromosome (99).

In addition to their roles as replication and silencing proteins, both Abf1p and Rap1p have also been found to bind upstream of a large number of yeast genes, where they function as transcriptional activators (Fig. 6d; reviewed in 100). Since many of the genes they activate are considered housekeeping genes, Abf1p and Rap1p are sometimes called 'general transcription factors'. Although Abf1p and Rap1p were the first transcription factors found to stimulate cellular replicators, transcription factors are also known to play important roles in the function of a number of viral replicators. This was first demonstrated for the eukaryotic DNA

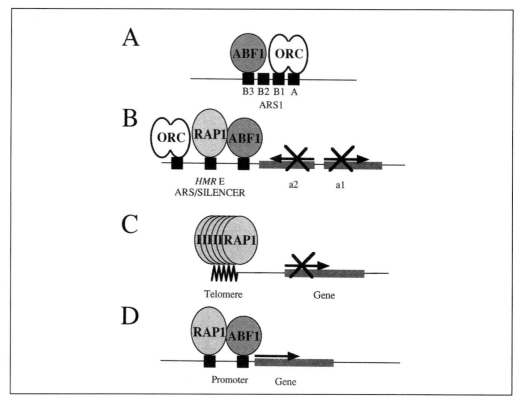

Fig. 6 Multiple roles for ORC, Abf1p, and Rap1p. (a) ORC binds to a bipartite recognition site comprising the A and B1 elements at ARS1 and Abf1p binds the B3 elements of chromosomal replicators (including the ARS1 replicator shown), stimulating replicator function. (b) At *HMR* E, ORC, Abf1p, and Rap1p binding has been implicated not only in *HMR* E replicator activation, but also in the transcriptional silencing of nearby genes (usually the a2 and a1 genes) through the establishment of an inactive chromatin domain. The ORC also binds to the silencer, but its binding site has not been shown to be bipartite. (c) Rap1P binds a subset of the repeats of each yeast telomere where the protein participates in telomere maintenance and also in the formation of an inactive chromatin domain that resembles the inactive chromatin at *HMR* E. (d) Both Abf1p and Rap1P binding the promoters of a variety of genes—sometimes to the same promoter—where both proteins function as transcriptional activators.

virus, polyomavirus, whose replicator includes a transcriptional enhancer. The enhancer is essential for replicator function and can be replaced by other enhancers (101). More recently, initiation of DNA replication in numerous other eukaryotic virus genomes was shown to be stimulated by transcriptional enhancers (see 102).

5.3 Abf1p and Rap1p may activate promoters and replicators by a common mechanism

Three lines of evidence indicate that transcriptional activators may stimulate eukaryotic virus replicators and eukaryotic transcription promoters by a similar mechanism. First, the *in vivo* activation of a mammalian viral promoter (MMTV-LTR) and the polyomavirus replicator by Gal4p-VP16 requires the VP16 activation region of this protein chimera. Mutations in the VP16 activating sequence (103–105) produce similar effects on promoter and replicator efficiency. Second, some transcription elements activate promoters and virus replicators in a distance- and orientation-independent manner (see 102). Third, under some circumstances, the stimulation *in vitro* of a eukaryotic promoter and the SV40 replicator by transcription factors requires the use of chromatin as a template (106–108).

Additional evidence suggests that transcriptional activators can stimulate yeast replicators and promoters by the same mechanism. First, the B3 element of *ARS1* was successfully replaced with a binding site for the Gal4p transcription factor (73). No loss of replicator function was observed in galactose where Gal4 is active as a transcription factor. In cells grown in glucose, however, where Gal4p is inactive as a transcription factor, stimulation of the replicator was not observed. Since Gal4p is thought to exist in the cell solely as a transcriptional regulator, the simplest explanation of these results is that the replicator is stimulated by a mechanism that normally stimulates transcriptional promoters. The *ARS1* replicator was also stimulated by the protein chimera LexA-Gal4p (73). This stimulation was dependent on the transcriptional activation domain of this protein. An attractive possibility is that the transcription activation region of some activators induces replicator function. Alternatively, transcription activation domains might facilitate the loading of DNA replication proteins to the replicator through direct protein–protein interactions (109). These two possibilities are not necessarily mutually exclusive.

ABf1p and Rap1p have each been ascribed three different biological functions which may not necessarily be performed by three distinct biochemical activities. It is an attractive possibility that the same activity is required for transcriptional activation and the stimulation of replicators while each protein utilizes a separate activity for transcriptional silencing. Indeed, several lines of evidence indicate that Rap1p performs its transcriptional activation and silencing functions through distinct domains in the protein (110). Interestingly, element A, an essential component of every replicator is thought to bind the initiator. Element A is also an important component of all four silencers at *HML* and *HMR*. A recent breakthrough in the study of eukaryotic replicators was the discovery of a protein that specifically binds the A element. This protein has been shown to play an important role in both DNA replication and transcriptional silencing.

6. The origin recognition complex

6.1 Identification of the origin recognition complex

The most recent milestone in the study of yeast DNA replication was the discovery of the origin recognition complex (ORC), a protein that is thought to be the yeast initiator protein. ORC was first detected as an activity in fractionated yeast nuclear extracts that specifically recognized the A element and protected the A and B1 elements of *ARS1* DNA from enzymatic cleavage by nuclease (Fig. 6a) (111). In addition to causing a region of protection, the interaction of ORC with *ARS1* created a series of sites of increased sensitivity to nuclease in the B domain, particularly over the B1 element. The purified activity consisted of six polypeptides of molecular weights 120, 72, 62, 56, 53, and 50 kDa. Point mutations in *ARS1* showed that ORC recognized the A element, which is essential for replicator function. Mutations in the A element that destroyed *ARS1* activity *in vivo* also eliminated ORC binding *in vitro*, whereas mutations in the A element that only reduced *ARS1* efficiency, similarly reduced ORC binding (111). The binding of yeast replicators by ORC appears to be a general property, as ORC forms a similar pattern of protection over every ARS that has been tested.

Three additional lines of evidence suggest that ORC recognizes *ARS1 in vivo*. First, nucleotide resolution genomic footprinting generated a protection pattern at *ARS1* that resembled the footprint of purified ORC (111, 112). Nucleotide resolution genomic footprinting examines the protection pattern seen at DNA sequences immediately after cells are lysed. The protection pattern seen is therefore thought to reflect protein–DNA interactions that also occur in the intact cell. Second, the *ORC2* gene encoding the second-largest subunit of ORC (called Orc2p) was recovered in a genetic screen for activities that interact with the A element *in vivo* (113). This screen made use of the observation that the A element blocks enhancer function when placed between an enhancer and a promoter. Mutant strains defective in this transcriptional block were recovered and mutations that allowed transcription were found to have defects in *ORC2*. Third, the *ORC6* gene encoding the smallest subunit of ORC (Orc6p) was recovered in a variation of the two-hybrid screen that selects protein chimeras that recognize the ACS of the *ARS1* A element *in vivo* (114). In this assay, a library of hybrid genes that contain protein-coding sequences fused to the Gal4p activation domain is expressed in cells that contain multiple copies of the ACS upstream of a *lacZ* reporter gene. A similar genetic screen has also been used to identify the Dbf4p as a protein that interacts indirectly with replicators (see below) (115).

More recently, it has been shown that ORC also recognizes the B1 element of yeast replicators (116, 117). Use of altered DNA binding conditions *in vitro* have revealed that ORC has a bipartite DNA recognition site that includes the A and B1 elements of the replicator. In this respect, ORC resembles the simian virus 40 (SV40) T antigen initiator protein (see Chapter 2). T antigen binds to a specific sequence (site II) and also binds a second inverted repeat (IR) sequence within the SV40 replicator (118). T antigen binding causes a structural distortion of the IR sequence in an ATP-dependent manner. Similarly, the dnaA initiator protein also binds two unrelated sequences (9-mers and 13-mers) at the *E. coli* replicator oriC (119). The 13-mers represent the site where oriC duplex opening first occurs (42). One role for ORC in the initiation of DNA replication may be to promote unwinding of the DNA, but this has not yet been demonstrated.

6.2 The role of ORC in DNA replication

Although it is widely assumed that ORC is the initiator protein for DNA replication in yeast, this remains to be proven. Several lines of evidence, however, strongly suggest that ORC is the key initiator protein. Yeast cells carrying a mutation in the *ORC2* gene arrest with a single large bud at the nonpermissive temperature, with the nucleus at the junction of the mother and daughter cells (62). This phenotype is also seen with other temperature-sensitive DNA replication mutants in *S. cerevisiae* (31). At the permissive temperature, yeast strains carrying mutant *orc2* mutant alleles were also found to lose the 2-μ plasmid and plasmids containing *ARS1* at greatly elevated frequencies (62, 113). In addition, fluorescent cytometric analysis of a temperature-sensitive *orc2* strain indicated that *orc2* gene product is required in S phase (61). A study in which the Orc6p subunit is overexpressed has also implicated ORC in DNA replication (114). The idea behind these experiments is that if ORC is involved in DNA replication, it is likely to interact with other replication proteins and overexpression of the protein or one of its subunits may therefore be disruptive to the DNA replication process. Indeed, overexpression of Orc6p was found to exacerbate the phenotypes of *cdc6* and *cdc7* mutant yeast cells, both of which are defective for entry into S phase (see below).

Interestingly, the binding of ORC to yeast replicators *in vitro* is seen only in the presence of ATP. ATP binds and plays an important role in the function of several viral and bacterial initiator proteins (26). One of these initiator proteins, the SV40 Large T antigen, has its DNA-binding properties affected by ATP (120–122). The *E. coli* DnaA protein, on the other hand, binds to oriC only when complexed with ATP, and hydrolysis of ATP to ADP prevents formation of an active initiation complex and reinitiation of DNA replication (123, 124).

6.3   The role of ORC in transcriptional silencing

ORC was found to be required for transcriptional silencing. The gene encoding the second largest ORC subunit, *ORC2*, was recovered in two genetic screens for mutant alleles that disrupted the function of silencers (62). In an independent study, a mutant *orc2* allele was recovered in a screen for proteins that bound to the *HMR* E silencer ACS and the mutant was shown to be defective in silencing (113). These studies are consistent with the observation that ORC binds to all four transcriptional silencers in yeast (Fig. 6b) (61).

It should be noted that establishment of transcriptional silencing requires passage through S phase of the cell cycle (82, 125). What then is the relationship between replicators and transcriptional silencers? Since the initiation of DNA replication is a fundamental process for all living things and silencing is not essential for yeast, it can be argued that the role of ORC in initiation of DNA replication probably represents its primary role. Silencing might be established through the activation of specialized replicators. Alternatively, the initiation of DNA replication and transcriptional silencing may be separable processes that utilize some of the same control elements.

Three lines of evidence argue that silencing does not require DNA replication. First, *HML* I and *HML* E are silencers without detectable replicator activity (52). Second, yeast telomeres can silence genes even after telomeric replicators are removed (58,

99, 126). Third, the *HMR* E locus can function as a transcriptional silencer in the absence of all defined replicator elements, including the A element, provided multiple copies of a protein involved in silencing, Sir1p, are tethered to the locus (127). This experiment was done by replacing the A, E, and B3 elements of *HMR* E with three Gal4p sites and expressing a Gal4p/Sir1p protein chimera in the cell. But alternative explanations can be issued for all three lines of evidence. *HML* I and *HML* E may be infrequently used replicators. Telomeres may contain replicator activity that remains undetected. Finally, additional elements remain to be defined at *HMR* E and these may contain ARS activity (81).

The alternative hypothesis, that replicator activation is essential for establishing transcriptional silencing, predicts that all proteins required for the initiation of DNA replication will also be required for transcriptional silencing. Cdc7p, a cell cycle-regulated kinase, has been implicated in both (128), but mutations in genes encoding other DNA replication proteins have not been recovered in genetic screens for silencing proteins. Additional studies are required to determine whether the other proteins thought to be involved in DNA replication are also involved in transcriptional silencing.

How might replicator activation participate in the formation of a particular chromatin structure? Replicator timing may be important. The mating-type silencers and telomeres replicate very late in S phase (129). Late activation may induce formation of inactive chromosome structures. One attractive hypothesis is that outward spread of a specific chromatin structure is linked to outward movement of the late activated replication forks. This hypothesis predicts that proteins required for replication fork movement, but not for replicator activation, are similarly required for the establishment of silent domains.

7. Other initiation proteins

7.1 Localization of a protein kinase subunit to replicators

Although no other proteins that have a clear role in DNA replication initiation have been identified that directly bind to yeast replicators (for example, to the B1 or B2 elements), it has recently been shown that a regulator of a cell cycle-regulated kinase is targeted to replicators. The kinase regulator, called Dbf4p, was discovered using a variation of the two-hybrid genetic screen that resembles the screen used to isolate the gene encoding the smallest subunit of the origin recognition complex, Orc6p (115). The screen that recovered the *DBF4* gene selected for proteins that recognize intact *ARS1 in vivo*. A library of hybrid genes that contain protein coding sequences fused to the Gal4p activation domain was expressed in cells that contain a reporter gene with the intact copy of *ARS1* in the place of an upstream activation sequence. The association of Dbf4p with a replicator appears to be meaningful as Dbf4p was localized to every replicator tested and a strong correlation was seen in the effect of mutations on replicator function and Dbf4p/Gal4p localization (115).

Dbf4p binds to and activates the Cdc7p protein kinase (reviewed in 130 and

Chapter 7) and therefore may localize Cdc7p activity to replication origins. Although the role of Cdc7p in DNA replication is not known, studies of temperature-sensitive *cdc7* mutants show that the protein is required for passage into S phase (131). Interestingly, overexpression of Cdc7p protein derepresses a silencer and mutant *cdc7* genes restore the silencing of some defective silencers (128), suggesting that this protein, like Abf1p, Rap1p, and ORC, plays a role in both DNA replication and silencing. A dominant mutation in an unknown gene called *BOB1* bypasses the need for both Cdc7p and Dbf4p in the cell (132). Bob1p might therefore represent a mutation in a key substrate of the Cdc7p/Dbf4p kinase. Identification of the *BOB1* gene will certainly generate insight into the role of the kinase in DNA replication and silencing.

7.2 Genomic footprinting at replicators

Recent genomic footprinting studies suggest that additional protein(s) bind yeast replicators (133). These studies elaborate on earlier observations, performed at lower resolution, that the chromatin structures at the chromosomal *ARS1* and *histone H4 ARS* loci change during the cell cycle (134). The more detailed footprinting studies examined protein complexes on plasmid-borne replicators. A footprint corresponding to the footprint obtained with purified ORC *in vitro* was present throughout most of the cell cycle (Fig. 7, postreplication state), including an obvious nuclease hypersensitive site in the B1 element (111, 112). This result suggests that ORC binds replicators soon after they have duplicated and remains bound throughout cell division until another round of initiation of DNA replication. Interestingly, the extent of nuclease protection over the B1 element changed during the cell cycle (Fig. 7, prereplication state). At the end of mitosis, a 'prereplicative complex' seems to be established and persists until the onset of the next S phase (133). These results

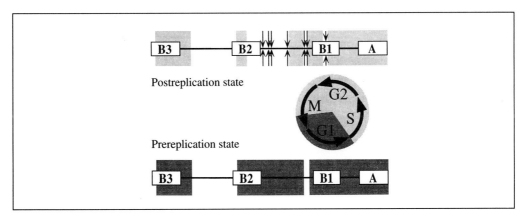

Fig. 7 The ARS1 replicator exists in distinct chromatin states before (late mitosis and G1 phases of the cell cycle; bottom) and after (S, G2 and early mitosis; top) replicator activation. Shaded boxes mark regions of the replicator protected from DNase I digestion by chromatin. Arrows indicate sites of enhanced DNase I cleavage. Adapted from Fig. 5D in Diffley *et al.* (133).

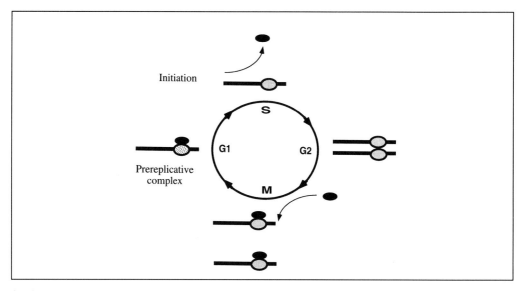

Fig. 8 Initiation of DNA replication and the cell cycle. A hypothetical model for the control of initiation of DNA replication during the eukaryotic cell cycle. The initiator protein ORC (gray ellipse) is associated with the DNA at all stages of the cell cycle. A protein or protein complex (black ellipse) associates with ORC during mitosis (133) and remains attached to the DNA during the G1 phase of the next cell cycle. At the G1 to S phase transition, the prereplicative complex becomes activated, either by removal of proteins (141) or by modification of ORC. This allows the initiation of DNA replication. If the prereplicative complex is destroyed at the G1 to S phase transition and is reset at mitosis, this could ensure one round of DNA replication per cell cycle.

suggest that at mitosis, changes occur at the replicator that facilitate initiation in the following S phase (Fig. 8). Although these changes may represent modifications that alter the DNA-binding properties of ORC, a more likely explanation is that additional proteins are bound to the replicator in G1.

The identity of the proteins that extend the genomic footprint at replicators is a matter of speculation. One possibility is that as yet unidentified proteins that recognize the B1 and/or B2 elements become associated with the replicator late in mitosis. Alternatively, the extended footprint could be caused by protein(s) that, like Dbf4p, associate with the replicator by protein–protein interactions. Attractive candidates for the latter scenario are a family of proteins that belong to the MCM family, Mcm2p, Mcm3p, and Mcm5p/Cdc46p. These proteins do not appear to bind DNA, but have been implicated in the initiation of DNA replication (135–138). All three genes were identified as mini chromosome maintenance (*mcm*) alleles that cause ARS plasmids to be lost at elevated frequencies (139). Consistent with the timing of formation of the 'prereplication complex', Mcm2p, Mcm3p, and the Mcm5p/Cdc46p all enter the nucleus late in mitosis and disappear from the nucleus either late in G1 or early in S phase (135, 136, 138). In addition, overexpression of one of the ORC subunits, Orc6p in *cdc46* mutant strains exacerbates the *cdc46/mcm5* phenotype, indicating that Mcm5p/Cdc46p and ORC might function in the initiation of DNA replication with one another (114).

7.3 A role for Cdc6p in replicator function

Cdc6p also appears to be involved in the initiation of DNA replication and may be one of the components of the prereplication complex. *cdc6* temperature-sensitive alleles arrest cell cycle progression late in G1, indicating that Cdc6p may be involved in DNA replication (131). Furthermore, the prereplicative complex appears to be absent in *cdc6* strains arrested at nonpermissive temperature (133). Finally, two lines of genetic evidence suggest that Cdc6p interacts with ORC. First, overexpression of Orc6p in *cdc6* mutant strains exacerbates the *cdc6* phenotype (114). Second, overexpression of Cdc6p suppresses a temperature-sensitive allele (*orc5-1*) of the gene encoding the fifth largest subunit of ORC, Orc5p (140). Results of 2D gel analyses show that like ORC, Cdc6p is also required for the initiation of DNA replication at each replicator in the chromosome (140). This suggests that Cdc6p cooperates with ORC to determine the frequency of initiation of DNA replication in the yeast genome. Moreover, it is possible that Cdc6p may be one of the ORC-interacting proteins represented as the black ellipse in Fig. 8. Ultimately, the proteins that bind to the replicator must cooperate to attract a DNA helicase and DNA primase to the ori so that the template DNA can unwind, DNA synthesis can be primed, and replication of DNA can occur (see 141).

8. Conclusion

Studies in yeast, with its powerful genetic system and reliable sequence-specific plasmid replication assay, have led attempts to understand the initiation of chromosomal DNA replication in eukaryotes. The identification of replicators in yeast has allowed the delineation of individual DNA elements that control the initiation of DNA replication, which in turn has led to the identification of important initiation proteins that bind some of these elements. Other important proteins have been discovered through protein–protein interactions and genetic screens for replication mutants. Yet more proteins important for replicator function will no doubt be discovered in the not too distant future.

The rapid strides made by studying yeast should facilitate the study of DNA replication in multicellular organisms. The 2D gel assays, developed to study oris in yeast, are proving to be very useful in the other systems (see Chapters 4 and 8). Insights obtained from the study of yeast artificial chromosomes will also prove to be of value in the development of mammalian artificial chromosomes. The cloning of genes for yeast replication proteins will allow the identification of homologues in multicellular organisms, including mammals. Fears that budding yeast, as a fast growing unicellular microorganism, may catalyze and control the replication of its genome in a fundamentally different manner from higher eukaryotes subsided somewhat as an increasing body of evidence in cell cycle and DNA replication fork studies suggests that yeast resembles higher eukaryotes in these fundamental processes. Mammalian replicator binding proteins such as mammalian ORC should aid our search for mammalian replicators. It would be interesting to determine whether

ORC homologues in *Drosophila* and mammals are involved in phenomena that resemble transcriptional silencing in yeast. Candidate phenomena include position effect variegation in *Drosophila* and gene imprinting and X-inactivation in mammals.

The research so far has provided insight and reagents that should facilitate a biochemical approach that allows us to delve into the molecular mechanisms of the initiation process. The development of cell-free DNA replication from both oriC from *E. coli* and the SV40 ori (see Chapters 1 and 2) led to a number of important advances, including the discovery of many new enzymatic functions, the identification of many mammalian proteins that are involved in SV40 DNA replication and probably also cell chromosomal replication, the delineation of the replication initiation process into multiple stages, insights into how eukaryotic genomes are replicated and repaired by multiple polymerases, and insights into how the DNA replication is controlled by the cell cycle machinery and damage control. The challenge now is to obtain a similar understanding of the initiation of eukaryotic chromosomal DNA replication.

References

1. Watson, J. D. and Crick, F. H. C. (1953) Genetical implications of the structure of deoxyribonucleic acid. *Nature*, **171**, 964.
2. Kornberg, A. (1989) Never a dull enzyme. *Annu. Rev. Biochem.*, **58**, 1.
3. Jacob, F. (1960) Genetic control of viral functions. *Harvey Lecture*, **54**, 1.
4. Morse, M. L., Lederberg, E. M., and Lederberg, J. (1956) Transduction of *E. coli*. *Genetics*, **41**, 142.
5. Jacob, F. and Monod, J. (1961) Genetic regulatory mechanisms in the synthesis of proteins. *J. Mol. Biol.*, **3**, 318.
6. Jacob, F., Brenner, S., and Cuzin, F. (1963) On the regulation of DNA replication in bacteria. *Cold Spring Harbor Symp. Quant. Biol.*, **28**, 329.
7. Stillman, B. (1993) Replicator renaissance. *Nature*, **366**, 506.
8. Wolfson, J., Dressler, D., and Magazin, M. (1972) Bacteriophage T7 DNA replication: a linear replicating intermediate. *Proc. Natl. Acad. Sci. USA*, **69**, 499.
9. Inman, R. B. (1966) A denaturation map of the lambda phage DNA molecule determined by electron microscopy. *J. Mol. Biol.*, **18**, 464.
10. Masters, M. and Broda, P. (1971) Evidence for the bidirectional replication of the *Escherichia coli* chromosome. *Nature New Biol.*, **232**, 137.
11. Bird, R. E., Louarn, J. M., Martuscelli, J., and Caro, L. (1972) Origin and sequence of chromosome replication in *Escherichia coli*. *J. Mol. Biol.*, **70**, 549.
12. Hiraga, S. (1976) Novel F prime factors able to replicate in *Escherichia coli* Hfr strains. *Proc. Natl. Acad. Sci. USA*, **73**, 198.
13. Rowley, A., Dowell, S. J., and Diffley, J. F. X. (1994) Recent developments in the initiation of chromosomal DNA replication: a complex picture emerges. *Biochim. Biophys. Acta*, **1217**, 239.
14. Hsiao, C.-L. and Carbon, J. (1979) High-frequency transformation of yeast by plasmids containing the cloned yeast ARG4 gene. *Proc. Natl. Acad. Sci. USA*, **76**, 3829.
15. Stinchcomb, D. T., Struhl, K., and Davis, R. W. (1979) Isolation and characterisation of a yeast chromosomal replicator. *Nature*, **282**, 39.

16. Clarke, L. and Carbon, J. (1980) Isolation of a yeast centromere and construction of functional small circular chromosomes. *Nature,* **287,** 504.

17. Dani, G. M. and Zakian, V. A. (1983) Mitotic and meiotic stability of linear plasmids in yeast. *Proc. Natl. Acad. Sci. USA,* **80,** 3406.

18. Murray, A. W. and Szostak, J. W. (1983) Construction of artificial chromosomes in yeast. *Nature,* **305,** 189.

19. Murray, A. W., Schultes, N. P., and Szostak, J. W. (1986) Chromosome length controls mitotic chromosome segregation in yeast. *Cell,* **45,** 529.

20. Huberman, J. A., Spotilla, L. D., Nawotka, K. A., El-Assouli, S. M., and Davis, L. R. (1987) The in vivo replication origin of the yeast 2µm plasmid. *Cell,* **51,** 473.

21. Brewer, B. J. and Fangman, W. L. (1987) The localization of replication origins on ARS plasmids in *S. cerevisiae. Cell,* **51,** 463.

22. Orr-Weaver, T. L. (1991) *Drosophila* chorion genes: crackling the eggshell's secrets. *BioEssays,* **13,** 97.

23. Hamlin, J. L. (1992) Mammalian origins of replication. *BioEssays,* **14,** 651.

24. Liang, C., Spitzer, J., D., Smith, H., S., and Gerbi, S. A. (1993) Replication initiates at a confined region during DNA amplification in *Sciara* DNA puff II/9A. *Genes Dev.,* **7,** 1072.

25. Bullock, P. A., Tevosian, S., Jones, C., and Denis, D. (1994) Mapping initiation sites for Simian Virus 40 DNA synthesis events in vitro. *Mol. Cell. Biol.,* **14,** 5043.

26. Kornberg, A. and Baker, T. A. (1992) *DNA replication,* 2nd ed. W. H. Freeman, New York.

27. Broach, J., Li, Y.-Y., Feldman, J., Jayaram, M., Abraham, J., Nasmyth, K. A., and Hicks, J. B. (1983) Localization and sequence analysis of yeast origins of DNA replication. *Cold Spring Harbor Symp. Quant. Biol.,* **47,** 1165.

28. Van Houten, J. V. and Newlon, C. S. (1990) Mutational analysis of the consensus sequence of a replication origin from yeast chromosome III. *Mol. Cell. Biol.,* **10,** 3917.

29. Newlon, C. S. and Theis, J. F. (1993) The structure and function of yeast ARS elements. *Curr. Opin. Genet. Dev.,* **3,** 752.

30. Newlon, C. S. (1988) Yeast chromosome replication and segregation. *Microbiol. Rev.,* **52,** 568.

31. Campbell, J. L. and Newlon, C. S. (1991) Chromosomal DNA replication. In *The molecular and cellular biology of the yeast Saccharomyces: genome dynamics, protein synthesis, and energetics.* Broach, J. R., Pringle, J. R., and Jones, E. W. (eds.), Cold Spring Harbor Laboratory Press, Cold Spring Harbor, p. 41.

32. Celniker, S. E., Sweder, K., Srienc, F., Bailey, J. E., and Campbell, J. L. (1984) Deletion mutations affecting autonomously replicating sequence ARS1 of *Saccharomyces cerevisiae. Mol. Cell. Biol.,* **4,** 2455.

33. Bouton, A. H. and Smith, M. M. (1986) Fine-structure analysis of the DNA sequence requirements for autonomous replication of *Saccharomyces cerevisiae* plasmids. *Mol. Cell. Biol.,* **6,** 2354.

34. Palzkill, T. G. and Newlon, C. S. (1988) A yeast replication origin consists of multiple copies of a small conserved sequence. *Cell,* **37,** 299.

35. Walker, S. S., Malik, A. K., and Eisenberg, S. (1991) Analysis of the interactions of functional domains of a nuclear origin of replication from *Saccharomyces cerevisiae. Nucleic Acids Res.,* **19,** 6255.

36. Walker, S. C., Francesconi, S. C., and Eisenberg, S. (1990) A DNA replication enhancer in *Saccharomyces cerevisiae. Proc. Natl. Acad. Sci. USA,* **87,** 4665.

37. Brewer, B. and Fangman, W. L. (1994) Initiation preference at a yeast origin of replication. *Proc. Natl. Acad. Sci. USA*, **91**, 3418.

38. Brewer, B. J., Diller, J. D., Friedman, K. L., Kolor, K. M., Raghuraman, M. K., and Fangman, W. L. (1993) The topography of chromosome replication in yeast. *Cold Spring Harbor Symp. Quant. Biol.*, **58**, 425.

39. Marahrens, Y. and Stillman, B. (1994) Replicator dominance in a eukaryotic chromosome. *EMBO J.*, **13**, 3395.

40. Simpson, R. T. (1990) Nucleosome positioning can affect the function of a *cis*-acting DNA element. *Nature*, **343**, 387.

41. Kowalski, D. and Eddy, M. J. (1989) The DNA unwinding element: a novel, *cis*-acting component that facilitates opening the *Escherichia coli* replication origin. *EMBO J.*, **8**, 4335.

42. Bramhill, D. and Kornberg, A. (1988) Duplex opening by dnaA protein at novel sequences in initiation of replication at the origin of the *E. coli* chromosome. *Cell*, **52**, 743.

43. Natale, D. A., Umek, R. M., and Kowalski, D. (1993) Ease of DNA unwinding is a conserved property of yeast replication origins. *Nucleic Acids Res.*, **21**, 555.

44. Huang, R.-Y. and Kowalski, D. (1993) A DNA unwinding element and an ARS consensus comprise a replication origin within a yeast chromosome. *EMBO J.*, **12**, 4521.

45. Shirahige, K., Iwasaki, T., Rashid, M. B., Ogasawara, N., and Yoshikawa, H. (1993) Location and characterization of replicating sequences from chromosome VI of *Saccharomyces cerevisiae*. *Mol. Cell. Biol.*, **13**, 5043.

46. Deshpande, A. and Newlon, C. (1992) The *ARS* consensus sequence is required for chromosomal origin function in *Saccharomyces cerevisiae*. *Mol. Cell. Biol.*, **12**, 4305.

47. Huberman, J. A., Zhu, J., Davis, L. R., and Newlon, C. S. (1988) Close association of a DNA replication origin and an ARS element on chromosome III of the yeast, *Saccharomyces cerevisiae*. *Nucleic Acids Res.*, **16**, 6373.

48. Linskens, M. H. K. and Huberman, J. A. (1988) Organization of replication of ribosomal DNA in *Saccharomyces cerevisiae*. *Mol. Cell. Biol.*, **8**, 4927.

49. Rivier, D. and Rine, J. (1992) An origin of DNA replication and a transcription silencer require a common element. *Science*, **256**, 659.

50. Zhu, J., Newlon, C. S., and Huberman, J. A. (1992) Localization of a DNA replication origin and termination zone on chromosome III of *Saccharomyces cerevisiae*. *Mol. Cell. Biol.*, **12**, 4733.

51. Greenfeder, S. and Newlon, C. (1992) A replication map of a 61-kb circular derivative of *Saccharomyces cerevisiae* chromosome *III*. *Mol. Biol. Cell*, **3**, 999.

52. Dubey, D., Davis, L. R., Greenfeder, S. A., Ong, L. Y., Zhu, J., Broach, J. R., Newlon, C. S., and Huberman, J. A. (1991) Evidence suggesting that the *ARS* elements associated with silencers of the yeast mating-type locus *HML* do not function as chromosomal DNA replication origins. *Mol. Cell. Biol.*, **11**, 5346.

53. Heck, M. M. S. and Spradling, A. C. (1990) Multiple replication origins are used during *Drosophila* chorion gene amplification. *J. Cell. Biol.*, **110**, 903.

54. Blumenthal, A. B., Kriegstein, H. J., and Hogness, D. S. (1973) The units of DNA replication in *Drosophila melanogaster* chromosomes. *Cold Spring Harbor Symp. Quant. Biol.*, **38**, 205.

55. McKnight, S. L. and Miller, O. L. J. (1977) Electron microscopic analysis of chromatin replication in the cellular blastoderm *Drosophila melanogaster* embryo. *Cell*, **12**, 795.

56. Collins, I. and Newlon, C. S. (1994) Chromosomal DNA replication initiates at the same origins in meiosis and mitosis. *Mol. Cell. Biol.*, **14**, 3524.

57. Newlon, C. S., Collins, I., Dershowitz, A., Deshpande, A. M., Greenfeder, S. A., Ong, L. Y., and Theis, J. F. (1993) Analysis of replication origin function on chromosome III of *Saccharomyces cerevisiae. Cold Spring Harbor Symp. Quant. Biol.*, **58**, 415.

58. Ferguson, B. M., Brewer, B. J., Reynolds, A. E., and Fangman, W. L. (1991) A yeast origin of replication is activated late in S phase. *Cell*, **65**, 507.

59. Ferguson, B. M. and Fangman, W. L. (1992) A position effect on the time of replication origin activation in yeast. *Cell*, **68**, 333.

60. Brewer, B. J. and Fangman, W. L. (1993) Initiation at closely spaced replication origins in a yeast chromosome. *Science*, **262**, 1728.

61. Bell, S. P., Kobayashi, R., and Stillman, B. (1993) Yeast origin recognition complex functions in transcription silencing and DNA replication. *Science*, **262**, 1844.

62. Foss, M., McNally, F. J., Laurenson, P., and Rine, J. (1993) Origin recognition complex (ORC) in transcriptional silencing and DNA replication in *S. cerevisiae. Science*, **262**, 1838.

63. Fangman, W. L. and Brewer, B. J. (1992) A question of time: replication origins of eukaryotic chromosomes. *Cell*, **71**, 363.

64. Gottschling, D. E., Aparicio, O. M., Billington, B. L., and Zakian, V. A. (1990) Position effect at *S. cerevisiae* telomeres: reversible repression of Pol II transcription. *Cell*, **63**, 751.

65. Karlsson, S. and Nienhuis, A. W. (1985) Developmental regulation of human β-globin genes. *Annu. Rev. Biochem.*, **54**, 1071.

66. Kitsberg, D., Selig, S., Keshet, I., and Cedar, H. (1993) Replication structure of the human β-globin gene domain. *Nature*, **366**, 588.

67. Lyon, M. F. (1993) Controlling the X chromosome. *Curr. Biol.*, **3**, 242.

68. Migeon, B. R. (1994) X-chromosome inactivation: molecular mechanisms and genetic consequences. *Trends Genet.*, **10**, 230.

69. Boggs, B. A. and Chinault, A. C. (1994) Analysis of replication timing properties of human X-chromosomal loci by fluorescence *in situ* hybridization. *Proc. Natl. Acad. Sci. USA*, **91**, 6083.

70. Torchia, B. S., Call, L. M., and Migeon, B. R. (1994) DNA replication analysis of FMR1, XIST, and Factor 8C loci by FISH shows nontranscribed X-linked genes replicate late. *Am. J. Hum. Genet.*, **55**, 96.

71. Dershowitz, A. and Newlon, C. S. (1993) The effect on chromosome stability of deleting replication origins. *Mol. Cell. Biol.*, **13**, 391.

72. McKnight, S. L. and Kingsbury, R. (1982) Transcriptional control signals of a eukaryotic protein-coding gene. *Science*, **217**, 316.

73. Marahrens, Y. and Stillman, B. (1992) A yeast chromosomal origin of DNA replication defined by multiple functional elements. *Science*, **255**, 817.

74. Rao, H., Marahrens, Y., and Stillman, B. (1994) Functional conservation of multiple elements in yeast chromosomal replicators. *Mol. Cell. Biol.*, **14**, 7643.

75. Theis, J. F. and Newlon, C. S. (1994) Domain B of *ARS307* contains two functional elements and contributes to chromosomal replication origin function. *Mol. Cell. Biol.*, **14**, 7652.

76. Zweifel, S. G. and Fangman, W. L. (1990) Creation of ARS activity in yeast through iteration of non-functional sequences. *Yeast*, **6**, 179.

77. Diffley, J. F. X. and Stillman, B. (1988) Purification of a yeast protein that binds to origins of DNA replication and a transcriptional silencer. *Proc. Natl. Acad. Sci. USA*, **85**, 2120.

78. Sweder, K. S., Rhode, P. R., and Campbell, J. L. (1988) Purification and characterization of proteins that bind to yeast ARSs. *J. Biol. Chem.*, **263**, 17 270.

79. Brand, A. H., Breeden, L., Abraham, J., Sternglanz, R., and Nasmyth, K. (1985) Characterization of a 'silencer' in yeast: a DNA sequence with properties opposite to those of a transcriptional enhancer. *Cell*, **41**, 41.
80. Rivier, D. H. and Pillus, L. (1994) Silencing speaks up. *Cell*, **76**, 963.
81. Brand, A. H., Micklen, G., and Nasmyth, K. (1987) A yeast silencer contains sequences that can promote autonomous plasmid replication and transcriptional activation. *Cell*, **51**, 709.
82. Laurenson, P. and Rine, J. (1992) Silencers, silencing, and heritable transcriptional states. *Microbiol. Rev.*, **56**, 543.
83. Rhode, P. R., Elsasser, S., and Campbell, J. L. (1992) Role of multifunctional autonomously replicating sequence binding factor 1 in the initiation of DNA replication and transcriptional control in *Saccharomyces cerevisiae*. *Mol. Cell. Biol.*, **12**, 1064.
84. Shore, D., Stillman, D. J., Brand, A. H., and Nasmyth, K. A. (1987) Identification of silencer binding proteins from yeast: possible roles in SIR control and DNA replication. *EMBO J.*, **6**, 461.
85. Shore, D. and Nasmyth, K. (1987) Purification and cloning of a DNA binding protein from yeast that binds to both silencer and activator elements. *Cell*, **51**, 721.
86. Diffley, J. F. X. and Stillman, B. (1989) Similarity between the transcriptional silencer binding proteins ABF1 and RAP1. *Science*, **246**, 1034.
87. Berman, J., Tachibana, C. Y., and Tye, B.-K. (1986) Identification of a telomere-binding activity from yeast. *Proc. Natl. Acad. Sci. USA*, **83**, 3713.
88. Shore, D. (1994) RAP1: a protein regulator in yeast. *Trends Genet.*, **10**, 408.
89. Blackburn, E. H. (1993) Telomerases. In *Reverse transcriptase*. Skalka, A. M. and Goff, S. P. (eds.), Cold Spring Harbor Laboratory Press, Cold Spring Harbor, p. 411.
90. Gilson, E., Roberge, M., Giraldo, R., Rhodes, D., and Gasser, S. M. (1993) Distortion of the DNA double helix by RAP1 at silencers and multiple telomeric binding sites. *J. Mol. Biol.*, **231**, 293.
91. Lustig, A. J., Kurtz, S., and Shore, D. (1990) Involvement of the silencer and UAS binding protein RAP1 in regulation of telomere length. *Science*, **250**, 549.
92. Kyrion, G., Boakye, K. A., and Lustig, A. J. (1992) C-terminal truncation of RAP1 results in the deregulation of telomere size, stability, and function in *Saccharomyces cerevisiae*. *Mol. Cell. Biol.*, **12**, 5159.
93. Hardy, C. F. J., Balderes, D., and Shore, D. (1992) Dissection of a carboxy-terminal region of the yeast regulatory protein RAP1 with effects on both transcriptional activation and silencing. *Mol. Cell. Biol.*, **12**, 1209.
94. Conrad, M. N., Wright, J. H., Wolf, A. J., and Zakian, V. A. (1990) RAP1 protein interacts with yeast telomeres *in vivo*: overproduction alters telomere structure and decreases chromosome stability. *Cell*, **63**, 739.
95. Buchman, A. R., Kimmerly, W. J., Rine, J., and Kornberg, R. (1988) Two DNA-binding factors recognize specific sequences at silencers, upstream activating sequences, and telomeres in *Saccharomyces cerevisiae*. *Mol. Cell. Biol.*, **8**, 210.
96. Abraham, J., Nasmyth, K. A., Strathern, J. N., Klar, A. J. S., and Hicks, J. B. (1984) Regulation of mating-type information in yeast: negative control requiring sequences both 5' and 3' to the regulated region. *J. Mol. Biol.*, **176**, 307.
97. Feldman, J. B., Hicks, J. B., and Broach, J. R. (1984) Identification of sites required for repression of a silent mating type locus in yeast. *J. Mol. Biol.*, **178**, 815.
98. Kurtz, S. and Shore, D. (1991) RAP1 protein activates and silences transcription of mating-type genes in yeast. *Genes Dev.*, **5**, 616.

99. Renauld, H., Aparicio, O. M., Zierath, P. D., Billington, B. L., Chhablani, S. K., and Gottschling, D. E. (1993) Silent domains are assembled continuously from the telomere and are defined by promotor distance and strength, and by SIR3 dosage. *Genes Dev.*, **7,** 1133.

100. Diffley, J. F. X. (1992) Global regulators of chromosome function in yeast. *Antonie Van Leeuwenhoek*, **62,** 25.

101. de Villiers, J., Schaffner, W., Tyndall, C., Lupton, S., and Kamen, R. (1984) Polyoma virus DNA replication requires an enhancer. *Nature*, **312,** 242.

102. DePamphilis, M. L. (1993) How transcription factors regulate origins of DNA replication in eukaryotic cells. *Trends Cell Biol.*, **3,** 1161.

103. Baru, M., Schissel, M., and Manor, H. (1991) The yeast GAL4 protein transactivates the polyomavirus origin of DNA replication in mouse cells. *J. Virol.*, **65,** 3496.

104. Bennet-Cook, E. R. and Hassell, J. A. (1991) Activation of polyomavirus DNA replication by yeast GAL4 is dependant on its transcriptional activation domains. *EMBO J.*, **10,** 959.

105. Sadowski, I., Ma, J., Triezenberg, S., and Ptashne, M. (1988) GAL4-VP16 is an unusually potent transcriptional activator. *Nature*, **335,** 563.

106. Cheng, L. and Kelly, T. J. (1989) Transcriptional activator nuclear factor 1 stimulates the replication of SV40 minichromosomes *in vivo* and *in vitro*. *Cell*, **59,** 541.

107. Workman, J. L., Abmayer, S. M., Cromlish, W. A., and Roeder, R. G. (1988) Transcriptional regulation by the immediate early protein of pseudorabies virus during in vitro nucleosome assembly. *Cell*, **55,** 211.

108. Workman, J. L., Taylor, I. C. A., and Kingston, R. E. (1991) Activation domains of stably bound GAL4 derivatives alleviate repression of promoters by nucleosomes. *Cell*, **64,** 533.

109. Mohr, I. J., Clark, R., Sun, S., Androphy, E. J., MacPherson, P., and Botchan, M. R. (1990) Targeting the E1 replication protein to the papillomavirus origin of replication by complex formation with the E2 transactivator. *Science*, **250,** 1694.

110. Sussel, L. and Shore, D. (1991) Separation of transcriptional activation and silencing functions of the RAP1-encoded repressor/activator protein 1: isolation of viable mutants affecting both silencing and telomere length. *Proc. Natl. Acad. Sci. USA*, **88,** 7749.

111. Bell, S. P. and Stillman, B. (1992) ATP-dependent recognition of eukaryotic origins of DNA replication by a multiprotein complex. *Nature*, **357,** 128.

112. Diffley, J. F. X. and Cocker, J. H. (1992) Protein–DNA interactions at a yeast replication origin. *Nature*, **357,** 169.

113. Micklem, G., Rowley, A., Harwood, J., Nasmyth, K., and Diffley, J. F. X. (1993) Yeast origin recognition complex is involved in DNA replication and transcriptional silencing. *Nature*, **366,** 87.

114. Li, J. J. and Herskowitz, I. (1993) Isolation of *ORC6*, a component of the yeast origin recognition complex by a one-hybrid system. *Science*, **262,** 1870.

115. Dowell, S. J., Romanowski, P., and Diffley, J. F. X. (1994) Interaction of Dbf4, the Cdc7 protein kinase regulatory subunit, with yeast replication origins *in vivo*. *Science*, **265,** 1243.

116. Rowley, A., Cocker, J. H., Harwood, J., and Diffley, J. F. X. (1995) Initiation complex assembly at budding yeast replication origins begins with the recognition of a bipartite sequence by limiting amounts of the initiator, ORC. *EMBO J.*, **14,** 2631.

117. Rao, H. and Stillman, B. (1995) The origin recognition complex interacts with a bipartite DNA binding site within yeast replicators. *Proc. Natl. Acad. Sci. USA*, **92,** 2224.

118. Parsons, R. E., Stenger, J. E., Ray, S., Welker, R., Anderson, M. E., and Tegtmeyer, P. (1991) Cooperative assembly of simian virus 40 T-antigen hexamers on functional halves of the replication origin. *J. Virol.*, **65**, 2798.

119. Yung, B. Y. and Kornberg, A. (1989) The dnaA initiator protein binds separate domains in the replication origin of *Escherichia coli*. *J. Biol. Chem.*, **264**, 6146.

120. Lorimer, H. E., Wang, E. H., and Prives, C. (1991) The DNA-binding properties of polyomavirus large T antigen are altered by ATP and other nucleotides. *J. Virol.*, **65**, 687.

121. Dean, F. B., Dodson, M., Echols, H., and Hurwitz, J. (1987) ATP-dependent formation of a specialized nucleoprotein structure by simian virus 40 (SV40) large tumor antigen at the SV40 replication origin. *Proc. Natl. Acad. Sci. USA*, **84**, 8981.

122. Deb, S. P. and Tegtmeyer, P. (1987) ATP enhances the binding of simian virus 40 large T antigen to the origin of replication. *J. Virol.*, **61**, 3649.

123. Sekimizu, K., Bramhill, D., and Kornberg, A. (1987) ATP activates dnaA protein in initiating replication of plasmids bearing the origin of the *E. coli* chromosome. *Cell*, **50**, 259.

124. Yung, B. Y. M., Crooke, E., and Kornberg, A. (1990) Fate of the DnaA initiator protein in replication at the origin of the *Escherichia coli* chromosome *in vitro*. *J. Biol. Chem.*, **265**, 1282.

125. Miller, A. M. and Nasmyth, K. A. (1984) Role of DNA replication in the repression of silent mating-type loci in yeast. *Nature*, **312**, 247.

126. Chan, C. S. M. and Tye, B.-K. (1983) A family of *Saccharomyces cerevisiae* repetitive autonomously replicating sequences that have very similar genomic environments. *J. Mol. Biol.*, **168**, 505.

127. Chien, C.-T., Buck, S., Sternglanz, R., and Shore, D. (1993) Targeting of SIR1 protein establishes transcriptional silencing at HM loci and telomeres in yeast. *Cell*, **75**, 531.

128. Axelrod, A. and Rine, J. (1991) A role for CDC7 in repression of transcription at the silent mating-type locus *HMR* in *Saccharomyces cerevisiae*. *Mol. Cell. Biol.*, **11**, 1080.

129. Reynolds, A. E., McCarroll, R. M., Newlon, C. S., and Fangman, W. L. (1989) Time of replication of ARS elements along yeast chromosome III. *Mol. Cell. Biol.*, **9**, 4488.

130. Sclafani, R. A. and Jackson, A. L. (1994) Cdc7 protein kinase for DNA metabolism comes of age. *Mol. Microbiol.*, **11**, 805.

131. Hartwell, L. H. (1976) Sequential function of gene products relative to DNA synthesis in the yeast cell cycle. *J. Mol. Biol.*, **104**, 803.

132. Jackson, A. L., Pahl, P. M. B., Harrison, K., Rosamond, J., and Sclafani, R. A. (1993) Cell cycle regulation of the yeast Cdc7 protein kinase by association with the Dbf4 protein. *Mol. Cell. Biol.*, **13**, 2899.

133. Diffley, J. F. X., Cocker, J. H., Dowell, S. J., and Rowley, A. (1994) Two steps in the assembly of complexes at yeast replication origins *in vivo*. *Cell*, **78**, 303.

134. Brown, J. A., Holmes, S. G., and Smith, M. M. (1991) The chromatin structure of *Saccharomyces cerevisiae* autonomously replicating sequences changes during the cell division cycle. *Mol. Cell. Biol.*, **11**, 5301.

135. Chen, Y., Hennessy, K. M., Botstein, D., and Tye, B.-K. (1992) CDC46/MCM5, a yeast protein whose subcellular localization is cell cycle-regulated, is involved in DNA replication at autonomously replicating sequences. *Proc. Natl. Acad. Sci. USA*, **89**, 10 459.

136. Hennessy, K. M., Clark, C. D., and Botstein, D. (1990) Subcellular localization of yeast CDC46 varies with the cell cycle. *Genes Dev.*, **4**, 2252.

137. Yan, H., Gibson, S., and Tye, B.-K. (1991) MCM2 and MCM3, two proteins important for ARS activity, are related in structure and function. *Genes Dev.*, **5**, 944.

138. Yan, H., Merchant, A. M., and Tye, B.-K. (1993) Cell cycle-regulated nuclear localisation

of MCM2 and MCM3, which are required for the initiation of DNA synthesis at chromosomal replication origins in yeast. *Genes Dev.*, **7**, 2149.

139. Maine, G. T., Sinha, P., and Tye, B.-K. (1984) Mutants of *S. cerevisiae* defective in the maintenance of minichromosomes. *Genetics*, **106**, 365.

140. Liang, C., Weinreich, M., and Stillman, B. (1995) ORC and Cdc6p interact and determine the frequency of initiation of DNA replication in the genome. *Cell*, **81**, 667.

141. Stillman, B. (1994) Initiation of chromosomal DNA replication in eukaryotes: Lessons from lambda. *J. Biol. Chem.*, **269**, 7047.

4 | Replication origins in metazoan chromosomes

MELVIN L. DEPAMPHILIS

1. Introduction

Central to understanding how animal cells regulate DNA replication is understanding the nature of DNA sites where replication begins. Based on analyses of the chromosomes of bacteria, bacteriophage, plasmids, yeast, animal viruses, and mitochondria, initiation of DNA replication begins with the binding of specific *trans*-acting proteins to specific *cis*-acting DNA sequences referred to as origins of replication (1–3). This event is followed by DNA unwinding at or near the protein-binding site in order to expose the two single-stranded, complementary DNA templates to DNA polymerases, their accessory proteins, and proteins that provide nucleotide primers for DNA synthesis. DNA synthesis is then initiated on one or both templates by a process that generally, but not always, results in replication bubble and fork structures such as those found in the chromosomes of prokaryotic and eukaryotic cells (Fig. 1).

Every genome analyzed so far contains at least one replication origin per chromosome, and the genomes of eukaryotic cells contain about one origin every 10 to 330 kb (4). Replication origins in 'simple' genomes are comprised of modular units acting in concert to determine where and when DNA replication will occur. In this sense, replication origins are equivalent to transcription promoters, except, of course, whereas each promoter is uniquely responsible for transcription of its associated gene, chromosomes containing many replication origins can afford to lose a few and still complete replication of the entire chromosome using replication forks that originated upstream and downstream of the origin-deficient region (5). Thus, replication origins ensure complete and efficient replication of a genome while at the same time allowing genetic flexibility.

Initiation of cellular DNA replication differs from initiation of viral or mitochondrial DNA replication in at least two important aspects. First, replication of cellular chromosomes is restricted to one phase of the cell proliferation cycle (S phase). Second, initiation at each of several thousand replication origins is restricted to once per S phase (6). Multiple initiation events at the same locus (gene amplification) can occur in tumors and transformed cell lines, but only rarely are genes amplified during normal animal development (7). In contrast, replication of mitochondrial DNA and

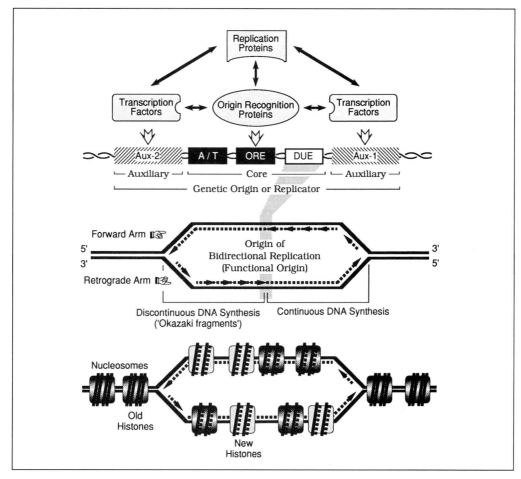

Fig. 1 Modular concept of origins of DNA replication found in simple genomes that function in eukaryotic cell nuclei (1). Replication is initiated by a combination of replication proteins and transcription factors interacting with specific DNA sequences and with each other. Bidirectional DNA replication originates at the DUE, resulting in a replication bubble with two replication forks traveling in opposite directions. The structure of replication forks (124), their organization into chromatin (125), and the process of chromatin assembly (107, 126) have been described in detail.

large viral DNA genomes such as Herpes and vaccinia are not dependent on the cell entering S phase. Mitochondrial DNA replicates randomly throughout the cell division cycle, and large viral genomes usurp the cell's machinery to provide their own replication components. Replication of small viral DNA genomes such as papovaviruses is restricted to S phase, but each genome copy may undergo two or more rounds of replication during a single S phase. Thus, it appears that mitochondrial and viral genomes were designed to escape the very controls required for cellular DNA replication.

2. Simple origins

Origins of DNA replication that function in eukaryotic cells can be divided into two groups: those found in 'simple' genomes such as animal viruses (papovaviruses, papillomaviruses, adenoviruses, herpes simplex viruses, Epstein–Barr viruses), mitochondria (human, mouse), protozoa (*Tetrahymena*), yeast (*S. cerevisiae, S. pombe*), and slime mold (*Physarum*), and those found in the more complex genomes of metazoa such as flies (*Drosophila, Sciara*), frogs (*Xenopus*), and mammals (rodents, human).

2.1 Characteristics of simple origins

● Simple origins are composed of unique, genetically required, sequences. These sequences occupy from 50 to ~1000 bp and are defined by *cis*-acting mutations that prevent DNA replication. They are anatomically similar to origins that function in prokaryotic cells and consist of at least four components, three of which are essential and therefore referred to as the origin core (Fig. 1) (1, 2). The first is an origin recognition element (ORE) that is the binding site for one or more origin recognition proteins that provide the helicase and polymerase activities required to initiate replication. For example, SV40 large tumor (T) antigen binds specifically to the SV40 origin recognition element and initiates DNA unwinding (Chapter 2). In addition, T antigen brings with it a single-stranded DNA-binding protein (RP-A) and the enzymes required to initiate RNA primed-DNA synthesis (DNA primase:DNA polymerase-α). The second component is an easily unwound DNA sequence ('DNA unwinding element', DUE) that is the site where DNA unwinding begins (8). The third component is an A/T-rich sequence consisting of a T-rich and an A-rich strand (1). In some origins (adenovirus, yeast), the A/T-rich element is part of the ORE. In other origins (SV40, polyomavirus), it appears to facilitate DUE activity, while in still others (herpes simplex virus) it may serve as the DUE element itself. Therefore, it is difficult to assign a specific role for the A/T-rich element. Moreover, the A/T-rich sequence is not evident in all origins (Epstein–Barr virus). The fourth component is one or more binding sites for transcription factors. These sites facilitate origin-core activity, but they are not required for replication under all conditions and do not affect the mechanism by which replication occurs. Therefore, they are auxiliary components. Auxiliary sequences stimulate origin-core activity in at least four ways: facilitating assembly of an initiation complex, DNA unwinding, RNA primer synthesis, and exclusion of chromosomal proteins from the origin region (9).

● The genetic origin is coincident with the functional origin. The functional origin is the site where replication actually begins.

● Simple origins act as autonomously replicating sequences (ARSs). ARS elements confer on other DNA molecules the ability to replicate when transferred to either cells or cell extracts containing the required replication proteins. So far, ARS elements have been demonstrated only with viral and yeast origins. Functional origins in the chromosomes of *S. cerevisiae* have been shown to correspond to individual ARS elements that are genetically required for origin activity (10–12) (Chapter 3). The same appears to be true for *S. pombe*, although sequence requirements for origins in *S. pombe* appear more diffuse and origins appear less efficient than in *S. cerevisiae* (13–15).

● Simple origins can function in a soluble cell-free DNA replication system. So far, this has been demonstrated only with viral origins, a problem that may be overcome when large amounts of purified replication proteins become available for other systems.

2.2 Cellular origins differ from viral and mitochondrial origins

Papillomaviruses (PV) and Epstein–Barr virus (EBV) DNA have been considered models for cellular DNA replication, because replication of these viral genomes is restricted to S phase, occurs in the nucleus, and maintains a low number of genome copies per cell. Moreover, early studies on PV DNA replication concluded that a complex interaction between positive and negative controls restricted initiation of replication to once per origin per S phase (130). However, later studies revealed that PV origins are remarkably similar to those in papovaviruses (1) and their replication is not restricted to once per S phase (16–18). EBV remains a candidate because its DNA replicates at the same rate as cellular chromosomes (19, 20), but it remains to be determined whether or not EBV DNA replication, like cellular DNA replication, does not reinitiate when cells are limited to a single S phase in the presence of a mitotic inhibitor such as nocodazole (18). This test should be applied to all putative mammalian ARS elements as well (see Table 2, below) (23, 24).

3. Metazoan origins

In comparison with simple genomes, origins of replication in multicellular animals (the metazoa) often appear paradoxical. Some data suggest that metazoan chromosomes utilize replication origins similar to those found in simpler genomes, whereas other data suggest that metazoan chromosomes lack specific DNA sequences that are required to initiate replication. As discussed below, mapping initiation sites of eukaryotic DNA replication *in situ* has revealed that metazoan chromosomes initiate DNA replication at specific sites that are genetically determined, although the nature of the sequences involved remain largely speculative. Early attempts to identify ori sequences in mammalian chromosomes by their ability to function as ARS elements were difficult to reproduce and therefore controversial (21, 22), although more recent attempts look promising (discussed below). Nevertheless, most large (>10 kb) DNA fragments from mammalian chromosomes can provide some ARS activity in mammalian cells (23, 24), suggesting that DNA length is more critical than DNA sequence. The same conclusion is reached when DNA is injected into the eggs of frogs (*Xenopus*) (25), sea urchin (26), and fish (27), or when DNA is added to extracts of *Xenopus* eggs (28) or *Drosophila* embryos (29) (see Chapter 5). DNA replication is initiated at a single randomly chosen site within virtually any DNA molecule (30–33). This lack of site-specific initiation also appears during chromosome replication in *Xenopus* (34) and *Drosophila* (35) embryos, suggesting that the lack of sequence requirements and site specificity observed when DNA is introduced into cultured cells, eggs, or egg extracts accurately reflects chromosome replication *in situ*.

The solution to this paradox appears to lie in the complexity of metazoan replication origins, a complexity that results from the interplay between specific DNA sequences, chromosome structure and nuclear organization. Simply stated, a metazoan origin of DNA replication consists of a broad initiation zone encompassing a smaller origin of bidirectional replication (OBR). While most initiation events occur at the OBR, some initiation events occur at other potential initiation sites distributed throughout the initiation zone. Potential initiation sites may exist wherever DNA can be unwound easily (at DUEs). An OBR may consist of a DUE associated with other sequence elements that facilitate DNA unwinding and initiation of DNA synthesis at a DUE (analogous to replication origins in simple genomes, Fig. 1). Perhaps the single most distinguishing feature of metazoan DNA replication is its requirement for nuclear structure to initiate replication at specific sites and to restrict initiation to once per S phase. Nuclear structure may repress the activity of some origins while facilitating the activity of others ('Jesuit model') (1, 2, 22, 36) (Chapter 5). Furthermore, nuclear structure may limit access of initiation factors ('licensing factor model') (25) (Chapter 6). Why such complexity appears in the chromosomes of metazoa but not single cell organisms may involve differences in their nuclear organization. For example, yeast nuclei do not dissociate during mitosis, and lack histone H1 and nuclear lamins.

3.1 Mapping origins of replication in metazoan chromosomes

Methods have been developed for identifying the actual sites where replication begins in eukaryotic cell chromosomes in the hope of revealing unique sequence characteristics and of correlating ARS activity with replication origin activity. These methods, which define the outer limits of an origin, fall into two categories: (a) analyses of nascent DNA strands, and (b) analyses of DNA structures (37). The first category locates sites where newly synthesized DNA originates by first incorporating radioactive and density substituted deoxyribonucleotides into nascent DNA and then isolating the labeled molecules and annealing them to sequence specific probes. Several approaches to locating the origins of these chains have proven successful. As cells enter S phase, DNA fragments closest to an origin of DNA replication are labeled first. Therefore, by digesting newly replicated DNA with various restriction endonucleases, one can identify the earliest labeled DNA fragment which should contain a replication origin. Newly formed replication bubbles also can be trapped near this origin by cross-linking the DNA templates with psoralen prior to initiation of replication in order to prevent migration of replication forks away from the origin region. However, these methods generally work best with highly amplified DNA sequences and when cells have been synchronized at the beginning of S phase. Thus, specific replication sites identified under these conditions could be either unique to amplified DNA sequences or an artifact of the cell synchronization conditions.

The ability to examine single copy sequences in exponentially proliferating cells in the absence of metabolic inhibitors is possible with the help of the DNA polymerase chain reaction (PCR). Labeled nascent DNA strands are separated from unreplicated DNA and then fractionated according to their length. Replication origins can be localized from either the length or abundance of nascent DNA strands passing through a specific genomic sequence (38, 39, 56). The closer a specific sequence lies relative to an OBR, the greater its abundance in shorter nascent DNA chains relative to longer nascent DNA chains. Alternatively, dividing the length of long, sequence specific, nascent DNA chains by two reveals the location of their replication origin. By examining several specific sequences on either side of a putative OBR, bidirectional replication events can be distinguished from unidirectional events, and the resolution of an OBR increases. Quantification of the number of nascent DNA molecules that contain a specific sequence (and therefore OBR resolution) is improved by competition between hybridization of the PCR primer to nascent DNA chains versus known amounts of competitor DNA (38).

If DNA replication proceeds by the replication fork mechanism (Fig. 1), then an OBR can be identified by determining replication fork polarity. DNA synthesis on the forward arm (leading strand template) is a continuous process, but DNA synthesis on the retrograde arm (lagging strand template) is a discontinuous process in which RNA-primed, nascent DNA chains approximately 135 nucleotides in length (Okazaki fragments) are repeatedly formed and joined into longer chains. Newly replicated DNA is then rapidly assembled into nucleosomes by randomly distributing to both arms of the fork old histone octamers from in front of the fork (40–42) and newly synthesized histone octamers from the cytoplasm. Therefore, since Okazaki fragments originate predominantly, perhaps exclusively, from the retrograde arm, the fraction of replication forks traveling in the same direction can be measured by hybridization of Okazaki fragments to the individual strands from a unique DNA sequence. The ratio of fragments annealed to the retrograde arm template versus the forward arm template provides a minimum estimate of the fraction of replication forks traveling in one direction. By examining fork polarity at several different chromosomal locations, one can locate the transition from continuous to discontinuous DNA synthesis that defines an OBR (Fig. 1). In SV40 and polyomavirus genomes, the ability to cut nascent DNA:template DNA hybrids at specific restriction sites allowed the OBR to be mapped to within 2 bp and 20 bp, respectively (43). In metazoan cells, limitations of material have so far prevented this strategy, so resolution has been limited to the sizes of restriction fragments that function as sequence-specific probes in the hybridization.

Fork polarity also can be measured in exponentially proliferating cells by inhibiting protein synthesis *in vivo*. Under these conditions, Okazaki fragment synthesis is preferentially inhibited, allowing accumulation of labeled long nascent DNA strands synthesized on the forward arms of replication forks. Again, the ratio of labeled strands that anneal to each of the two template strands of a specific sequence identifies fork polarity. In this procedure, however, nascent DNA should preferentially anneal to the template complementary to that recognized by Okazaki fragments. Initial studies using this mapping protocol assumed that it depended on preferential segregation of prefork histone octamers to the forward arm of replication forks in the absence of histone synthesis, and therefore employed micrococcal nuclease to digest nascent DNA on retrograde arms (44). Subsequent studies demonstrated that this mapping protocol did not depend on chromatin structure and therefore did not require micrococcal nuclease (40, 45).

The second category of methods for identifying replication origins is based on fractionating DNA structures by two-dimensional gel electrophoresis (46, 47). Total DNA is first digested with specific restriction endonucleases, then fractionated by gel electrophoresis, and DNA from specific genomic regions is detected by blotting–hybridization with sequence-specific probes. This provides a sensitive method for identifying replication intermediates at specific genomic sites, and is particularly useful in systems such as yeast where incorporation of labeled nucleotides into nascent DNA is difficult. The presence of replication bubbles (Fig. 1) in a particular genomic region is recognized by a distinctive migration pattern when DNA is fractionated under two different neutral pH conditions. The direction in which replication forks travel can be determined by fractionating them according to size at neutral pH, and then fractionating their nascent DNA strands according to length at alkaline pH. The closer to an OBR, the greater the abundance of short nascent DNA strands containing that sequence. This origin-mapping strategy is equivalent to analysis of the lengths of labeled nascent DNA strands described above. 2D neutral and alkaline gel fractionations of genomic DNA can be run sequentially to determine more precisely the sizes of forks and bubbles at specific genomic locations (48). These methods have localized replication origins in papillomavirus, yeast and slime mold to specific sites of 0.2 to 1 kb (1).

3.2 Metazoan chromosomes initiate DNA replication at specific sites

Mapping the location of 18 different replication origins in differentiated metazoan cells (Table 1) has revealed four general characteristics of initiation of DNA replication in metazoan chromosomes.

● DNA synthesis does not initiate randomly throughout cellular chromosomes, but at specific DNA sites. Therefore, at some point during animal development, specific replication origins are formed. For example, at the beginning of *Xenopus* development, initiation events appear randomly distributed throughout the rRNA gene repeats, but by the time gastrulation is complete, initiation events are confined to the intergenic regions (131). These sites, however, appear more complicated than anticipated from studies on simple genomes.

● From 80% to 95% of DNA synthesis occurs bidirectionally from specific genomic loci referred to as OBRs. This conclusion is based on the fraction of replication forks traveling in the same direction in 2D neutral/alkaline gels (35, 48, 49), the ratio of Okazaki fragments that hybridize to the separated strands of a specific

Table 1 Metazoan chromosomes initiate DNA replication at specific sites

Organism	Location	OBR[a]	Initiation zone[b]	Reference
Hamster	DHFR gene (ori-β)	0.5–3 kb	55 kb	36, 40, 44, 50, 52, 60, 118, 119
Human	rRNA genes	4	31	56, 59, 132
Mouse	rRNA genes	3		E. Gogel and F. Grummt (unpublished data), 132
Drosophila	DNA polymerase α gene	5	10	57
Drosophila	Chorion gene	1[c]	12	58, 120
Sciara	Chromosome-2, locus 9	1[c]	6	48, 49
Human	*hsp70* gene	0.4		68
Human	Lamin B2 gene	0.5		38
Mouse	Ig heavy chain gene	0.6		121
Mouse	ADA gene (late S phase)	2		67
Human	*c-myc* gene	2		54, 122
Hamster	Ribosomal protein S14 gene	2		52
Human	cDNA 343 (early S phase)	2		66
Human	β-Globin gene	2		45
Mouse	CAD gene group	5		55
Hamster	Rhodopsin gene	5		123
Hamster	DHF gene (ori-γ)	8		44
Mouse	ADA gene (early S phase)	11		51

[a] Mapped by labeling of nascent DNA strands.
[b] Mapped by 2D gel analyses of replication bubbles.
[c] Origins of gene amplification that were mapped by 2D gel analyses of replication fork direction.

DNA sequence (50–55), the ratio of long nascent DNA strands synthesized in the presence of emetine that hybridize to the two strands of a specific DNA sequence (40, 44, 45, 55), and the amount of nascent DNA that anneals preferentially to the OBR region (38, 53, 56).

● An OBR is contained within as little as 0.5 kb to as much as 3 kb. This conclusion is based on 11 different OBRs that appear to lie within a 2 kb region and on the DHFR ori-β locus where five different nascent DNA strand methods have been applied with remarkable agreement (reviewed in 1, 36). Thus, metazoan replication origins appear to be 3 to 10 times larger than replication origins in simple genomes (0.05 to 1 kb) (1). The fact that 18 OBRs (Table 1) have been identified by independent investigators using a variety of different methods gives confidence that site-specific initiation is not an artifact of the experimental conditions used to map them. Similar results were obtained with synchronized and unsynchronized cells, with cells containing single-copy sequences and with cells containing amplified multi-copy sequences, with untreated cells, and with cells treated with metabolic inhibitors, and with different methods for detecting specific DNA sequences.

Mapping origins of replication by analyses of nascent DNA strands has revealed that replication begins bidirectionally within an 0.4 kb site in the human hsp70 promoter (Table 1), a 0.5 kb site at the 3′-end of the human lamin B2 gene, a 0.6 kb region upstream of the mouse immunoglobulin heavy chain gene, a 2 kb region at the 5′-end of the human c-*myc* gene, a 2 kb site 160 kb downstream of the mouse adenosine deaminase (ADA) gene, 2 kb site within or immediately adjacent to the

3'-end of the hamster ribosomal protein S14 gene, a 2 kb site within the transcribed region of an undefined human gene in cDNA343, and a 2 kb site between the human δ- and β-globin genes. Neutral/alkaline 2D gel electrophoresis has been used to map an OBR to 1 kb at an amplification locus in *Scira* (48, 49), giving credence to an earlier interpretation of 2D gel electrophoresis mapping data that 80% of replication forks at the chorion gene amplification locus in *Drosophila* originate from a specific 1 kb site (58). Neutral/alkaline 2D gel electrophoresis also identified an S-phase OBR 15–20 kb downstream of the *Drosophila* DNA polymerase–α gene (57).

Some OBRs lie within larger regions such as a 5 kb site at the 5'-end of the hamster rhodopsin gene, a 5 kb region in the mouse CAD gene, an 8 kb region located ~40 kb downstream of the hamster DHFR gene (ori–γ), a 3 kb (E. Gogel and F. Grummt, unpublished data) to 8 kb (56) region at the 5'-end of mammalian rRNA gene transcription units, and an 11 kb region ~29 kb upstream of the mouse ADA gene. These estimates are for the maximum size of an OBR; resolution is limited by the difficulty in preparing probes large enough to give a strong hybridization signal with radio-labeled nascent DNA chains, but that lack any repetitive sequences. Future refinements in origin-mapping techniques will likely resolve these OBRs to a smaller locus.

● Replication bubbles are detected throughout a larger 'initiation zone' of 6–55 kb that includes the OBR. This conclusion is based on analyses of DNA structures by 2D gel neutral/neutral electrophoresis at five different genomic loci (Table 1) (48, 49, 58, 59, 60, 119), two of which (*Drosophila* chorion gene, *Sciara* locus 9) are developmentally programmed amplification origins. Four of these 'initiation zones' encompass an OBR that was detected either by nascent strand analyses (rRNA genes, DHFR ori–β) or by measuring the direction of fork movement using neutral/alkaline 2D gel electrophoresis (*Sciara* locus 9, *Drosophila* Polα gene). One study on the *Drosophila* chorion gene locus (58) concluded that although multiple initiation sites may exist within a 12 kb locus, a model in which a single origin is preferred 70–80% of the time could explain their neutral/neutral 2D gel electrophoresis data.

● Metazoan replication origins consist of a primary initiation site (OBR) surrounded by many secondary initiation sites distributed 'randomly' throughout a larger DNA region. Results of neutral/neutral 2D gel analyses are consistent with newly synthesized DNA analyses and most neutral/alkaline gel analyses if one assumes that the frequency of initiation events at the OBR is much greater than the frequency of initiation events outside the OBR. In fact, replication bubbles detected by neutral/neutral 2D gel analyses appear more abundant in the 12 kb region containing the DHFR ori–β OBR (60), and in the 8 kb region at the 5'-end of the rRNA transcription unit (59) where subsequent nascent DNA strand analysis revealed a >10-fold excess of newly synthesized DNA relative to other sites within the initiation zone (56). In practice, the relative number of initiation events in different DNA segments is difficult to quantify by 2D gel analysis because of concerns over variable loss of replication bubbles and other technical problems (24, 59, 60), whereas analysis of labeled nascent DNA chains lends itself readily to quantification and thus reveals the preference for one site relative to another. For example, the ratio of

DNA synthesis between the two templates of a specific DNA fragment automatically provides the minimum fraction of replication forks moving in the same direction through this region. Initiation events distributed randomly outside the OBR simply contribute to the background level in these mapping protocols.

3.3 Metazoan origins of replication are genetically determined

The simple fact that metazoan origins map to specific sites that replicate at specific times during S phase (63, 64) demonstrates that origins of replication are inherited from one cell division to the next. This conclusion is reinforced by reports that the same OBR identified in cells containing two copies per diploid genome are also identified in cells containing 1000 (hamster DHFR gene) to 30 000 (mouse ADA gene) tandem copies of either chromosomal or extrachromosomal (episomal) sequences (Table 2). Therefore, each copy of the amplified region that initiates replication must use the same OBR; otherwise, initiation would appear to occur at many different sites within the same DNA locus.

Direct evidence that metazoan replication origins are genetically determined comes from reports that the DHFR ori–β (44) and the chorion gene amplification origin (65) retain their activity when translocated to other chromosomal sites. Conversely, an 8 kb deletion between the human δ- and β-globin genes that includes the only OBR found within a 135 kb region eliminates bidirectional replication from this site; all replication forks now move in one direction through this 135 kb region (45). These data demonstrate that metazoan origins of replication are determined by as yet undefined DNA sequences. Nevertheless, genetically required DNA sequences that function as ARS elements have been difficult to identify.

To date, five reports of ARS elements that function in mammalian cells and cell extracts have been documented in detail and shown to correspond to sites where

Table 2 Metazoan origins of replication are genetically determined

Organism	Location	Same origin in single copy and multicopy genomes*	Translocated to other sites	Deletion in OBR	ARS activity	APE activity
Hamster	DHFR gene (ori-β)	Chromosomal	Active (44)			Yes (127)
Mouse	rRNA gene					Yes (74)
Drosophila	Chorion gene		Active (65)			
Human	hsp70 gene				Yes (68)	
Mouse	IgH gene enhancer				Yes (120, 121)	
Mouse	ADA gene (late S phase)	Episomal			Yes (67)	
Human	c-myc gene				Yes (54)	
Human	cDNA 343				Yes (66)	
Hamster	CAD gene group	Episomal				
Human	β-globin gene			Inactive (45)		
Mouse	ADA gene (early S phase)	Episomal				

*Amplified gene copies are found in either chromosomal or episomal locations. OBR (origin of bidirectional replication), ARS (autonomously replicating sequence), APE (amplification promoting element).

replication occurs in mammalian chromosomes (Table 2). In other plasmid assays, replication appears to depend on the distribution of as yet undefined sequence signals over a large area (>10 kb), signals that are more prevalent in human DNA than in bacterial DNA (24). Sequences have been identified in human DNA that stimulate plasmid replication ~3-fold and are present at a 2 kb OBR mapped in chromosomal DNA (23). These results suggest that replication is stimulated by simple sequence features that occur frequently in mammalian DNA and therefore may promote initiation events throughout the initiation zone.

ARS activity in mammalian cells may depend on several variables. For example, some OBR regions may exhibit stronger ARS activity than others. Incubation of negatively supercoiled plasmid DNA with DNA primase:DNA polymerase–α, RP-A, T-antigen helicase, and DNA gyrase resulted in site-specific initiation of DNA replication at the strong yeast origin, ARS1, and at the c-*myc* OBR (69). These conditions employ the energy derived from negative superhelical turns to initiate DNA replication at DUEs that can be unwound by T antigen in the presence of RP-A. However, in the DHFR ori–β region where ARS activity has not been detected (50), preference for the OBR region was observed, but it was less pronounced than with the other two origins.

Other studies on plasmid DNA replication in human cells (13) or in *Xenopus* eggs and egg extracts (53) failed to observe either preferential replication of plasmids containing the DHFR ori–β region or site-specific initiation within ori–β in those plasmids that contained this sequence. However, when nuclei from G1 phase hamster cells were incubated in *Xenopus* egg extract, then DNA replication was initiated specifically at or near the same ori–β OBR utilized by hamster cells (53). Therefore, site-specific initiation of DNA replication in metazoan chromosomes involves nuclear structure, a requirement that may be difficult to fulfill with plasmid DNA. For example, 'matrix (scaffold) attachment regions' (70) or 'locus control regions' can increase transcription rates for integrated but not episomal templates, demonstrating that some potential components of replication origins function only in the context of cell chromosomes. Conversely, many sequences that can function as ARS elements in plasmids, do not function as replication origins in chromosomes (5, 71). Therefore, plasmid DNA replication may not be an appropriate model for metazoan cellular DNA replication.

Finally, the sequence context of an origin can strongly affect its activity. When two or more yeast ARS elements are in close proximity (~6 kb), the efficiency of each is reduced, and only one is activated in each cell cycle (72). This phenomenon has been demonstrated in *S. pombe* chromosomes where initiation zones have been shown to be composed of two or three independent origins (14, 15). Other sequences in the neighborhood also can affect origin activity, making one yeast ARS element preferred over its neighbor (5, 73). Thus, one could imagine that a metazoan initiation zone is composed of many 'simple origins' of the type found in yeast, for example, and that the resulting interference patterns from neighboring origins and extraneous sequences would give the impression that initiation occurs at many sites distributed throughout a broad initiation zone, while some site or cluster of sites is favored and appears as an OBR. Moreover, the anatomical complexity observed for metazoan initiation zones could vary considerably as a function of the number and arrangement of the simple origins that comprise them.

These considerations suggest that the ability to detect ARS activity in mammalian cells may depend on a number of factors, among which are negative superhelical density in the extrachromosomal DNA, sequence context of the cellular OBR, number and proximity of initiation signals that comprise a replication origin, size of the extrachromosomal DNA, and the relative strengths of various OBRs. In addition, detection of ARS activity may require stringent selection conditions (67). Detection of ARS in *S. pombe*, for example, are more difficult than in *S. cerevisiae*, because virtually every DNA sequence, even vector DNA, is capable of replicating to a limited extent (13–15). Furthermore, if replication sites in nuclei are limited, only a small number of extrachromosomal origins will be accommodated and detection may require a sensitive PCR based assay (68).

An alternative assay for *cis*-acting sequences that initiate DNA replication is to look for an 'amplification promoting element' (APE) that promotes formation of large numbers of integrated copies of a DNA sequence, rather than replication of extrachromosomal DNA sequences. A 370 bp APE has been identified in the non-transcribed spacer region of mouse rRNA gene (74) which mediates a 40- to 800-fold amplification of the vector DNA in transformed cells. This DNA segment also contains an OBR that maps from 0.5 to 3.5 kb upstream of the transcription initiation site for mouse rRNA gene (E. Gogel and F. Grummt, unpublished data), in agreement with the OBR at the 5'-end of human rRNA genes (56, 59). The 4.5 kb OBR region in the hamster DHFR ori–β also acts as an APE and contains homologies to the APE found in rRNA genes (Fig. 2) (127). APE activity may provide a more reproducible assay for metazoan replication origin sequences, if these sequences function efficiently only in the context of a large chromosome.

3.4 DNA features of a metazoan replication origin

Metazoan replication origins contain a number of structural features that may be related to their role in DNA replication. Many of these features are found in the DHFR ori–β region (Fig. 2), although none have so far been shown to be required for initiation of chromosome replication (1, 75, 83, 84). ori–β is flanked by two matrix attachment regions commonly associated with newly replicated DNA and, in some cases, origins of cellular replication. The OBR is flanked by two Alu repeats, sequences often associated with DNA amplification and autonomously replicating plasmid DNA. A segment of bent DNA is also present as are several close matches to the yeast ARS consensus sequence that is required for ARS activity. PUR is a protein that binds single-stranded DNA at a specific purine-rich element conserved in origins of DNA replication (45, 78). Transcription factor binding sites are frequently present, but only those for c-Myc protein (68) and octamer binding transcription factors (120) have been reported to contribute to ARS activity in mammalian cells. Replication and transcription sites are colocalized in mammalian

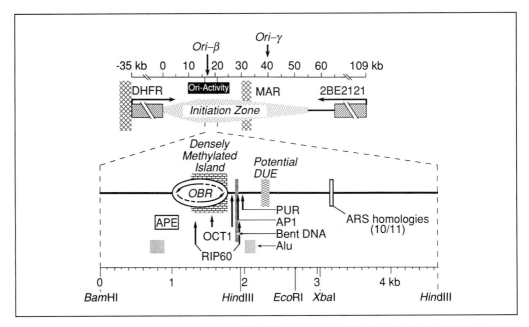

Fig. 2 Notable features of metazoan replication origins. The most extensively characterized replication origin is found between the hamster DHFR and 2BE2121 transcription units (arrow indicates direction of transcription). Map positions are indicated for ori-β, ori-γ, the 16 kb segment that can initiate replication when translocated to other genomic locations (ori-activity), the initiation zone mapped by 2D gel electrophoresis, the origin of bidirectional replication (OBR), a potential DUE, homologies to the *S. cerevisiae* ARS core consensus sequence, homologies to the amplification promoting element (APE) found in the mouse rRNA gene OBR region, nuclear matrix associated region (MAR), densely methylated island (DMI), Alu repeats, bent DNA, and binding sites for transcription factors AP1 and OCT1, RIP60, and PUR. See Table 1 and text for references.

nuclei (89), but transcription through cellular replication origins does not appear to be required for replication since only three of the 18 OBRs in Table 2 are located within a transcribed region (cDNA 343, rS14 gene, CAD gene). More likely is that both processes are facilitated by an open chromatin structure that provides access to initiation factors and negative superhelical energy that facilitates DNA unwinding. These features may be provided by a particular nuclear structure.

One feature likely to be shared by all OBRs is the presence of one or more DUEs (75). Although a DUE is not a unique DNA sequence, it is determined by base stacking interactions, and therefore is not simply a function of AT content, but depends on nucleotide sequence (76). It seems unlikely that a DUE serves as a protein-binding site, because DUEs are interchangeable, and some easily unwound DNA sequences that are not components of replication origins can be substituted for ones that are. DUEs have been demonstrated in *E. coli* oriC (77), yeast ARS elements (76), yeast chromosomal replication origins (12), and SV40 (8). In *E. coli* oriC, the DUE corresponds to the site where dnaA-dependent DNA unwinding begins (77). In SV40 and polyomavirus origins, the nucleotide location of the DUE is coincident with the site where T antigen-dependent DNA unwinding begins, and with the OBR (Fig. 1) (1, 8). Therefore, the DUE most likely determines where DNA replication begins. Any mechanism that promotes DNA unwinding at a specific DUE will also promote site-specific initiation of DNA replication. In yeast and *E. coli*, the DUE function is position and orientation dependent with respect to the ORE, consistent with DNA unwinding in response to protein binding at the ORE. Proteins that bind to specific sequences in single-stranded DNA (78) and matrix (scaffold) attachment regions may operate in conjunction with a DUE to stabilize specific single-stranded DNA regions (70, 79).

Not all easily unwound DNA sequences are part of replication origins, but the fact that such sequences can substitute for known DUEs qualifies them as 'potential DUEs'. Potential DUEs exist at or close to several OBRs, including DHFR ori–β (75). Since the energetic, length, and spacing requirements of true DUEs remain to be defined, it is difficult to estimate the frequency at which potential DUEs occur in natural DNA. For purposes of comparison, D. Natale (personal communication) suggests that a potential DUE be given a definition that is likely to exclude even some true DUEs, such as a 100 bp sequence whose helical stability is 20 or more kcal/mol below the average for the sequence analyzed. Such a potential DUE would be expected once every 3.2 kb in a random sequence of 60% A + T content, suggesting that potential DUEs occur much more frequently than replication origins in yeast (1 origin/36 kb) and mammalian (1 origin/100 kb) DNA. Thus, the ability of an easily unwound sequence to function as a DUE likely depends on conditions such as its proximity to other origin elements (Fig. 1), the concentration of initiation factors, the influence of chromatin structure, and the amount of negative superhelical energy available.

Another feature is a 'densely methylated island' (DMI) that consists of 127 bp (RPS14 gene OBR) to 512 bp (DHFR ori–β) of DNA in which all dC residues are methylated on both strands, regardless of the adjacent nucleotide (80). This unusual methylation pattern has been observed only in association with replication origins, and then only in proliferating cells. Intriguingly, DNA methyltransferase, the enzyme responsible for converting hemimethylated sites to methylated sites in nascent DNA, becomes associated with replication foci during S phase (82). DMIs might act in a positive way by providing a binding site for a replication-specific factor, by altering DNA structure to promote unwinding at a neighboring DUE, or by altering chromatin structure to increase accessibility to initiation factors. In fact, the DMI overlapping the DHFR ori–β OBR is flanked by binding sites for RIP60, a protein that copurifies with helicase activity (83) and that can link the two sites to form a 736 bp DNA loop (128) encompassing the DMI and flanking a potential DUE. By analogy to *E. coli* dnaA, bacteriophage λ O protein, and SV40 T antigen, binding of origin recognition proteins can impose superhelical tension that causes untwisting of DNA in near-by DUEs (3). Alternatively, by analogy to *E. coli oriC* (81), the DMI may help to limit initiation to once per S phase. *OriC* is methylated on both strands at 11 sites. When these sites become hemimethylated as a result of replication, *oriC* associates with an outer membrane component, delaying rebinding of its origin recognition protein, dnaA, and thus delaying reinitiation.

When sufficient negative superhelical energy is provided, palindromic sequences in double-stranded DNA can collapse into a cruciform structure, and these structures may promote initiation of DNA replication at specific sites. Cruciform extrusion at the origin of *E. coli* plasmid pT181 is promoted by the plasmid-encoded initiator protein RepC (85), and antibodies directed against DNA cruciforms can stimulate overall DNA synthesis and copy number of specific genes in permeabilized mammalian cells (86). Moreover, staining mammalian cells with anti-cruciform antibodies suggests that cruciform structures accumulate as cells prepare to enter S phase (87). Alternatively, cruciform structures can be a direct result of aberrant DNA replication involved in gene amplification (88).

4. The role of nuclear structure in metazoan DNA replication

Prokaryotic genomes and animal virus genomes can all replicate in the presence of purified soluble proteins and cofactors; no requirement for a cellular structure has been observed, although there exists a transient interaction between *E.coli oriC* and the outer membrane that regulates the rate at which reinitiation can occur at *oriC* (81). Whether or not the same is true for replication origins in simple eukaryotic organisms remains to be seen. However, one of the most striking requirements for initiation of DNA replication in metazoan chromosomes is that of nuclear structure (see Chapter 5).

● DNA replication in metazoan chromosomes occurs at discrete nuclear foci. It has long been observed that the density of initiation sites (origin to origin distance) in metazoan chromosomes varies in different cell types and under different experimental conditions, leading to the conclusion that many more replication sites are available than are used during a normal S phase, and that the number of active replication origins depends on the ratio of initiation factors to DNA and accessibility of origins as determined by higher order chromatin structure (4). Furthermore, clusters of replication origins initiate replication synchronously (4), giving rise to discrete 'replication complexes' that contain from 100 to 300 replication forks (25, 89). Formation of these replication complexes accounts for the many observations that newly synthesized DNA is preferentially bound to components of nuclear structure generally referred to as 'nuclear matrix' or 'nuclear scaffold' (90).

These replication complexes appear to be assembled in an energy-dependent process prior to S phase at the sites where replication begins (Fig. 4). Replication protein-A (RP-A), a heterotrimeric single-strand DNA-binding protein that is required for replication of metazoan chromosomes (91), is bound at discrete foci in nuclei prior to DNA unwinding and DNA synthesis (92). High levels of cyclin B/cdc2 protein kinase, an enzyme that is required for entrance into mitosis, prevents the appearance of these RP-A foci, consistent with their absence in mitotic chromosomes (92). Cyclin A-dependent cdk2 protein kinase, an enzyme that is required for entrance into S phase (93), colocalizes with RP-A (94). Proliferating cell nuclear antigen (PCNA), a cofactor for DNA polymerase δ, and DNA polymerase α, an enzyme required for synthesis of Okazaki fragments, are also found at replication foci in S-phase nuclei (95). Whether or not they are prebound to these foci before replication begins is not clear. Presumably, 'licensing factor' (25), a cytoplasmic initiation factor that gains access to replication origins only when the nuclei become permeable during mitosis, also binds to preinitiation complexes.

● Initiation of DNA replication in metazoan chromosomes requires an intact nuclear structure. Replication of DNA introduced into either *Xenopus* eggs or egg extracts does not occur unless DNA is first assembled into chromatin and then organized into a nuclear structure that includes lamin B3 and functional nuclear pores (96–99). In addition, the nuclear envelope is instrumental in regulating the onset of S phase, apparently by regulating access of chromosomal DNA to one or more initiation factors ('licensing factor') present in the cytoplasm (reviewed in 25 and Chapter 6).

● Initiation site specificity requires an intact nuclear structure. *Xenopus* egg extract can initiate DNA replication in purified DNA molecules only after the DNA is organized into a pseudo-nucleus, but under these conditions, DNA replication is independent of DNA sequence and begins at many sites distributed throughout the molecules. However, *Xenopus* egg extract can initiate DNA replication at specific sites in mammalian chromosomes, but only when the DNA is presented in the form of an intact nucleus from differentiated cells (53). Initiation of DNA synthesis in nuclei isolated from G1 phase hamster cells is distinguished from continuation of DNA synthesis at preformed replication forks in S-phase nuclei by a delay that precedes DNA synthesis, a dependence on soluble *Xenopus* egg factors, sensitivity to the protein kinase inhibitor 6-dimethylaminopurine (DMAP), and complete labeling of nascent DNA chains. Initiation sites for DNA replication were mapped downstream of the amplified DHFR gene region by: (i) identification of the earliest labeled DNA fragments (100); (ii) quantitative hybridization of newly synthesized DNA to double-stranded DNA probes to reveal genomic loci where DNA synthesis began; and (iii) quantitative hybridization of Okazaki fragments to single-stranded DNA probes to reveal the transition between continuous and discontinuous DNA synthesis on each template within this initiation locus. When bare DNA substrates are used, then *Xenopus* eggs or egg extracts do not distinguish between prokaryotic DNA, hamster DNA that does not contain a replication origin, and hamster DNA that does contain a replication origin. Moreover, initiation events were distributed equally throughout a 30 kb cosmid containing the DHFR ori–β region. When nuclei are used, *Xenopus* egg extract continues DNA synthesis in S-phase nuclei at sites that had been initiated in hamster cells (e.g., DHFR ori–β). When the integrity of the nuclear membrane is preserved, *Xenopus* egg extract initiates DNA replication in G1-phase nuclei specifically at or near the origin of bidirectional replication (ori–β) utilized by hamster cells. When nuclear integrity is damaged, preference for initiation at ori–β is significantly reduced or eliminated. Therefore, initiation sites for DNA replication in mammalian cells are established prior to S phase by some component of differentiated nuclear structure, and this replication origin can be recognized by soluble initiation factors present in *Xenopus* eggs. Subsequent studies (133) have revealed that *Xenopus* egg extract initiates replication at many sites throughout the DHFR gene region in nuclei isolated from early G1-phase hamster cells, whereas the same extract initiates specifically at ori-β in nuclei isolated from mid to late G1-phase hamster cells. Therefore, specific origins of replication in mammalian chromosomes are reestablished during each cell division cycle several hours after nuclear assembly occurs.

5. What is a metazoan replication origin and how does it work?

There are three basic models for understanding the cumulative data. The first is the 'strand separation model' in which extensive DNA unwinding is followed by initiation of RNA-primed DNA synthesis at many sites on both templates (84). This model was based largely on reports of single-stranded DNA in *Xenopus* and *Drosophila* embryos, but the single-stranded DNA bubbles that one would expect as replication intermediates have not been detected by 2D gel analyses of DNA replication in *Xenopus* eggs (32, 34) or *Drosophila* embryos (48), or by electron microscopy of DNA replicating in mammalian cells (101, 102). On the contrary, the fact that replication forks and bubbles are detected by these methods, and the fact that replication fork polarity is observed by hybridization of either Okazaki fragments or long nascent DNA chains to complementary DNA templates (see Section 3.1) provides compelling evidence that DNA synthesis occurs at replication forks in both simple and complex genomes.

To accommodate replication forks, one could imagine that a prereplication complex assembles at the OBR, but then migrates to other positions on either side of the OBR before initiating DNA unwinding and DNA synthesis. This would generate a Gaussian distribution of initiation events with the OBR at its apex. However, one would expect the OBR sequence to strongly stimulate DNA replication under all conditions, a prediction that is difficult to reconcile with the lack of DNA sequence preference observed in the eggs and cleavage stage embryos of frogs, flies, fish, and sea urchins.

Perhaps the simplest model is suggested by the Jesuit dictum that 'many are called, but few are chosen' (Fig. 3) (1, 2, 22, 36). While naked DNA contains many potential sites where replication can begin, nuclear structure may lead to suppression of some of these sites while activating others. For example, transcription can suppress ARS activity in mammalian cells (115), and chromatin structure can interfere with origin activity in yeast, flies, and mammalian cells (103–106). Therefore, initiation events outside the initiation zone could be strongly suppressed by transcription or higher order chromatin structure (4), two features that are generally absent in rapidly cleaving embryos (107). On the other hand, organization of chromatin into a nuclear structure can promote DNA unwinding (79). This could determine which potential sites become OBRs. Thus, *in vivo*, most initiation events

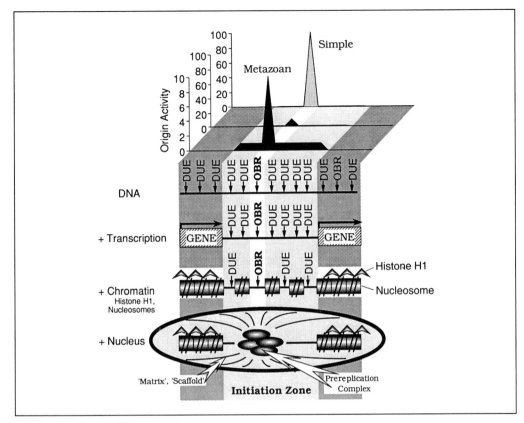

Fig. 3 The 'Jesuit model' for metazoan replication origins and their relationship to simple replication origins. An origin of bidirectional replication (OBR) in simple genomes appears much more efficient than an OBR in metazoan genomes, thus reducing the relative intensity of secondary initiation events observed in the initiation zone surrounding the OBR. 'Quiet zones' are sequences where no initiation events are detected. Eukaryotic DNA contains many potential origins, some may be simply DNA unwinding elements (DUE) while other, more efficient replication origins, may be analogous to those in simple genomes (Fig. 1). The number of potential origins that can become active origins is restricted by actively transcribed DNA regions ('Genes') and chromatin structure, which is required for initiation of metazoan DNA replication, activates selected origins.

will occur at a primary origin (the OBR), but some initiation events will occur at one of many other secondary origins (the initiation zone). Thus, the 'Jesuit model' (Fig. 3) can account for all of the characteristics of metazoan DNA replication using the paradigm provided by DNA replication in *Xenopus* egg (Fig. 4) (Chapter 6).

Naked DNA contains many potential sites where DNA replication can begin. These sites most likely correspond to easily unwound DNA sequences that are components of replication origins in simple (76) as well as metazoan (75) genomes, and that can promote site-specific replication in plasmid DNA (69). When naked DNA is introduced into *Xenopus* eggs or egg extract, chromatin is assembled in the absence of histone H1, which is required for compaction of DNA into 30 nm fibers (107). In addition, nuclei assembled in *Xenopus* eggs are less compact than nuclei in

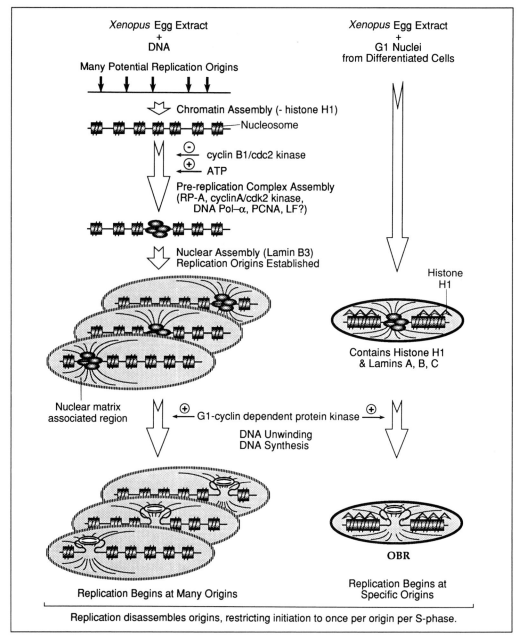

Fig. 4 Acquisition of site-specific DNA replication. *Xenopus* eggs or egg extracts assemble bare DNA or sperm chromatin into a relaxed nuclear structure that permits initiation of DNA replication within many sequences, allowing the early cleavage stage amphibian embryo to replicate its genome. In contrast, preformed nuclei from G1 phase differentiated mammalian cells initiate DNA replication under the same conditions at or near a site-specific OBR that was selected by the mammalian cell to be used as a replication origin during its subsequent S phase. Selection of initiation sites may be restricted by chromatin structure masking some potential origins, and nuclear matrix (scaffold) associated regions stabilizing DNA unwinding at other potential origins. At some point during animal development, changes occur in nuclear organization that restricts the number of sites that can be used as origins of replication. In *Xenopus* (and presumably other animals whose embryos undergo rapid cell cleavages) this transition appears to occur after the mid-blastula transition (see text). 'LF' is licensing factor.

differentiated cells. This may be due partially to fact that the composition of nuclear lamin proteins, a component of nuclear structure required for DNA replication (97, 108), is simpler in *Xenopus* embryos than in *Xenopus* differentiated cells (109, 110). This relaxed environment may allow assembly of prereplication complexes at any one of the many potential origins located within a large DNA region, thus facilitating rapid replication of the *Xenopus* embryonic genome. The fact that replication does not initiate at more than one site within a single plasmid molecule (30–33) suggests that replication in embryos undergoing rapid nuclear division cycles (e.g., frogs, flies, fish, sea urchin) occurs at only one of many potential sites within a large DNA region. Formation of prereplication complexes does not require nuclear membrane formation (92), but initiation of DNA replication does, suggesting that some component of nuclear structure interacts directly with each replication origin. Moreover, activation of prereplication complexes appears to require the action of G1-cyclin-dependent protein kinase cdk2. General inhibitors of protein kinases (e.g., DMAP) prevent initiation of DNA replication in either sperm chromatin or G1 nuclei, but do not inhibit active replication forks (53, 111, 112). Cip1 protein, a specific inhibitor of cdk2 kinase, exhibits a similar effect, and this inhibition can be overcome with cyclin E (129). Cip1 inhibition occurs after formation of preinitiation complexes but before DNA unwinding has begun (92).

It has long been observed that the density of initiation sites (origin to origin distance) in metazoan chromosomes varies in different cell types and under different experimental conditions, leading to the conclusion that many more replication sites are available than are used during a normal S phase, and that the number of active replication origins depends on the ratio of initiation factors to DNA and accessibility of origins as determined by higher order chromatin structure (4). As animal development progresses and cells begin to undergo differentiation, a 5- to 10-fold reduction can occur in the frequency of initiation sites (6, 35, 113, 114). In fact, while initiation events in the *Xenopus* rRNA gene region are distributed randomly throughout these sequences in cleavage stage embryos, initiation events become confined to the intergenic regions as development progresses beyond the blastula stage (131). Changes in chromatin and nuclear structure that occur at the mid-blastula transition in *Xenopus*, for example, may repress initiation of DNA replication at some loci while facilitating it at others (Fig. 3) (53). In yeast, for example, at least half of the ARS elements on chromosome III are not active as chromosomal replication origins (5). Factors that likely contribute to these changes in metazoa include the onset of transcription and cell differentiation, and the appearance of histone H1 and lamins A and C. Thus, some of the initiation sites chosen in fertilized *Xenopus* eggs may remain as origins throughout development, while other sites may be eliminated.

6. Summary and perspective

It is now clear that all eukaryotic cells initiate replication at specific DNA sites, although in the case of metazoan chromosomes these sites appear more complex and the sequences that determine them remain unclear. Nevertheless, metazoan

OBRs may turn out to be less efficient versions of OBRs in simple genomes. If we compare site specificity in a metazoan genome with that in a simple genome, initiation events outside the metazoan OBR would go undetected unless the scale is expanded to accommodate the lower activity of a metazoan replication origin (Fig. 3). For example, SV40 and polyomavirus replication origins are at least 100-fold more efficient at initiating replication than neighboring DNA sequences, but initiation events can still occur at other sites in the DNA molecule, although they are difficult to detect (61, 62). Similarly, bacteriophages T4 and T7 both contain a primary replication origin as well as several secondary ones (3).

As for now, what is clear is that initiation sites for DNA replication in metazoan chromosomes consists of an OBR and an initiation zone (Fig. 4). Most initiation events occur bidirectionally within a 0.5–3 kb locus that contains the OBR. In differentiated cells, where most potential replication origins are suppressed, favored sites may consist of an easily unwound DNA sequence in combination with other origin components such as densely methylated islands, origin recognition elements, and transcription factor binding sites (Figs 1 and 3). These features, in addition to sequences that may attach to nuclear matrix (scaffold) in order to stabilize unwound DNA at the OBR, would comprise the site-specific, heritable replication origins that have been mapped in metazoan chromosomes (Table 1). Additional initiation events are detected throughout a 6 to 55 kb region that encompasses the OBR (the initiation zone). Therefore, not every cell initiates replication at the OBR; some cells choose one of the many potential origins (DUEs?) that remain accessible to replication factors. Alternatively, replication may sometimes occur by another mechanism (see Section 5). The prominence of secondary initiation sites in metazoan replication origins compared with simple replication origins may simply reflect the low efficiency of primary replication origins in metazoans (Fig. 3).

Site specificity requires an intact nuclear structure, consistent with the concept of a replication origin as a specific nucleoprotein complex that interacts with soluble replication factors, rather than simply a specific sequence that is recognized by soluble factors. This requirement can account for three important features of metazoan DNA replication. First, it provides a way to change the number of origins per chromosomal locus at different stages in animal development, thus accommodating the need for shorter S phases in the early developmental stages of some animals. Second, it provides a way to limit the number of origins per genome without simultaneously limiting the cell's capacity to sustain genetic alternations. A large number of less efficient origins that demand less sequence specificity allows more flexibility for rearranging genetic information than a small number of highly efficient origins that require a great deal of sequence specificity. Finally, it provides a way to restrict initiation at each origin to once per S phase. Initiation at each replication origin would be restricted to once per cell cycle if the nucleoprotein complex that defines a replication origin in G1 nuclei is disrupted by the act of replication. In that event, reinitiation at the same site will be prevented, even in the presence of excess soluble initiation factors.

Metazoan replication origins in G1 nuclei of differentiated mammalian cells may be analogous to the six-protein origin replication complex assembled at yeast ARS sites (116) (Chapter 3), or the EBNA-1 protein/origin core DNA complex in Epstein–Barr virus episomes (117), both of which appear to be stable DNA–protein complexes that exist throughout the cell cycle. Thus, a licensing factor (25) may recognize this DNA/protein platform, permitting it to be activated by a G1-cyclin-dependent protein kinase. Licensing factor is inactivated by the replication process and then replaced when the nuclear membrane becomes permeable to cytoplasmic factors during mitosis (25).

Progress in this field is moving rapidly. What appeared several years ago as a paradox (50, 118), has now resolved itself into a novel view of DNA replication that brings together the realms of biochemistry and cell biology. Further progress is likely to come from three approaches: identification of soluble factors that are required for initiation of S phase, identification of common features among replication origins, and identification of structural features of nuclei that are required for DNA replication.

References

1. DePamphilis, M. L. (1993) Eukaryotic DNA replication: Anatomy of an origin. *Annu. Rev. Biochem.*, **62,** 29.
2. DePamphlis, M. L. (1993) Origins of DNA replication that function in eukaryotic cells. *Curr. Opin. Cell Biol.*, **5,** 434.
3. Kornberg, A. and Baker, T. (1992) *DNA replication.* W. H. Freeman, New York.
4. Hand, R. (1978) Eucaryotic DNA: organization of the genome for replication. *Cell,* **15,** 317.
5. Newlon, C., Collins, I., Dershowitz, A., Deshpande, A. M., Greenfeder, S. A., Ong, L. Y., and Theis, J. F. (1994) Analysis of replication origin function on chromosome III of *Saccharomyces cerevisiae. Cold Spring Harbor Symp. Quant. Biol.*, **58,** 415.
6. Blumenthal, A. B., Kreigstein, H. J., and Hogness, D. S. (1973) The units of DNA replication in *Drosophila melanogaster* chromosomes. *Cold Spring Harbor Symp. Quant. Biol.*, **38,** 205.
7. Tlsty, T. D. (1990) Normal diploid human and rodent cells lack a detectable frequency of gene amplification. *Proc. Natl. Acad. Sci. USA,* **87,** 3132.
8. Lin, S. and Kowalski, D. (1994) DNA helical instability facilitates initiation at the SV40 replication origin. *J. Mol. Biol.*, **235,** 496.
9. DePamphilis, M. L. (1993) How transcription factors regulate origins of DNA replication in eukaryotic cells. *Trends Cell Biol.*, **3,** 161.
10. Rivier, D. H. and Rine, J. (1992). An origin of DNA replication and a transcription silencer require a common element. *Science,* **256,** 659.
11. Deshpande, A. M. and Newlon, C. S. (1992) The ARS consensus sequence is required for chromosomal origin function in *S. cerevisiae. Mol. Cell. Biol.*, **12,** 4305.
12. Huang, R.-Y. and Kowalski, D. (1993) A DNA unwinding element and an ARS consensus comprise a replication origin within a yeast chromosome. *EMBO J.*, **12,** 4521.
13. Caddle, M. S. and Calos, M. P. (1994) Specific initiation at an origin of replication from *Schizosaccharomyces pombe. Mol. Cell. Biol.*, **14,** 1796.
14. Dubey, D. D., Zhu, J., Carlson, D. L., Sharma, K., and Huberman, J. A. (1994) Three ARS elements contribute to the ura4 replication origin region in the fission yeast, *Schizosaccharomyces pombe. EMBO J.*, **13,** 3638.

15. Wohlgemuth, J. G., Bulboaca, G. H., Moghadam, M., Caddle, M. S., and Calos, M. P. (1994) Physical mapping of origins of replication in the fission yeast *Schizosaccharomyces pombe. Mol. Biol. Cell.*, **5**, 839.

16. Gilbert, D. M. and Cohen, S. N. (1987) BPV plasmids replicate randomly in mouse fibroblasts throughout S-phase of the cell cycle. *Cell*, **50**, 59.

17. Ravnan, J.-B., Gilbert, D. M., Kelly, G., Hagen, T. and Cohen, S. N. (1992) Random-choice replication of extrachromosomal BPV molecules in heterogeneous, clonally derived BPV-infected cell lines. *J. Virol.*, **66**, 6946.

18. Nallaseth, F. S. and DePamphilis, M. L. (1994) Papillomavirus contains *cis*-acting sequences that can suppress but not regulate origins of DNA replication. *J. Virol.*, **68**, 3051.

19. Haase, S. B. and Calos, M. P. (1991) Replication control of autonomously replicating human sequences. *Nucleic Acids Res.*, **19**, 5053.

20. Yates, J. L. and Guan, N. (1991) Epstein–Barr virus-derived plasmids replicate only once per cell cycle and are not amplified after entry into cells. *J. Virol.*, **65**, 483.

21. Gutierrez, C., Guo, Z.-S., Burhans, W., DePamphilis, M. L., Farrell-Towt, J., and Ju, G. (1988) Is c-Myc protein directly involved in DNA replication? *Science*, **240**, 1202.

22. Burhans, W. C. and Huberman, J. A. (1994) DNA replication origins in animal cells: a question of context? *Science*, **263**, 639.

23. Masukata, H., Satoh, H., Obuse, C., and Okazaki, T. (1993) Autonomous replication of human chromosomal DNA fragments in human cells. *Mol. Cell. Biol.*, **4**, 1121.

24. Krysan, P. J., Smith, J. G., and Calos, M. P. (1993) Autonomous replication in human cells of multimers of specific human and bacterial DNA sequences. *Mol. Cell. Biol.*, **13**, 2688.

25. Coverly, D. and Laskey, R. A. (1994) Regulation of eukaryotic DNA replication. *Annu. Rev. Biochem.*, **63**, 745.

26. McMahon, A. P., Flytzanis, C. N., Hough-Evans, B. R., Katula, K. S., Britten, R. J., and Davidson, E. H. (1985) Introduction of cloned DNA into sea urchin egg cytoplasm: replication and persistence during embryogenesis. *Dev. Biol.*, **108**, 420.

27. Stuart, G. W., McMurray, J. V., and Westerfield, M. (1988) Replication, integration and stable germ-line transmission of foreign sequences injected into early zebrafish embryos. *Development*, **103**, 403.

28. Blow, J. J. and Laskey, R. A. (1986) Initiation of DNA replication in nuclei and purified DNA by a cell-free extract of *Xenopus* eggs. *Cell*, **47**, 577.

29. Crevel, G. and Cotterill, S. (1991) DNA replication in cell-free extracts from *Drosophila melanogaster*. *EMBO J.*, **10**, 4361.

30. Mechali, M. and Kearsey, S. (1984) Lack of specific sequence requirement for DNA replication in *Xenopus* eggs compared with high sequence specificity in yeast. *Cell*, **38**, 55.

31. Hines, P. J. and Benbow, R. M. (1982) Initiation of replication at specific origins in DNA molecules microinjected into unfertilized eggs of the frog *Xenopus laevis*. *Cell*, **30**, 459.

32. Mahbubani, H. M., Paull, T., Elder, J. K., and Blow, J. J. (1992) DNA replication initiates at multiple sites on plasmid DNA in *Xenopus* egg extracts. *Nucleic Acids Res.*, **20**, 1457.

33. Hyrien, O. and Méchali, M. (1992) Plasmid replication in *Xenopus* eggs and egg extracts: A 2D gel electrophoretic analysis. *Nucleic Acids Res.*, **20**, 1463.

34. Hyrien, O. and Méchali, M. (1993) Chromosomal replication initiates and terminates at random sequences but at regular intervals in the ribosomal DNA of *Xenopus* early embryos. *EMBO J.*, **12**, 4511.

35. Shinomiya, T. and Ina, S. (1991) Analysis of chromosomal replicons in early embryos of *Drosophila melanogaster* by two-dimensional gel electrophoresis. *Nucleic Acids Res.*, **19**, 3935.

36. DePamphilis, M. L. (1993) Origins of DNA replication in metazoan chromosomes. *J. Biol. Chem.*, **268**, 1.

37. Vassilev, L. T. and DePamphilis, M. L. (1992) Guide to identification of origins of DNA replication in eukaryotic cell chromosomes. *Crit. Rev. Biochem. Mol. Biol.*, **27**, 445.

38. Giacca, M., Zentilin, L., Norio, P., Diviacco, S., Dimitrova, D., Contreas, G., Biamonti, G., Perini, G., Weighardt, F., Riva, S., and Falaschi, A. (1994) Fine mapping of a replication origin of human DNA. *Proc. Natl. Acad. Sci. USA*, **91**, 7119.

39. Vassilev, L. T., Burhans, W. C., and DePamphilis, M. L. (1990) Mapping an origin of DNA replication at a single copy locus in exponentially proliferating mammalian cells. *Mol. Cell. Biol.*, **10**, 4685.

40. Burhans, W. C., Vassilev, L. T., Wu, J., Sogo, J. M., Nallaseth, F. N., and DePamphilis, M. L. (1991) Emetine allows identification of origins of mammalian DNA replication by imbalanced DNA synthesis, not through conservative nucleosome segregation. *EMBO J.*, **10**, 4351.

41. Krude, T. and Knippers, R. (1991) Transfer of nucleosomes from parental to replicated chromatin. *Mol. Cell. Biol.*, **11**, 6257.

42. Sugasawa, K., Ishimi, Y., Eki, T., Hurwitz, J., Kikuchi, A. and Hanaoka, F. (1992) Non-conservative segregation of parental nucleosomes during SV40 chromosome replication in vitro. *Proc. Natl. Acad. Sci. USA*, **89**, 1055.

43. DePamphilis, M. L., Martínez-Salas, E., Cupo, D. Y., Hendrickson, E. A., Fritze, C. E., Folk, W. R., and Heine, U. (1988) Initiation of polyomavirus and SV40 DNA replication, and the requirements for DNA replication during mammalian development. In *Eukaryotic DNA replication. Cancer Cells*. Stillman, B. and Kelly, T. (eds.). Cold Spring Harbor Laboratory Press, Cold Spring Harbor, NY, Vol. 6, pp. 165–175.

44. Handeli, S., Klar, A., Meuth, M., and Cedar, H. (1989) Mapping replication units in animal cells. *Cell*, **57**, 909.

45. Kitsberg, D., Selig, S., Keshet, I., and Cedar, H. (1993) Replication structure of the human β-globin gene domain. *Nature*, **366**, 588.

46. Fangman, W. L. and Brewer, B. J. (1991) Activation of replication origins within yeast chromosomes. *Annu. Rev. Cell Biol.*, **7**, 375.

47. Huberman, J. A. (1994) Analysis of DNA replication origins and directions by two-dimensional gel electrophoresis. In: *The cell cycle: A practical approach*. Fantes, P. and Brooks, R. F. (eds.). Oxford University Press, Oxford, pp. 213–234.

48. Liang, C. and Gerbi, S. A. (1994) Analysis of an origin of DNA amplification in *Sciara coprophila* by a novel three dimensional gel method. *Mol. Cell. Biol.*, **14**, 1520.

49. Liang, C., Spitzer, J. D., Smith, H. S., and Gerbi, S. A. (1993) Replication initiates at a confined region during DNA amplification in *Sciara* DNA puff II/9A. *Genes Dev.*, **7**, 1072.

50. Burhans, W. C., Vassilev, L. T., Caddle, M. S., Heintz, N. H., and DePamphilis, M. L. (1990) Identification of an origin of bidirectional DNA replication in mammalian chromosomes. *Cell*, **62**, 955.

51. Carroll, S. M., DeRose, M. L., Kolman, J. L., Nonet, G. H., Kelly, R. E., and Wahl, G. M. (1993) Localization of a bidirectional DNA replication origin in the wild type and in episomally amplified murine ADA loci. *Mol. Cell. Biol.*, **13**, 2971.

52. Tasheva, E. S. and Roufa, D. J. (1994). A mammalian origin of bidirectional DNA replication within the Chinese hamster RPS14 locus. *Mol. Cell. Biol.*, **14**, 5628.

53. Gilbert, D. M., Miyazawa, H., and DePamphilis, M. L. (1995). Site-specific initiation of DNA replication in *Xenopus* egg extract requires nuclear structure. *Mol. Cell. Biol.*, **15**, 2942.

54. Berberich, S., Trivedi, A., Daniel, D. C., Johnson, E. M., and Leffak, M. (1994). In vitro replication of plasmids containing human c-myc DNA. *J. Mol. Biol.*, **245**, 92.

55. Kelly, R. E., DeRose, M. L., Draper, B. W., and Wahl, G. M. (1994) Identification of an origin of bidirectional replication within the coding region of the ubiquitously expressed CAD gene, *Mol. Cell. Biol.*, **15**, 4136.

56. Yoon, Y., Sanchez, J. A., Brun, C. and Huberman, J. A. (1994) Mapping of replication initiation sites in human ribosomal DNA by nascent strand abundance analysis, *Mol. Cell. Biol.*, **15**, 2482.

57. Shinomiya, T. and Ina, S. (1994) Mapping an initiation region of DNA replication at a single-copy chromosomal locus in *Drosophila melanogaster* cells by two-dimensional gel methods and PCR-mediated nascent-strand analysis: multiple replication origins in a broad zone. *Mol. Cell. Biol.*, **14**, 7394.

58. Heck, M. M. S. and Spradling, A. C. (1990) Multiple replication origins are used during *Drosophila* chorion gene amplification. *J. Cell Biol.*, **4**, 903.

59. Little, R. D., Platt, T. H. K., and Schildkraut, C. L. (1993) Initiation and termination of DNA replication in human rRNA genes. *Mol. Cell. Biol.*, **13**, 6600.

60. Dijkwel, P. A. and Hamlin, J. L. (1992) Initiation of DNA replication in the DHFR locus is confined to the early S-period in CHO cells synchronized with the plant amino acid mimosine. *Mol. Cell. Biol.*, **12**, 3715.

61. Tack, L. C. and Proctor, G. N. (1987) Two major replicating SV40 chromosome classes: synchronous replication fork movement is associated with bound large T-ag during elongation. *J. Biol. Chem.*, **262**, 6339.

62. Martin, R. G. and Setlow, V. P. (1980) Initiation of SV40 DNA synthesis is not unique to the replication origin. *Cell*, **20**, 381.

63. Gilbert, D. M. (1986) The temporal order of replication of *Xenopus laevis* 5S ribosomal RNA genes in somatic cells. *Proc. Natl. Acad. Sci. USA*, **83**, 2924.

64. Hatton, K. S., Dhar, V., Brown, E. H., Iqbal, M. A., Stuart, S., Didamo, V. T., and Schildkraut, C. L. (1988) Replication program of active and inactive gene families in mammalian cells. *Mol. Cell. Biol.*, **8**, 2149.

65. Orr-Weaver, T. L. (1991) Drosophila chorion genes: cracking the eggshell's secrets. *BioEssays*, **13**, 97.

66. Wu, C., Zannis-Hadjopoulos, M., and Price, G. B. (1993) In vivo activity for initiation of DNA replication resides in a transcribed region of the human genome. *Biochim. Biophys. Acta*, **1174**, 258.

67. Virta-Pearlman, V. J., Gunaratne, P. H., and Chinault, A. C. (1993) Analysis of a replication initiation sequence from the adenosine deaminase region of the mouse genome. *Mol. Cell. Biol.*, **13**, 5931.

68. Taira, T., Iguchi-Ariga, S. M. M., and Ariga, H. (1994) A novel DNA replication origin identified in the human heat shock protein 70 gene promoter. *Mol. Cell. Biol.*, **14**, 6386.

69. Ishimi, Y., Matsumoto, K., and Ohba, R. (1994) DNA replication from initiation zones of mammalian cells in a model system. *Mol. Cell. Biol.*, **14**, 6489.

70. Schlake, T., Klehr-Wirth, D., Yoshida, M., Beppu, T., and Bode, J. (1994) Gene expression within a chromatin domain: the role of core histone hyperacetylation. *Biochemistry*, **33**, 4197.

71. Kipling, D. and Kearsey, S. E. (1990) Reversion of autonomously replicating sequence mutations in *S. cerevisiae*: creation of a eucaryotic replication origin within procaryotic vector DNA. *Mol. Cell. Biol.*, **10**, 265.

72. Brewer, B. J. and Fangman, W. L. (1993) Initiation at closely spaced replication origins in a yeast chromosome. *Science*, **262**, 1728.

73. Brewer, B. J. and Fangman, W. L. (1994) Initiation preference at a yeast origin of replication. *Proc. Natl. Acad. Sci. USA*, **91**, 3418.

74. Hermann, C., Gartner, E., Weidle, U. H., and Grummt, F. (1994) High copy expression vector based on amplification promoting sequences. *DNA Cell. Biol.*, **13**, 437.

75. Dobbs, D. L., Shaiu, W.-L., and Benbow, R. M. (1994) Modular sequence elements associated with origin regions in eukaryotic chromosomal DNA. *Nucleic Acids Res.*, **22**, 2479.

76. Natale, D. A., Umek, R. M., and Kowalski, D. (1993) Ease of DNA unwinding is a conserved property of yeast replication origins. *Nucleic Acids Res.*, **21**, 555.

77. Kowalski, D. and Eddy, M. J. (1989) The DNA unwinding element: a novel, *cis*-acting component that facilitates opening of the *E. coli* replication origin. *EMBO J.*, **8**, 4335.

78. Bergemann, A. D. and Johnson, E. M. (1992) HeLa Pur factor binds single-stranded DNA at a specific element conserved in gene flanking regions and origins of DNA replication. *Mol. Cell. Biol.*, **12**, 1257.

79. Bode, J., Kohwi, Y., Dickinson, L., Joh, T., Klehr, D., Mielke, C., and Kohwi-Shigematsu, T. (1992) Biological significance of unwinding capability of nuclear matrix-associating DNAs. *Science*, **55**, 195.

80. Tasheva, E. S. and Roufa, D. J. (1994) Densely methylated DNA islands in mammalian chromosomal replication origins. *Mol. Cell. Biol.*, **14**, 5636.

81. Herrick, J., Kern, R., Guha, S., Landulsi, A., Fayet, O., Malki, A., and Kohiyama, M. (1994) Parental stand recognition of the DNA replication origin by the outer membrane in *E. coli*. *EMBO J.*, **13**, 4695.

82. Leonhardt, H., Page, A. W., Weier, H. U., and Bestor, T. H. (1992) A targeting sequence directs DNA methyltransferase to sites of DNA replication in mammalian nuclei. *Cell*, **71**, 865.

83. Held, P. G. and Heintz, N. H. (1992) Eukaryotic replication origins. *Biochim. Biophys. Acta*, **1130**, 235.

84. Benbow, R. M., Zhao, J., and Larson, D. D. (1992) On the nature of origins of DNA replication in eukaryotes. *BioEssays*, **14**, 661.

85. Noirot, P., Bargonetti, J., and Novick, R. P. (1990) Initiation of rolling circle replication in pT181 plasmid: initiator protein enhances cruciform extrusion at the origin. *Proc. Natl. Acad. Sci. USA*, **87**, 8560.

86. Zannis-Hadjopoulos, M., Frappier, L., Khoury, M., and Price, G. B. (1988) Effect of anti-cruciform DNA monoclonal antibodies on DNA replication. *EMBO J.*, **7**, 1837.

87. Ward, G. K., Shihab-el-Deen, A., and Zannis-Hadjopoulos, M. (1991) DNA cruciforms and the nuclear supporting structure. *Exp. Cell Res.*, **195**, 92.

88. Cohen, S., Hassin, D., Karby, S., and Lavi, S. (1994) Hairpin structures are the primary amplification products: a novel mechanism for generation of inverted repeats during gene amplification. *Mol. Cell. Biol.*, **14**, 7782.

89. Hassan, A. B., Errington, R. J., White, N. S., Jackson, D. A., and Cook, P. R. (1994) Replication and transcription sites are colocalized in human cells. *J. Cell Sci.*, **107**, 425.

90. Nakayasu, H. and Berezney, R. (1989) Mapping replication sites in the eukaryotic cell nucleus. *Exp. Cell Res.*, **165**, 291.

91. Fang, F. and Newport, J. (1993) Distinct roles of cdk2 and cdc2 in RP-A phosphorylation during the cell cycle. *J. Cell Sci.*, **106**, 983.

92. Adachi, Y. and Laemmli, U. K. (1994) Study of the cell cycle-dependent assembly of the DNA pre-replication centres in *Xenopus* egg extract. *EMBO J.*, **13**, 4153.

93. Fang, F. and Newport, J. (1991) Evidence that the G1–S and G2–M transitions are controlled by different cdc2 proteins in higher eukaryotes. *Cell*, **66**, 731.

94. Cardoso, M. C., Leonhardt, H., and Nadal-Ginard, B. (1993) Reversal of terminal differentiation and control of DNA replication: cyclin A and cdk2 specifically localize at subnuclear sites of DNA replication. *Cell*, **74**, 979.

95. Kill, I. R., Bridges, J. M., Campbell, K. H. S., Maldonado-Codina, G., and Hutchison, C. J. (1991) The timing of the formation and usage of replicase clusters in S-nuclei of human diploid fibroblasts. *J. Cell Sci.*, **100**, 869.

96. Blow, J. J. and Sleeman, A. M. (1990) Replication of purified DNA in *Xenopus* egg extracts is dependent on nuclear assembly. *J. Cell Sci.*, **95**, 383.

97. Newport, J. W., Wilson, K. L., and Dunphy, W. G. (1990) A lamin-independent pathway for nuclear envelope assembly. *J. Cell Biol.*, **111**, 2247.

98. Meier, J., Campbell, K.-H. S., Ford, C. C., Stick, R., and Hutchison, C. J. (1991) Role of lamin LIII in nuclear assembly and DNA replication in cell-free extracts of *Xenopus* eggs. *J. Cell Sci.*, **98**, 271.

99. Cox, L. (1992) DNA replication in cell-free extracts from *Xenopus* eggs is prevented by disrupting nuclear envelope function. *J. Cell Sci.*, **101**, 43.

100. Gilbert, D. M., Miyazawa, H., Nallaseth, F. S., Ortega, J. M., Blow, J. J., and DePamphilis, M. L. (1994) Site-specific DNA replication in metazoan chromosomes and the role of nuclear organization. *Cold Spring Harbor Symp. Quant. Biol.*, **58**, 475.

101. Hamlin, J. L., Dijkwel, P. A., and Vaughn, J. P. (1992) Initiation of replication in the Chinese hamster DHFR domain. *Chromosoma*, **102** (Suppl. 1), 17.

102. Gruss, C., Wu, J., and Sogo, J. M. (1994) Disruption of nucleosomes at the replication fork. *EMBO J.*, **12**, 4533.

103. Forrester, W. C., Epner, E., Driscoll, M. C., Enver, T., Brice, M., Papayannopoulou, T., and Groudine, M. (1990) A deletion of the human β-globin locus activation region causes a major alteration in chromatin structure and replication across the entire β-globin region. *Genes Dev.*, **4**, 1637.

104. Simpson, R. T. (1990) Nucleosome positioning can affect the function of a *cis*-acting DNA element in vivo. *Nature*, **343**, 387.

105. Karpen, G. H. and Spradling, A. C. (1992) Reduced DNA polytenization of a mini-chromosome region undergoing position effect variegation in *Drosophila*. *Cell*, **63**, 97.

106. Ferguson, B. M. and Fangman, W. L. (1992) A position effect on the time of replication origin activation in yeast. *Cell*, **68**, 333.

107. Wolffe, A. P. (1994) The role of transcription factors, chromatin structure and DNA replication in 5S RNA gene regulation. *J. Cell Sci.*, **107**, 2055.

108. Jenkins, H., Holman, T., Lyon, C., Lane, B., Stick, R., and Hutchison, C. (1993) Nuclei that lack a lamina accumulate karyophilic proteins and assemble a nuclear matrix. *J. Cell Sci.*, **106**, 275.

109. Benavante, R., Krohne, G., and Franke, W. W. (1985) Cell type-specific expression of nuclear lamina proteins during development of *Xenopus laevis*. *Cell*, **41**, 177.

110. Stick, R. and Hausen, P. (1985) Changes in the nuclear lamina composition during early development of *Xenopus laevis*. *Cell*, **41**, 191.

111. Kubota, Y. and Takisawa, H. (1993) Determination of initiation of DNA replication before and after nuclear formation in *Xenopus* egg cell-free extracts. *J. Cell Biol.*, **123**, 1321.

112. Blow, J. J. (1993) Preventing re-replication of DNA in a single cell cycle: evidence for a replication licensing factor. *J. Cell Biol.*, **122**, 993.

113. McKnight, S. L. and Miller, O. L. (1977) Electron microscopic analysis of chromatin replication in the cellular blastoderm *Drosophila melanogaster* embryo. *Cell*, **12**, 795.

114. Buongiorno-Nardelli, M., Micheli, G., Carri, M. T., and Marilley, M. (1982) A relationship between replicon size and supercoiled loop domains in the eukaryotic genome. *Nature*, **298**, 100.

115. Haase, S. B., Heinzel, S. S., and Calos, M. P. (1994) Transcription inhibits the replication of autonomously replicating plasmids in human cells. *Mol. Cell. Biol.*, **14**, 2516.

116. Diffley, J. F. X., Cocker, J. H., Dowell, S. J., and Rowley, A. (1994) Two steps in the assembly of complexes at yeast replication origins in vivo. *Cell*, **78**, 303.

117. Hsieh, D.-J., Camiolo, S. M., and Yates, J. L. (1993) Constitutive binding of EBNA1 protein to the Epstein–Barr virus replication origin oriP, with distortion of DNA structure during latent infection. *EMBO J.*, **12**, 4933.

118. Vaughn, J. P., Dijkwell, P., and Hamlin, J. L. (1990) Initiation of DNA replication occurs in a broad zone in the DHFR gene locus. *Cell*, **61**, 1075.

119. Delidakis, C. and Kafatos, F. C. (1989) Amplification enhancers and replication origins in the autosomal chorion gene cluster of *Drosophila*. *EMBO J.*, **8**, 891.

120. Iguchi-Ariga, S. M. M., Ogawa, N., and Ariga, H. (1993) *Biochim. Biophys. Acta*, **1172**, 73.

121. Ariizumi, K., Wang, Z., and Tucker, P. W. (1993) Immunoglobulin heavy chain enhancer is located near or in an initiation zone of chromosomal DNA replication. *Proc. Natl. Acad. Sci. USA*, **90**, 3695.

122. Vassilev, L. T. and Johnson, E. M. (1990) An initiation zone of chromosomal DNA replication located upstream of the c-myc gene in proliferating HeLa cells. *Mol. Cell. Biol.*, **10**, 4899.

123. Gale, J. M., Tobey, R. A., and D'Anna, J. A. (1992) Localization and DNA sequence of a replication origin in the rhodopsin gene locus of chinese hamster cells. *J. Mol. Biol.*, **224**, 343.

124. DePamphilis, M. L. and Wassarman, P. M. (1980) Replication of eukaryotic chromosomes: A close-up of the replication fork. *Annu. Rev. Biochem.*, **49**, 627.

125. Cusick, M. E., Wassarman, P. M., and DePamphilis, M. L. (1989) Application of nucleases to visualizing chromatin organization at replication forks. In *Methods in Enzymology – nucleosomes.* Wassarman, P. M. and Kornberg, R. D. (eds.). Vol. 170, pp. **290**.

126. Gruss, C. and Sogo, J. M. (1992) Chromatin replication. *BioEssays*, **14**, 1.

127. Stolzenburg, F., Gerwig, R., Dinkl, E., and Grummt, F. (1994) Structural homologies and functional similarities between mammalian origins of replication and amplification promoting sequences. *Chromosoma*, **103**, 209.

128. Mastrangelo, I. R., Held, P. G., Dailey, L., Wall, J. S., Hough, P. V. C., Heintz, N., and Heintz, N. H. (1993) RIP60 dimers and multimers of dimers and assemble link structures at an origin of bidirectional replication in the DHFR amplicon of CHO cells. *J. Mol. Biol.*, **232**, 766.

129. Strausfeld, U. P., Howell, M., Rempel, R., Maller, J. L., Hunt, T., and Blow, J. J. (1994) Cip1 blocks the initiation of DNA replication in *Xenopus* extracts by inhibition of cyclin-dependent kinases, *Curr. Biol.*, **4**, 876.

130. Roberts, J. M. and Weintraub, H. (1988). *Cis*-acting negative control of DNA replication in eukaryotic cells. *Cell*, **52**, 397.

131. Hyrien, O., Maric, C., and Méchali, M. (1995). Transition in specification of embryonic metazoan DNA replication origins. *Science*, **270**, 994.

132. Gencheva, M., Anachkova, B. and Russev, G. (1996) Mapping the sites of initiation of DNA replication in rat and human rRNA genes. *J. Biol. Chem.*, **271**, 2608.

133. Wu, J.-R. and Gilbert, D. M. (1996) A distinct G1-phase skp required to specify a mammalian replication origin. *Science*, in press.

5 | Role of nuclear structure in DNA replication

PAVEL HOZÁK, DEAN A. JACKSON, and PETER R. COOK

1. Introduction

The idea that DNA polymerases track along the template as they duplicate the DNA pervades our thinking (Figs 1a–c) (1, 2; and this volume). We imagine that it must be the small polymerase that moves along the very much larger template and that each polymerase acts relatively independently of others. However, recent evidence suggests many active polymerases are concentrated together into large nuclear structures – 'factories' – and that replication occurs as a number of templates slide past the many enzymes fixed in the factory; the template moves whilst the

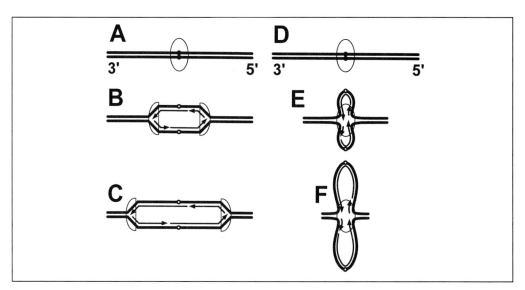

Fig. 1 Models for replication involving mobile (a–c) and immobile (d–f) polymerases. (a–c). The conventional model involves (a) binding of the polymerizing complex (oval) to the origin (black circles), before the complex splits into two; (b, c) the two halves then track along the template (thick lines) as nascent DNA (thin grey lines) is made. Arrows indicate direction of growth of nascent chains. (d–f) The alternative model involves passage of the template through a fixed complex and extrusion of nascent DNA. The origin is shown here detaching from the complex, but it may remain attached. Redrawn from Hozák and Cook (79).

enzyme remains stationary, rather than vice versa (Figs 1d–f). Then, higher-order structure dictates how and when DNA is replicated. Discussion concentrates on eukaryotic enzymes, but as the catalytic domains of all polymerases are structurally related (3), prokaryotic enzymes probably work in the same way.

Jacob *et al.* (4) first suggested in their 'replicon' model that polymerases might be attached to a larger structure; attachment of the bacterial chromosome to the membrane would regulate initiation of replication and specific growth of the membrane between the two attached progeny chromosomes could ensure that they segregated correctly to daughter cells. Although membranes do play several roles in these processes, they are not those that were initially envisaged (5, 6). Nevertheless, this model provoked a search for analogous membrane attachments in eukaryotes and this eventually led to the discovery that nascent DNA was attached, not to the membrane, but to various subnuclear structures, including 'matrices' and 'cages'. (For reviews, see 7–9.) As nascent DNA is sticky and as polymerases sometimes aggregate during extraction in the unphysiological salt concentrations used during isolation (10), it was easy to dismiss these results as reflecting nothing more than the artefactual precipitation of soluble polymerases on to an underlying structure (11). And as, at about this time, protein chemists were isolating relatively pure polymerases and finding that they worked *in vitro* in the absence of any larger structures, they included no such structures in their models. And, of course, it is these models that fill our textbooks.

2. Artefacts

Isotonic salt concentrations were not initially used during fractionation, or during polymerase assay, because they cause chromatin to aggregate into an unworkable mess. Therefore biochemists used more tractable conditions, often isolating nuclei and chromatin in (at least) one-tenth the physiological salt concentration. But this destroys the 30 nm chromatin fibre, extracts most DNA polymerases, and doubles the number of attachments of the chromatin fibre to the substructure. Often residual aggregation is suppressed by adding 'stabilizing' cations but these generate further artefactual attachments. Therefore it is not surprising that the slightly different conditions used to isolate 'matrices', 'scaffolds', and 'cages' ensures that each has its own characteristic set of sequences associated with a different subset of proteins. For example, matrix-attached regions or MARs are bound to various different proteins depending on the precise method of isolation, scaffold-attached regions or SARs are often specifically associated with topoisomerase II, and transcribed sequences are bound to 'cages'. Sceptics point to these differences and naturally suggest that some, or all, of these various complexes are artefacts and have no counterparts *in vivo* (11, 12).

Against this background, it was not surprising that convincing evidence for a role of larger structures in replication was only obtained when more physiological conditions were used.

3. Attached polymerases

The practical problems caused by the aggregation of chromatin at isotonic salt concentrations can be overcome if cells are encapsulated in agarose microbeads (50–150 µm diameter) before lysis. As agarose is permeable to small molecules such as nutrients, encapsulated cells grow in standard tissue culture media. They can also be permeabilized with a mild detergent in a 'physiological' buffer (13); then, most soluble cytoplasmic proteins and RNA diffuse out to leave the cytoskeleton surrounding the nucleus (Figs 2a–c). These cell remnants are protected by the agarose coat; importantly, they can be manipulated freely without aggregation whilst remaining accessible to probes like antibodies and enzymes. While no isolate is free of all artefacts, it seems unlikely that polymerases could have aggregated artefactually in this one: template integrity is retained, essentially all replicative activity of the living cell is preserved, and this activity increases 50-fold from low levels as cells pass from G1 into S phase.

Models involving tracking or immobile (i.e., attached) polymerases can be distinguished by cutting this encapsulated chromatin into fragments of <10 kb with an

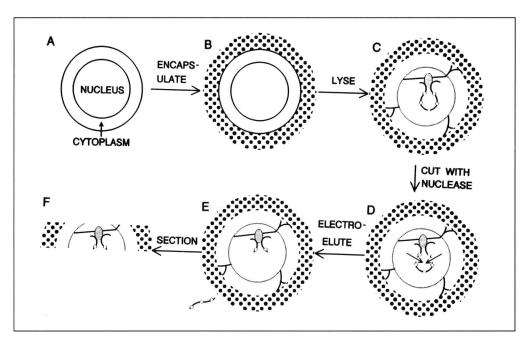

Fig. 2 Procedure for visualizing nucleoskeletons and associated structures. Cells (a) are encapsulated in agarose (dots) (b) and lysed (c) to leave a cytoskeleton, nuclear lamina (dotted circle), and nucleoskeleton (straight line) to which is attached a replication 'factory' (grey oval) and a DNA loop covered with nucleosomes (open circles). Dense chromatin obscures the nucleoskeleton but can be removed by (d) cutting the chromatin fibre with a restriction enzyme, and then (e) removing any unattached fragments electrophoretically. (f) After cutting thick sections, the nucleoskeleton and associated factory can now be seen in the electron microscope. From Cook (76) with permission of ICSU Press.

endonuclease and then removing electrophoretically any unattached fragments (Fig. 2). If polymerizing complexes are attached to a large structure (e.g., the grey oval that is attached in turn to the skeleton in Fig. 2c), they should remain in the agarose bead after electroelution; if unattached, they should electroelute from the bead with the eluting chromatin fragment in Fig. 2e. Cutting chromatin in unsynchronized and permeabilized HeLa cells, followed by electroelution of 75% of the chromatin, hardly reduced DNA polymerizing activity. As very large chromatin fragments (i.e., containing 150 kb DNA) can escape from beads, this polymerizing activity must be attached. Nascent DNA, whether labelled *in vivo* or *in vitro* by short incubations with [^{3}H]thymidine or [^{32}P]TTP for 0.5 or 2.5 min, respectively, also resisted elution (13–15). These results are simply explained if polymerases are attached, directly or indirectly, to a large structure like a skeleton that cannot be removed electrophoretically.

4. Visualization of replication sites

4.1 Light microscopy

Seeing is believing; recently discrete sites of replication have been visualized. Rat fibroblasts in S phase were incubated with bromodeoxyuridine, then sites where the analogue had been incorporated were labelled using fluorescently tagged antibodies directed against the analogue. These sites were not diffusely spread throughout nuclei but concentrated in ~150 foci (16, 17, 36). Early during S phase the foci are small and discrete; later they became larger (18–21) when centromeric and other heterochromatic regions are being replicated (22). Permeabilized mammalian cells (Fig. 3) (23, 24) or demembranated frog sperm in egg extracts (25–28) incorporate biotin-labelled dUTP into analogous foci, visualized in this case with fluorescently labelled streptavidin or appropriate antibodies. These foci are not fixation artefacts because similar foci are seen after incorporation of fluorescein–dUTP into permeabilized, but unfixed, cells (29). The foci remain even when most chromatin is removed (17, 24).

Surprisingly, *Xenopus* extracts will organize pure DNA from phage lambda into foci and replicate their DNA (30) (see Chapter 6); this must mean that the extract contains all the necessary signals. Quite unexpectedly, glycogen, present in high concentrations in the extracts, plays some role in focus formation (31).

A single replication fork could not incorporate sufficient labelled analogue under these conditions to allow detection, so many forks must be active in each focus. As we know the number of foci, the rate of fork progression, the spacing between forks, the size of the genome, and the length of S phase, we can calculate that ~40 forks must be active in each of the small foci found early in S phase. If polymerases track, the 40 loops would have to be confined in some way to one small part of the nucleus; if fixed, the enzymes must be locally concentrated on an underlying structure. Indeed, particles containing the requisite number of polymerases can be extracted by lightly disrupting somatic nuclei (32), whilst more vigorous procedures give the simpler isolates usually studied by biochemists (33).

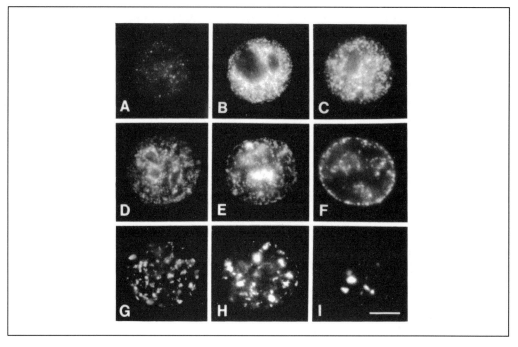

Fig. 3 Fluorescence micrographs of replication patterns found at different stages of S phase. Synchronized HeLa cells were encapsulated in agarose, permeabilized, incubated with biotin–dUTP and incorporation sites indirectly immunolabelled. Fluorescence marks replication sites, which change in number and distribution as cells progress from (A) early to (I) late S phase. Bar: 5 μm. From Hozák *et al.* (40) with permission of the Company of Biologists Ltd.

4.2 Electron microscopy

An early S-phase focus is ~200 nm in diameter, a size that is at the limit of resolution by light microscopy. Many attempts have been made over the years to localize replication sites by electron microscopy. For example, early autoradiographic studies showed that living cells incorporated [³H]thymidine into dispersed chromatin, close to variably sized masses of condensed chromatin (34, 35). But as the pathlength of β-particles is many tens, even hundreds, of nanometers long, autoradiography does not provide sufficient resolution to allow precise localization. Moreover, nascent DNA could well move away from its synthetic site during the relatively long labelling times required to incorporate sufficient label to be detected. The same is true of experiments in which synthetic sites are immunolabelled after incubating living cells for 5 min or more with bromodeoxyuridine (22, 26).

4.2.1 Replication factories

If permeabilized cells are incubated with biotin–dUTP under suboptimal conditions which limit elongation to a few nucleotides, replication sites can be immunolabelled with gold particles to a higher resolution; the centres of gold particles are now connected through an antibody bridge to the incorporated biotin and lie within 20 nm of it. In the first such experiments, most obscuring chromatin was also removed as described in Fig. 2 before 400 nm thick (resinless) sections were viewed in the electron microscope (Fig. 4). Residual clumps of chromatin could be seen attached to a 'diffuse skeleton' that ramified throughout the nucleus. (Although this network is morphologically complex, it contains lamin proteins, which are members of the intermediate-filament family; its 'core filaments' have the axial repeat typical of the family (37) and its nodes can be immunolabelled with anti-lamin antibodies (38). The lamins were named because it was originally thought that they were confined to the nuclear periphery; however, it now seems that the dense chromatin normally obscures the internal lamins.) Electron-dense bodies were scattered along this diffuse skeleton. They are present in the same numbers as the foci seen by light microcopy and during early S phase they are relatively constant in size (100–300 nm diameter). After elongating nascent DNA by ~500 nucleotides, gold particles were associated mainly with these electron-dense bodies (Fig. 4). As the incubation time was progressively increased, longer pieces of DNA were made and gold particles were found progressively further away from the dense bodies. This implies that nascent DNA is extruded from the dense body as templates pass through it.

As cells progress through S phase, these dense structures change in numbers, size, shape, and distribution just like the foci seen by light microscopy. Each contains ~40 active forks and associated leading- and lagging-strand polymerases, so it seems appropriate to call them replication 'factories'.

4.2.2 Factories are a subset of nuclear bodies

The factories seen in thick resinless sections correspond to a subset of the nuclear 'bodies' that have been seen over the years in conventional (thin) sections (34, 39); nuclear bodies are sites where biotin–dUTP is incorporated, they contain PCNA and they change in number, shape, and distribution just like factories (40).

4.3 Extra-factory replication

Although most replication takes place in factories, there is some extra-factory synthesis which increases as cells progress through S phase (40). There are special topological problems associated with replicating the last few basepairs between two replicons (41, 42) and this is true even if the two replicating forks are not immobilized (43). It is attractive to suppose that extra-factory labelling reflects a 'tidying-up' duplication of this hitherto unreplicated DNA.

5. Molecular content of replication factories

5.1 Attachment of origins

Ever since the replicon model was proposed, there has been discussion as to whether origins might be attached permanently to some underlying structure (44, 45).

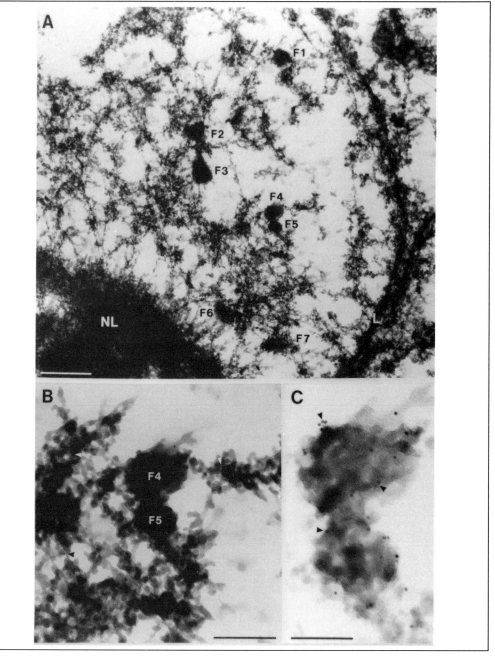

Fig. 4 Replication factories. Encapsulated cells were permeabilized with streptolysin, incubated with biotin–dUTP for 2.5 min, treated with nucleases, ~90% chromatin eluted, and sites of biotin incorporation immunolabelled with 5 nm gold particles. From Hozák *et al.* (24) with permission of Cell Press. (a) Seven replication factories (F1–7). NL: nucleolus. L: nuclear lamina. Of the 180 gold particles in the nuclear region (not visible at this magnification) 72% were in factories, indicating that factories were the site of DNA synthesis. Bar: 0.5 μm. (b) Higher-power view of F4 and F5. Three arrowheads point to the only extra-factory particles. Bar: 0.2 μm. (c) Underexposure and further 2 × magnification of F4 and F5 to show labelling; three arrowheads indicate some of the 30 gold particles. Bar: 0.1 μm.

However, experiments addressing this problem used unphysiological conditions which generate artefactual attachments of the chromatin fibre to the underlying skeleton (46); therefore, we still await a decisive answer to this question.

5.2 Proteins

Various proteins have been found in replication foci/factories including those specifically involved in synthesis as well as others that might be involved in regulation.

- DNA polymerase α (19, 24);
- PCNA (19, 23, 24);
- RP-A (47);
- cyclin A, cdk2, and RPA70, but not cyclin B1, cdc2, or RPA34 (48, 49). Cyclins are of interest because they may play a role in initiation (Chapters 6, 7, and 8).
- lamin B (50);
- DNA methyltransferase (51). This enzyme methylates DNA, so affecting gene expression. The presence of the enzyme at the replication site suggests that a methylation pattern, and so a pattern of gene expression, might be inherited by progeny duplexes. This transferase also contains a sequence that targets it to replication foci so similar sequences may target other proteins to replication sites.
- Various viral and cellular proteins have been found in the analogous 'factories' involved in viral replication; examples involving herpes virus include the virally encoded single-strand binding protein, ICP8 (52, 53), the cellular single-strand binding protein, as well as PCNA, Rb, p53, DNA ligase I, and DNA polymerase α, but not snRNP, c-Myc, p68, nucleophosmin (B23), or Ki67 (54).

6. Replication and transcription

Several tantalizing pieces of evidence suggest that RNA polymerases play a role in initiation (55).

Eukaryotic origins are always closely associated with transcription units and they are rich in binding sites for transcription factors which directly influence initiation (56, 57).

A temperature-sensitive mutant of BHK cells (i.e., tsAF8) arrests in G1 at the nonpermissive temperature, but microinjection of RNA polymerase II allows progression into S phase (58). A phenotypically similar mutant can also be rescued by transfection of a cDNA encoding the cell cycle gene 1 (CCG1) which turns out to encode a transcription factor associated with the TATA-binding protein, $TAF_{II}250$ (59–61). Another mutant can be rescued from G1 arrest by a human cell cycle gene, BN51; this is homologous with the yeast RPC53 gene that encodes a subunit of RNA polymerase III (62). It can hardly be fortuitous that so many cell cycle mutants are deficient in polymerases and transcription factors unless transcription is critically involved in replication.

Discrete sites of transcription, which are analogous to replication sites, can be immunolabelled after incubation with Br-UTP (63, 64). There are ~300 such sites in a HeLa nucleus during G1 phase, each one containing ~40 active RNA polymerases. On entry into S phase, the ~300 sites aggregate to colocalize with the ~150 sites of replication and even late during S phase replication sites remain transcriptionally active (65, but see also 66). Again it seems unlikely that sites of replication and transcription would be so closely associated unless transcription plays some role during replication.

7. Repair replication

The repair of damage induced in DNA by ultraviolet light involves excision of the damage and then repair synthesis to fill the gap. Early work suggested that such repair synthesis both did, and did not, take place on an underlying skeleton (67–69). Sites of repair synthesis have now been immunolabelled after incorporation of biotin–dUTP; again they are not diffusely spread throughout nuclei but concentrated in discrete foci (70, 71). The repair activity seems not to be as closely associated with the nucleoskeleton as the S-phase activity; after treatment with an endonuclease as in Fig. 2, most repaired DNA and the repair foci are removed from beads with the chromatin fragments. However, as electroelution destroys repair activity (but not the S-phase activity), repaired DNA might be attached *in vivo* through a polymerase that was removed by the procedure. So this approach has not allowed us to determine decisively whether repair sites are associated with a skeleton *in vivo*.

8. Models for replication by immobile polymerases

We are all familiar with models involving tracking polymerases and at first sight those with fixed enzymes seem more complicated. However, they are not more so, it is only that we are used to seeing the enzymes move rather than the template. Figs 5–7 illustrate three views of chromosome duplication by fixed enzymes. (See 72–74 for some earlier models.)

8.1 Low-power view (Fig. 5)

During G1 phase, we imagine that the chromatin fibre is looped by attachment to transcription factories that are, in turn, attached to a nucleoskeleton; euchromatin is usually organized into smaller loops than heterochromatin. Loops are tied either through transcription factors at promoters/enhancers, or through active RNA polymerases to the factories (75, 76). Late during G1, some as yet undefined signal (mediated by a kinase?) triggers assembly of a replication factory around about two transcription factories (55, 65). Here the structures involved in transcription play a role in replication by nucleating assembly of replication factories.

Two extreme models for the evolution of factories can be envisaged. In one (Fig. 5, left), small replication factories (1–6) might quickly assemble at the beginning of S phase around transcription factories, which would generally be on the same chromosome, and immediately become active. As one factory replicates many loops, not all origins need to fire simultaneously. By mid S phase, some small replication factories (e.g., 1) would have become redundant; they would be disassembled and their components incorporated into medium-sized factories that arise by growth and fusion of smaller factories (e.g., 2, 3). By late S phase, disassembly of most factories and growth/fusion of a few would generate large factories (i.e., 4–6) which replicate most heterochromatin; they will always contain nascent DNA as they grow from smaller active factories. Factories might apparently 'move' along the chromosome as now-redundant components at one (inactive) end become soluble and then become incorporated into the other (active) end.

In a second model, factories might be built *ab initio* at new sites on the chromatin fibre at different stages during S phase (Fig. 5, right). For example, the small factories A1, A2, and A3 might be built before the medium-sized factory B, which in turn is built before the large factory C, and some signal must trigger assembly at the appropriate time. If assembly of large factories takes time, we should see (but do not) partially built, but inactive, factories (not shown). Although the kinetics of appearance and activity of factories supports the (simple) first alternative involving factory growth and fusion, it is probably premature to exclude the second (40).

8.2 Medium-power view

Once a replication factory has been built, an origin within a loop will now lie close to polymerases on the surface of a factory and so will be likely to attach to one of them (Fig. 6; stage 1). Initiation occurs as strands within the origin separate (stage 2); then daughter duplexes are extruded (arrows), as parental duplexes slide (arrows) through the fixed polymerization site in the factory (stage 3). Replication continues (stage 4) as parental loops shrink (arrows) and daughter loops grow (arrows), until most DNA is duplicated (stage 5). Even the smallest factory replicates ~25 loops simultaneously. 'Tidying-up' replication of unreplicated DNA probably takes place later outside factories. Note that the origin is shown here to detach from the factory after initiation, but it may remain attached throughout the cycle (43).

8.3 High-power view

Figure 7 illustrates a model for the way that leading- and lagging-strand synthesis can be coordinated (see Chapter 1 for a model involving polymerases that are not fixed). DNA associated with a polymerizing complex (like the one illustrated in Fig. 6, stage 3), is shown in (a); it contains two replication forks. Parental duplexes slide into the symmetrical complex from each side (arrows). We will follow the path of the two grey segments behind the small open circles through the right-hand side of the complex. For clarity, the left-hand side has been omitted in panels (b)–(g), where only one fork is shown.

Continuous (leading-strand) synthesis is straightforward (illustrated at the bottom

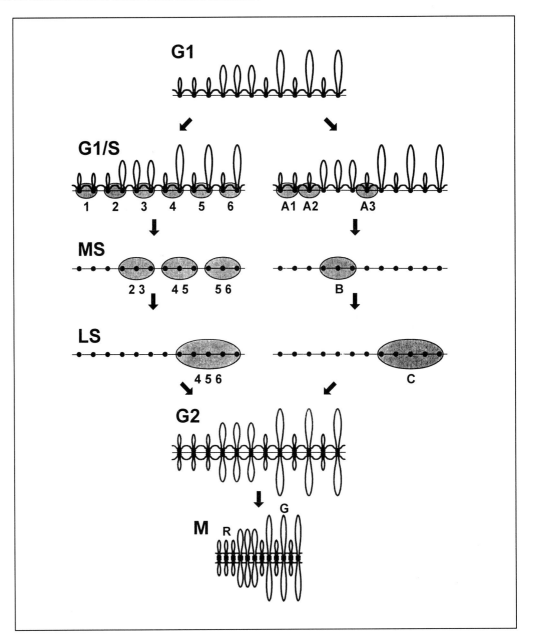

of each panel); nascent DNA is extruded as the template slides past the fixed polymerase. (a) The 3′ end of the grey segment of the parental strand (marked by the small open circle) became single stranded as it slid past the helicase/topoisomerase; it has just reached the polymerizing site (bottom oblong). (b) The grey segment now slides past the polymerase (arrow) as nascent DNA is extruded into the bottom loop, enlarging it. (c) Copying of the grey segment has been completed.

Fig. 5 Models for chromosome duplication (low-power view). The G1 chromatin fibre is shown looped by attachment to transcription factories (circles) on a skeleton (horizontal line). Small and large loops represent euchromatin and heterochromatin respectively. For the sake of clarity, only one of the ~50 loops attached to each transcription factory during G1, G1/S, and G2, and none of those during mid and late S phase (MS and LS), are shown. Replication factories (ovals) are assembled around transcription factories and then most DNA is synthesized as templates slide through the factories, although some 'tidying-up' replication occurs outside factories to give the duplicated G2 fibre. On entry into mitosis (M), the skeleton disassembles (without changing the contour length of loops) and residual transcription factories (plus associated euchromatin and heterochromatin) collapse on to the chromosome axis to generate R and G bands. Two extreme models for the evolution of factories are illustrated. Left: Large factories grow from small factories. At the G1/S border, small replication factories (1–6) quickly assemble around pairs of transcription factories (generally on the same chromosome) which immediately become active. As one factory replicates many loops, not all origins need to fire simultaneously. By mid S phase, some small replication factories (e.g., 1) have become redundant; on disassembly their components are incorporated into medium-sized factories that arise by growth and fusion (e.g., 2, 3). By late S phase, disassembly of most factories and growth/fusion of a few generates large factories (i.e., 4–6) which replicate most heterochromatin; they are always labelled as they grow from smaller active factories. Factories might apparently 'move' along the chromosome as now-redundant components at one (inactive) end become soluble and then become incorporated into the other (active end). Right: Factories of different sizes are created *ab initio* at new sites at different times (first A1, A2, A3, then B, then C). If assembly of large factories takes time, we should see (but do not) partially built, but inactive, factories (not shown). Redrawn from Hozák *et al.* (40).

Lagging-strand synthesis is more complicated (illustrated at the top of each panel); because synthesis occurs $5' \to 3'$ on strands of opposite polarity, the two strands must move locally past the relevant synthetic sites in opposite directions, which gives rise to a 'lagging-strand shuffle'. (a) The leading (5') end of the grey parental strand (marked by the small open circle) became single stranded as it slid past the helicase/topoisomerase. (b) The 5' end remains stationary as more of the grey segment slides into the complex (arrow), forming a loop. (c) A large loop has now formed. (d) The lagging (3') end of the grey segment attached to the lagging-strand polymerase/primase (top oblong) and then slid 'backwards' (arrow) past it as an RNA primer was extruded. Both ends of the primer remain attached, the 3' to the polymerase and the 5' end to another site (arbitrarily shown here as part of the helicase/topoisomerase). DNA synthesis now begins as the 'backwards' flow continues (arrow). (e) A loop of nascent DNA is extruded as the segment continues to slide 'backwards' (arrow) past the polymerase (f) Copying of the grey segment to give the nascent Okazaki fragment is complete. (g) The loop containing the grey segment and its Okazaki fragment plus its primer rearranges, sliding past the 'processing' site (where gaps are sealed and primers removed) to enlarge the top loop. The cycle now repeats.

Models for initiation and termination involving fixed polymerases can also be drawn (43).

9. Conclusions

A wide body of results now suggests that replication occurs in eukaryotes as each template slides through a fixed polymerizing site and that many sites are organized into complex structures called 'factories'. It is easy to imagine how gently disrupting a cell might first release factories, before further disruption yields simpler polymerizing

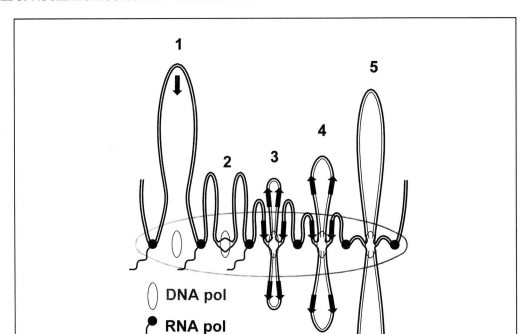

Fig. 6 A macroscopic model for replication (medium-power view). Various stages (1–5) in the replication of a chromatin loop, which is attached to a replication factory (large oval) through RNA polymerase/transcription factor complexes (filled circles); see text for details.

activities. Such pure enzymes might well track along templates *in vitro* but this does not necessarily mean that they do so *in vivo*. Immobilization in simple 'organisms' might be achieved by membrane attachment (e.g., in bacteria and mitochondria) or dimerization (e.g., in a virus), with a complex at one fork attached to, and immobilized by, its sister at the other. And if one kind of polymerase is immobile, it may be that other polymerases like those involved in RNA synthesis (77), reverse transcription (78), telomere synthesis, and recombination are also immobilized.

Immobilization of replicative enzymes in factories immediately suggests new and different mechanisms of control including:

(1) control of factory construction and activation;

(2) control of attachment of origins to the factory;

(3) restriction of subsequent attachments of the template to the factory to ensure that the template is replicated once, and only once, during the cell cycle; then structures, rather than soluble components, might 'license' only one cycle of replication (26).

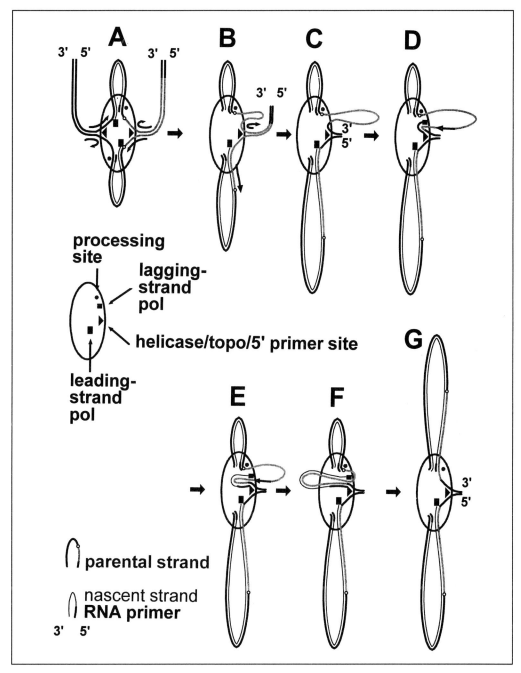

Fig. 7 A model for elongation (high-power view). Various stages during elongation are shown (see text for details). Redrawn from Cook (43).

Acknowledgements

We thank the Cancer Research Campaign, the Wellcome Trust, and The British Council for support.

References

1. Alberts, B., Bray, D., Lewis, J., Raff, M., Roberts, K. and Watson, J. D. (1983) *Molecular biology of the cell*. Garland, New York.
2. Darnell, J., Lodish, H., and Baltimore, D. (1986) *Molecular cell biology*. Scientific American Books, New York.
3. Braithwaite, D. K. and Ito, J. (1993) Compilation, alignment, and phylogenetic relationships of DNA polymerases. *Nucleic Acids Res.*, **21**, 787.
4. Jacob, F., Brenner, S., and Cuzin, F. (1963) On the regulation of DNA replication in bacteria. *Cold Spring Harbor Symp. Quant. Biol.*, **28**, 329.
5. Kornberg, A. and Baker, T. (1992) *DNA replication*, 2nd ed. W. H. Freeman, New York.
6. Funnell, B. E. (1993) Participation of the bacterial membrane in DNA replication and chromosome partition. *Trends Cell Biol.*, **3**, 20.
7. Jackson, D. A., McCready, S. J., and Cook, P. R. (1984) Replication and transcription depend on attachment of DNA to the nuclear cage. *J. Cell Sci. Suppl.*, **1**, 59.
8. Nelson, W. G., Pienta, K. J., Barrack, E. R., and Coffey, D. S. (1986) The role of the nuclear matrix in the organization and function of DNA. *Annu. Rev. Biophys. Chem.*, **15**, 457.
9. Berezney, R. (1991) The nuclear matrix: a heuristic model for investigating genomic organization and function in the cell nucleus. *J. Cell. Biochem.*, **47**, 109.
10. Martelli, A. M., Gilmour, R. R., Falcieri, E., Manzoli, F. A., and Cocco, L. (1990) Temperature-dependent association of DNA polymerase α activity with the nuclear matrix. *Exp. Cell Res.*, **190**, 227.
11. Cook, P. R. (1988) The nucleoskeleton: artefact, passive framework or active site? *J. Cell Sci.*, **90**, 1.
12. Jack, R. S. and Eggert, H. (1992) The elusive nuclear matrix. *Eur. J. Biochem.*, **209**, 503.
13. Jackson, D. A., Yuan, J., and Cook, P. R. (1988) A gentle method for preparing cyto- and nucleo-skeletons and associated chromatin. *J. Cell Sci.*, **90**, 365.
14. Jackson, D. A. and Cook, P. R. (1986) Replication occurs at a nucleoskeleton. *EMBO J.*, **5**, 1403.
15. Jackson, D. A. and Cook, P. R. (1986) A cell-cycle-dependent DNA polymerase activity that replicates intact DNA in chromatin. *J. Mol. Biol.*, **192**, 65.
16. Nakamura, H., Morita, T., and Sato, C. (1986) Structural organisation of replicon domains during DNA synthetic phase in the mammalian nucleus. *Exp. Cell Res.*, **165**, 291.
17. Nakayasu, H. and Berezney, R. (1989) Mapping replication sites in the eukaryotic cell nucleus. *J. Cell Biol.*, **108**, 1.
18. Fox, M. H., Arndt-Jovin, D. J., Jovin, T. M., Baumann, P. H., and Robert-Nicoud, M. (1991) Spatial and temporal distribution of DNA replication sites localized by immunofluorescence and confocal microscopy in mouse fibroblasts. *J. Cell Sci.*, **99**, 247.
19. Kill, I. R., Bridger, J. M., Campbell, K. H. S., Maldonado-Codina, G., and Hutchison, C. J. (1991) The timing of the formation and usage of replicase clusters in S-phase nuclei of human diploid fibroblasts. *J. Cell Sci.*, **100**, 869.

20. Humbert, C. and Usson, Y. (1992) Eukaryotic DNA replication is a topographically ordered process. *Cytometry*, **13**, 603.

21. Manders, E. M. M., Stap, J., Brakenhoff, G. J., van Driel, R., and Aten, J. A. (1992) Dynamics of three-dimensional replication patterns during the S-phase, analysed by double labelling of DNA by confocal microscopy. *J. Cell Sci.*, **103**, 857.

22. O'Keefe, R. T., Henderson, S. C., and Spector, D. L. (1992) Dynamic organization of DNA replication in mammalian cell nuclei: spatially and temporally defined replication of chromosome-specific α-satellite sequences. *J. Cell Biol.*, **116**, 1095.

23. Bravo, R. and Macdonald-Bravo, H. (1987) Existence of two populations of cyclin/proliferating cell nuclear antigen during the cell cycle: association with DNA replication sites. *J. Cell Biol.*, **105**, 1549.

24. Hozák, P., Hassan, A. B., Jackson, D. A., and Cook, P. R. (1993) Visualization of replication factories attached to a nucleoskeleton. *Cell*, **73**, 361.

25. Blow, J. J. and Laskey, R. A. (1986) Initiation of DNA replication in nuclei and purified DNA by a cell-free extract of *Xenopus* eggs. *Cell*, **47**, 577.

26. Blow, J. J. and Laskey, R. A. (1988) A role for the nuclear envelope in controlling DNA replication within the cell cycle. *Nature*, **332**, 546.

27. Hutchison, C. J., Cox, R., Drepaul, R. S., Gomperts, M., and Ford, C. C. (1987) Periodic DNA synthesis in cell-free extracts of *Xenopus* eggs. *EMBO J.*, **6**, 2003.

28. Hutchison, C. J., Cox, R., and Ford, C. C. (1988) The control of DNA replication in a cell-free extract that recapitulates a basic cell cycle *in vitro*. *Development*, **103**, 553.

29. Hassan, A. B. and Cook, P. R. (1993) Visualization of replication sites in unfixed human cells. *J. Cell Sci.*, **105**, 541.

30. Cox, L. S. and Laskey, R. A. (1991) DNA replication occurs at discrete sites in pseudo-nuclei assembled from purified DNA *in vitro*. *Cell*, **66**, 271.

31. Hartl, P., Olson, E., Dang, T., and Forbes, D. J. (1994) Nuclear assembly with λ DNA in fractionated *Xenopus* egg extracts: an unexpected role for glycogen in formation of a higher order chromatin intermediate. *J. Cell. Biol.*, **124**, 235.

32. Tubo, R. A. and Berezney, R. (1987) Identification of 100 and 150S DNA polymerase alpha-primase megacomplexes solubilized from the nuclear matrix of regenerating rat liver. *J. Biol. Chem.*, **262**, 5857.

33. Waga, S., Bauer, G., and Stillman, B. (1994) Reconstitution of complete SV40 DNA replication with purified replication factors. *J. Biol. Chem.*, **269**, 10923.

34. Bouteille, M., Laval, M., and Dupuy-Coin, A. M. (1974) Localization of nuclear functions as revealed by ultrastructural autoradiography and cytochemistry. In *The cell nucleus.* Busch, H. (ed.). Academic Press, New York, Vol. 1, p. 5.

35. Fakan, S. (1978) High resolution autoradiographic studies on chromatin functions. In *The cell nucleus.* Busch, H. (ed.). Academic Press, New York, Vol. 5, p. 3.

36. Mazzotti, G., Rizzoli, R., Galanzi, A., Papa, S., Vitale, M., Falconi, M., Neri, L. M., Zini, N., and Maraldi, N. M. (1990) High-resolution detection of newly-synthesized DNA by anti-bromodeoxyuridine antibodies identifies specific chromatin domains. *J. Histochem. Cytochem.*, **38**, 13.

37. Jackson, D. A. and Cook, P. R. (1988) Visualization of a filamentous nucleoskeleton with a 23 nm axial repeat. *EMBO J.*, **7**, 3667.

38. Hozák, P., Sasseville, A. M.-J., Raymond, R., and Cook, P. R. (1994) Lamin proteins form an internal nucleoskeleton as well as a peripheral lamina in human cells. *J. Cell Sci.*, **108**, 635.

39. Brasch, K. and Ochs, R. L. (1992) Nuclear bodies (NBs): a newly 'rediscovered' organelle. *Exp. Cell Res.*, **202**, 211.

40. Hozák, P., Jackson, D. A., and Cook, P. R. (1994) Replication factories and nuclear bodies: the ultrastructural characterization of replication sites during the cell cycle. *J. Cell Sci.*, **107**, 2191.

41. Sundin, O. and Varshavsky, A. (1980) Terminal stages of SV40 DNA replication proceed via multiply intertwined catenated dimers. *Cell*, **21**, 103.

42. Sundin, O. and Varshavsky, A. (1981) Arrest of segregation leads to accumulation of highly intertwined catenated dimers: dissection of the final stages of SV40 DNA replication. *Cell*, **25**, 659.

43. Cook, P. R. (1991) The nucleoskeleton and the topology of replication. *Cell*, **66**, 627.

44. Dijkwel, P. A., Wenink, P. W., and Poddighe, J. (1986) Permanent attachment of replication origins to the nuclear matrix in BHK-cells. *Nucleic Acids Res.*, **14**, 3241.

45. Dijkwel, P. A., Vaughn, J. P., and Hamlin, J. L. (1991) Mapping of replication initiation sites in mammalian genomes by two-dimensional analysis: stabilization and enrichment of replication intermediates by isolation of the nuclear matrix. *Mol. Cell. Biol.*, **11**, 3850.

46. Jackson, D. A., Dickinson, P., and Cook, P. R. (1990) The size of chromatin loops in HeLa cells. *EMBO J.*, **9**, 567.

47. Adachi, Y. and Laemmli, U. K. (1992) Identification of nuclear pre-replication centres poised for DNA synthesis in *Xenopus* egg extracts: immunolocalization study of replication protein A. *J. Cell Biol.*, **119**, 1.

48. Cardoso, M. C., Leonhardt, H., and Nadal-Ginard, B. (1993) Reversal of terminal differentiation and control of DNA replication: cyclin A and cdk2 specifically localize at subnuclear sites of DNA replication. *Cell*, **74**, 979.

49. Sobczak-Thepot, J., Harper, F., Florentin, Y., Zindy, F., Brechot, C., and Puvion, E. (1993) Localization of cyclin A at sites of cellular DNA replication. *Exp. Cell Res.*, **206**, 43.

50. Moir, R. D., Montag-Lowy, M., and Goldman, R. D. (1994) Dynamic properties of nuclear lamins: lamin B is associated with sites of DNA replication. *J. Cell Biol.*, **125**, 1201.

51. Leonhardt, H., Page, A. W., Weier, H.-U., and Bestor, T. H. (1992) A DNA targeting sequence directs DNA methyltransferase to sites of DNA replication in mammalian nuclei. *Cell*, **71**, 865.

52. Quinlan, M. P., Chen, L. B., and Knipe, D. M. (1984) The intranuclear location of a herpes simplex virus DNA-binding protein is determined by the status of viral DNA replication. *Cell*, **36**, 857.

53. Kops, A. B. and Knipe, D. M. (1988) Formation of viral DNA replication structures in herpes virus-infected cells requires a viral DNA binding protein. *Cell*, **55**, 857.

54. Wilcock, D. and Lane, D. P. (1991) Localization of p53, retinoblastoma and host replication proteins at sites of viral replication in herpes-infected cells. *Nature*, **349**, 429.

55. Hassan, A. B. and Cook, P. R. (1994) Does transcription by RNA polymerase play a direct role in the initiation of replication? *J. Cell Sci.*, **107**, 1381.

56. Heintz, H. H. (1992) Transcription factors and the control of DNA replication. *Curr. Opin. Cell Biol.*, **4**, 459.

57. DePamphlis, M. L. (1993) Eukaryotic DNA replication: anatomy of an origin. *Annu. Rev. Biochem.*, **62**, 29.

58. Waechter, D. E., Avignolo, C., Freund, E., Riggenbach, C., Mercer, W., McGuire, P. M.,

and Baserga, R. (1984) Microinjection of RNA polymerase II corrects the temperature-sensitive defect of tsAF8 cells. *Mol. Cell. Biochem.*, **60**, 77.

59. Ruppert, S., Wang, E. H., and Tjian, R. (1993) Cloning and expression of human TAF$_{II}$250: a TBP-associated factor implicated in cell-cycle regulation. *Nature*, **362**, 175.

60. Hisatake, K., Hasegawa, S., Takada, R., Nakatani, Y., Horikoshi, M., and Roeder, R. G. (1993) The p250 subunit of native TATA box-binding factor TFIID is the cell-cycle regulatory protein CCG1. *Nature*, **362**, 179.

61. Kokubo, T., Gong, D.-W., Yamashita, S., Horikoshi, M., Roeder, R. G., and Nakatani, Y. (1993) *Drosophila* 23-kD TFIID subunit, a functional homologue of the human cell cycle gene product, negatively regulates DNA binding of the TATA box-binding subunit of TFIID. *Genes Dev.*, **7**, 1033.

62. Mann, C., Micouin, J. Y., Chiannilkulchai, N., Treich, I., Buhler, J. M., and Sentenac, A. (1992) RPC53 encodes a subunit of *Saccharomyces cerevisiae* RNA polymerase C (III) whose inactivation leads to a predominantly G1 arrest. *Mol. Cell. Biol.*, **12**, 4314.

63. Jackson, D. A., Hassan, A. B., Errington, R. J., and Cook, P. R. (1993) Visualization of focal sites of transcription within human nuclei. *EMBO J.*, **12**, 1059.

64. Wansink, D. G., Schul, W., van der Kraan, I., van Steensel, B., van Driel, R., and de Jong, L. (1993) Fluorescent labelling of nascent RNA reveals transcription by RNA polymerase II in domains scattered throughout the nucleus. *J. Cell Biol.*, **122**, 283.

65. Hassan, A. B., Errington, R. J., White, N. S., Jackson, D. A., and Cook, P. R. (1994) Replication and transcription sites are colocalized in human cells. *J. Cell Sci.*, **107**, 425.

66. Wansink, D. G., Manders, E. E. M., van der Kraan, I., Aten, J. A., van Driel, R., and de Jong, L. (1994) RNA polymerase II transcription is concentrated outside replication domains throughout S-phase. *J. Cell Sci.*, **107**, 1449.

67. McCready, S. J. and Cook, P. R. (1984) Lesions induced in DNA by ultra-violet light are repaired at the nuclear cage. *J. Cell Sci.*, **70**, 189.

68. Harless, J. and Hewitt, R. R. (1987) Intranuclear localization of UV-induced DNA repair in human VA13 cells. *Mut. Res.*, **183**, 177.

69. Mullenders, L. H. F., van Leeuwen, A. C. van K., van Zeeland, A. A., and Natarajan, A. T. (1988) Nuclear matrix associated DNA is potentially repaired in normal human fibroblasts, exposed to a low dose of ultraviolet light but not in Cockayne's syndrome fibroblasts. *Nucleic Acids Res.*, **16**, 10 607.

70. Jackson, D. A., Balajee, A. S., Mullenders, L., and Cook, P. R. (1994) Sites in human nuclei where DNA damaged by ultra-violet light is repaired: visualization and localization relative to the nucleoskeleton. *J. Cell Sci.*, **107**, 1745.

71. Jackson, D. A., Hassan, A. B., Errington, R. J., and Cook, P. R. (1994) Sites in human nuclei where damage induced by ultra-violet light is repaired: localization relative to transcription sites and concentrations of proliferating cell nuclear antigen and the tumour suppressor protein, p53. *J. Cell Sci.*, **107**, 1753.

72. Dingman, C. W. (1974) Bidirectional chromosome replication: some biological considerations. *J. Theor. Biol.*, **43**, 187.

73. Pardoll, D. M., Vogelstein, B., and Coffey, D. S. (1980) A fixed site of DNA replication in eukaryotic cells. *Cell*, **19**, 527.

74. McCready, S. J., Godwin, J., Mason, D. W., Brazell, I. A., and Cook, P. R. (1980) DNA is replicated at the nuclear cage. *J. Cell Sci.*, **46**, 365.

75. Jackson, D. A. and Cook, P. R. (1993) Transcriptionally-active minichromosomes are attached transiently in nuclei through transcription units. *J. Cell Sci.*, **105**, 1143.

76. Cook, P. R. (1994) RNA polymerase: structural determinant of the chromatin loop and the chromosome. *BioEssays*, **16**, 425.

77. Cook, P. R. (1989) The nucleoskeleton and the topology of transcription. *Eur. J. Biochem.*, **185**, 487.

78. Cook, P. R. (1993) A model for reverse transcription by a dimeric enzyme. *J. Gen. Virol.*, **74**, 691.

79. Hozák, P. and Cook, P. R. (1994) Replication factories. *Trends Cell Biol.*, **4**, 48.

6 | Chromosome replication in *Xenopus* egg extracts

J. JULIAN BLOW

1. Introduction

Chromosomal DNA replication must be coordinated with other events of the cell division cycle. This means that:

(1) DNA replication should only occur when the cell is legitimately committed to undergoing cell division;

(2) the entire genome must be fully replicated in each S phase, as any section of DNA left unreplicated at mitosis would lead to chromosome breakage;

(3) no DNA should be replicated more than once in a single cell cycle, as this is likely to lead to an irreversible amplification of a section of the genome;

(4) S phase and M phase must occur in strict alternation to maintain the ploidy of the cell.

Cell-free extracts of eggs of the South African clawed toad *Xenopus laevis* support the major transitions of the eukaryotic cell cycle, including complete chromosome replication under normal cell cycle control (1). Apart from a similar system derived from embryos of the fruit fly *Drosophila melanogaster* (2), it is currently the only eukaryotic cell-free system that supports efficient chromosome replication *in vitro*. Since replication in the *Xenopus* system is regulated in the four ways mentioned above, it offers the opportunity to study the way that DNA replication is coordinated within the cell cycle.

2. Overview of the *Xenopus* system

When *Xenopus* eggs are lysed by gentle centrifugation (approximately $10\,000 \times g$), the resultant cell-free extracts ('low-speed supernatants') continue to support almost all of the intracellular activities of the intact egg. Low-speed supernatants contain abundant particulate material including ribosomes and membrane vesicles, and can support both translation and the assembly of exogenous DNA into functional interphase nuclei (3). The ability of the extract to assemble nuclei allows it to support the efficient initiation of DNA replication on nuclear templates, leading to the precise duplication of added DNA molecules (1).

2.1 The *Xenopus* egg

The *Xenopus* egg is arrested at second meiotic metaphase with a diploid set of chromosomes aligned on the spindle, awaiting fertilization. Transcription does not occur in the egg, and does not resume until later in development, although translation of preexisting mRNA continues. As expected of a mitotic cell, DNA introduced into the egg is not replicated so long as the metaphase arrest is maintained (4–7). On fertilization, when a sperm pronucleus pierces the egg plasma membrane, a calcium wave is generated which releases the egg from its metaphase arrest and allows it to progress into interphase of the first cell cycle. If cell-free extracts are made from unactivated eggs, they can be released from metaphase arrest by addition of low levels of calcium (8).

2.2 Early embryonic cell cycles

After fertilization, the activated egg undergoes 12 synchronous rounds of cell division in about 8 hours. It oscillates between S phase and mitosis, taking about 15 min for each, with virtually no G1 or G2 phases. During each interphase, chromosomal DNA is assembled into a nucleus and is replicated precisely once; during mitosis the nuclear envelope breaks down whilst the chromosomes condense and are segregated to the daughter cells. No protein synthesis is required for the early embryo to progress from mitosis into interphase and undergo a complete S phase (4, 9). However, protein synthesis is eventually required for the embryo to progress from interphase back into mitosis.

The *Xenopus* embryo appears to lack certain checkpoint functions which coordinate cell cycle events in most other cell types (Fig. 1). *Xenopus* embryos treated with aphidicolin to block DNA synthesis still undergo mitosis and cell division on schedule (10). Similarly the basic oscillation between S phase and mitosis is not disturbed if chromosome segregation or cytokinesis have been blocked by inhibitors (10–12). This lack of checkpoint control may be due to the very low ratio of DNA to cytoplasm in the early *Xenopus* embryo (at fertilization, there is only a single diploid set of chromosomes in approximately 0.5 µl cytoplasm) which may be unable to restrain cell cycle activities present throughout the cytoplasm. In support of this idea, when egg extracts are loaded with large quantities of DNA they become capable of arresting entry into mitosis if DNA replication is not complete (13–15). Since the fertilized egg embarks on a series of preprogrammed cell divisions, the cell cycle of the early embryo is also free of two other controls that exist in most other cell types: the need to regulate cell division with cell growth, and the need to regulate cell division with extracellular signals provided by growth factors or nutrients. The cell cycles of the early *Xenopus* embryo thus appears to be stripped down to its essential components.

Two important cell cycle dependencies are nevertheless maintained in the early embryo (Fig. 1). Firstly, the initiation of DNA replication only occurs during interphase, not in mitosis (7). Secondly, if embryos or extracts are blocked in G2, DNA

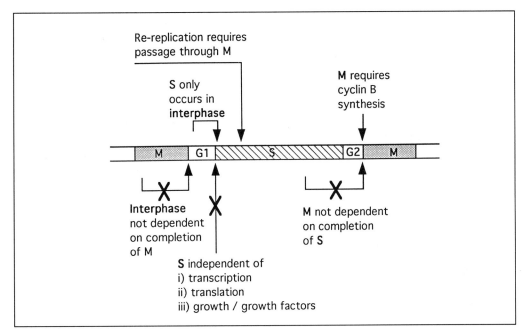

Fig. 1 Checkpoint controls functioning in the early *Xenopus* embryo. Above the outline cell cycle are shown cell cycle dependencies that operate in the *Xenopus* system. Below are three cell cycle dependencies that do not operate in the *Xenopus* system. See text for more details.

does not undergo further rounds of replication and the ploidy of the cells is maintained (4, 9).

3. The role of nuclear assembly in DNA replication

3.1 Nuclear assembly

When DNA is added to *Xenopus* low-speed supernatants a characteristic series of morphological events take place leading to the assembly of interphase nuclei (Fig. 2). Sperm chromatin, which is initially highly condensed, undergoes a very rapid decondensation (<5 min) as sperm protamines are removed and replaced by histones present in the egg extract. This chromatin remodelling process requires the presence of nucleoplasmin, a molecular chaperone capable of binding histones (16, 17). Membrane vesicles are then seen assembling around the periphery of this decondensed chromatin (Fig. 2, 20–25 min) (3, 18, 19). Nuclear pores are also assembled onto the surface of the chromatin. The result is a double unit envelope studded with nuclear pores, the typical structure of the eukaryotic nuclear envelope (3, 7, 19–22). With a complete nuclear envelope around the DNA, selective nuclear protein accumulation rapidly occurs, leading to a significant swelling of the nuclear interior (Fig. 2, 30–40 min) (20, 23). An early consequence of this import is that a nuclear lamina is laid down inside the nuclear envelope (21, 24–26).

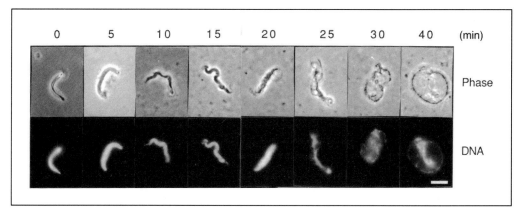

Fig. 2 Nuclear assembly in the *Xenopus* cell-free system. Photomicrographs of demembranated *Xenopus* sperm nuclei incubated in the cell-free system for the indicated periods of time, and viewed by phase contrast (upper panel) or by UV fluorescence to show total DNA (lower panel). Only after nuclear assembly is complete, evidenced by acquisition of a complete phase-dense nuclear envelope (as seen at 40 min), does the initiation of DNA replication occur. Scale bar 10 μm. Reproduced from Blow (60).

3.2 Dependence of replication on nuclear assembly

In eukaryotic cells, chromosome replication occurs within the nucleus. This appears to be functionally important, as the *Xenopus* system requires template DNA to be assembled into nuclei before initiation can occur. When low-speed supernatants are centrifuged hard to remove particulate material, the resultant high-speed supernatants fail both to assemble interphase nuclei and to initiate DNA replication (7, 18, 21, 22). Both these activities can be restored to the high-speed supernatants by adding back pelleted membrane material to the supernatants. However, high-speed supernatants can support replication of intact tissue culture nuclei (27). When naked DNA is microinjected into *Xenopus* eggs or is incubated in low-speed supernatants, it is assembled into structures resembling normal interphase nuclei, termed 'pseudo-nuclei' (1, 20, 21, 28). Only a fraction of the total DNA is incorporated into pseudo-nuclei: DNA that is not assembled into nuclei does not undergo the initiation of DNA replication, whilst DNA in pseudo-nuclei does (7). These results strongly suggest that some aspect of nuclear assembly is required before DNA replication occurs in the *Xenopus* system. A similar dependence of initiation on nuclear assembly is seen in extracts of *Drosophila* eggs (2). Since high-speed supernatants support the elongation stage of DNA replication (1, 29–31), it appears that nuclear assembly is specifically required for the initiation of DNA replication.

Why is the initiation of DNA replication dependent on nuclear assembly in the *Xenopus* system? One plausible explanation is that it permits the selective nuclear accumulation of proteins involved in initiation. Consistent with this, wheat germ agglutinin, a lectin that blocks nuclear protein accumulation (32, 33), renders extracts unable to support the initiation of DNA replication although they are capable of elongating previously initiated replication forks (30).

In addition, nuclear assembly may provide structural components of the nucleus required for DNA replication. A number of proteins thought to be needed for the structural integrity of nuclei are required for DNA replication in the *Xenopus* system. The lamina is assembled within the nucleus once nuclear envelope assembly is complete (21). If extracts are immunodepleted of lamin precursors, subsequent DNA replication does not occur, although nuclear envelopes are assembled (24–26). The involvement of the lamina in the initiation of DNA replication is unexpected, since its position against the nuclear envelope places it a long way away from the sites of DNA replication in the nuclear interior (34). In somatic cells, a small proportion of lamin proteins have been colocalized to the nuclear interior (35), which could potentially give them a direct role in DNA replication (36). RCC1, a chromatin-associated protein which is involved in the regulation of chromosome condensation, is also required for DNA replication in the *Xenopus* system (37), and appears to be involved in the maintenance of normal interphase nuclear architecture (38).

3.3   Subnuclear organization of DNA replication

Replication forks are not scattered randomly throughout the nucleus of higher eukaryotic cells, but are clustered into discrete 'replication foci' or 'replication factories' (see Chapter 5). This clustering of active replication forks is also seen in the *Xenopus* cell-free system (34) (Fig. 3). When nuclei assembled from sperm chromatin are pulse-labelled with biotinylated dNTPs, DNA synthesis is localized to approximately 100–200 discrete foci in the nuclear interior. With 100–200 thousand active replication forks in each of these nuclei, each replication focus must contain in the order of a thousand individual replication forks. It seems quite likely that the need to assemble replication forks into these structures is part of the reason why the initiation of DNA replication is dependent on nuclear assembly in the *Xenopus* system.

Replication proteins can be observed in replication foci in the *Xenopus* system. RP-A, a single-stranded DNA-binding protein required for DNA replication (see Chapters 1 and 2), binds rapidly to chromatin incubated in *Xenopus* extract (39, 40). RP-A binding occurs in 'prereplication foci' which become sites of replication once

Fig. 3 Replication foci in a nucleus replicating in the *Xenopus* cell-free system. Sperm chromatin replicating in the *Xenopus* cell-free system was pulse labelled with biotinylated dUTP. Sites of biotin incorporation were labelled with fluorescent streptavidin, and then analysed by confocal microscopy. A single optical section is shown. Each bright spot contains several hundred replication forks that remain clustered together throughout S phase. Reproduced from Mills *et al.* (34).

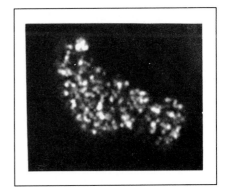

initiation occurs. The binding of RP-A to prereplication foci is relatively weak, and does not assume the character of association with single-stranded DNA until initiation occurs. This suggests that the unwinding of DNA presumably required for initiation does not occur prior to nuclear assembly in the *Xenopus* system. Binding of RP-A to prereplication foci does not occur during mitosis (40), and this may be one of the reasons why initiation does not occur at this stage of the cell cycle (7). PCNA, the DNA polymerase δ auxiliary factor (see Chapter 1) is also found tightly bound to chromatin at the replication foci (26, 41). PCNA is only seen in foci after nuclear assembly has occurred, but there is a short period of time when PCNA can be seen in foci that have not synthesized significant quantities of DNA (41). These PCNA-containing foci may be at an early stage in the process of initiation.

3.4 Nuclear structure and replication origin usage

Xenopus eggs and egg extracts can replicate a wide variety of DNA templates, including naked DNA templates such as small plasmid molecules (1, 4, 7, 21, 42–45). Replication of naked DNA templates is less efficient than the replication of sperm chromatin, since the DNA is only inefficiently assembled into pseudo-nuclei (7, 46). However, once nuclear assembly has occurred, replication appears to be very similar to that occurring with sperm chromatin (1, 7, 21, 43), with replication occurring in discrete foci (47). The wide range of different DNA molecules replicated in *Xenopus* means that special DNA sequences are required neither for the initiation of DNA replication nor for preventing rereplication of DNA in a single cell cycle. In this respect the *Xenopus* embryo appears to behave more like somatic cells of higher eukaryotes than lower eukaryotes such as yeast (see Chapters 3 and 4).

The lack of sequence specificity of initiation has been confirmed by neutral/neutral 2D gel mapping techniques (44, 45) and electron microscopy (48). These studies showed single initiation bubbles at many different sites on different copies of replicating plasmid DNA molecules. Similar results were obtained by 2D gel analysis of replicating sperm chromatin, showing initiation bubbles scattered throughout the rDNA of sperm chromatin (49). Despite the lack of sequence specificity for initiation, apparently normal replicative intermediates were observed. About half of the nascent DNA transiently appeared in Okazaki-sized fragments which were rapidly ligated into higher molecular weight DNA (1). These results suggest a 'classical' replication process in the *Xenopus* embryo, with each initiation event generating a single replication bubble that grows only from its ends, so that with the exception of short-lived Okazaki fragments, it contains a continuous strand of nascent DNA on each side of the bubble (Fig. 4a).

S phase duration in the early *Xenopus* embryo is about 15 min, and with a replication fork rate of 10 nt/s (44, 50), this means that adjacent origins should be spaced no more than 15 kb apart (44, 49, 51). However, if the 100 000 initiation events were to occur at completely random sites on the genome, there would be a statistical distribution of replicon sizes, with some excessively large replicons (44). Since the early embryo is not proficient at delaying mitosis if S phase is incomplete, and an

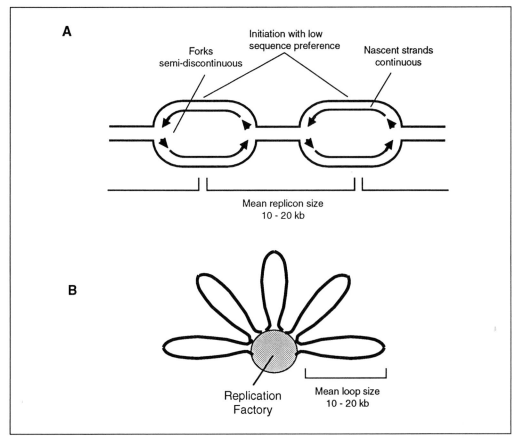

Fig. 4 Replication origin usage in the *Xenopus* system. (a) Cartoon summarizing experiments analysing replication origin usage and the structure of replicative intermediates in the *Xenopus* system. (b) Cartoon to show a possible role for chromosome looping in generating a relatively constant replicon size in the *Xenopus* system. See text for more details.

unreplicated stretch of DNA is likely to be broken if mitosis goes ahead, the *Xenopus* embryo must have an alternative mechanism to regulate the replicon size.

Instead of being dictated by DNA sequence, replicon size in the *Xenopus* early embryo may be mediated by nuclear structure (Fig. 4b). In interphase cells, chromosomal DNA appears to be organized into supercoiled loops, the size of which correlate well with the average replicon size as they both increase during *Xenopus* development (52). In particular, each copy of rDNA appeared to comprise one supercoiled loop (53), which supports only a single initiation event (44, 49). When intact hamster nuclei were transferred to the *Xenopus* extract, they continued to use the dihydrofolate reductase (DHFR) origin of replication, although naked DNA containing the DHFR gene showed no preferential initiation at this site (54). This further suggests that initiation sites are determined by some higher-order structure of the eukaryotic nucleus (see Chapter 4).

3.5 The nucleus as a fundamental unit of DNA replication

The dependence of initiation on an intact nuclear envelope means that the *Xenopus* system can be used to study the way that cytoplasmic signals induce nuclei to initiate DNA replication. When many sperm nuclei are added to *Xenopus* extract, they are not all assembled into nuclei at the same rate. Flow cytometry was used to measure the replication kinetics of individual nuclei under such conditions (51). Different nuclei started to replicate at different times once nuclear assembly was complete. As they entered S phase, each nucleus underwent a burst of approximately 50 000 near-synchronous initiation events. Even though nuclei continued to enter S phase over a significant period of time, no nuclei replicated more than once. This pattern of replication is shown in Fig. 5. Nuclei act as individual 'units' each receiving a signal to replicate from the cytoplasm, which causes them to undergo once and only once a burst of initiation events (51). Further experiments indicated that the feature that defines this 'unit' of replication is indeed the nuclear envelope (55). When demembranated chicken erythrocyte nuclei were incubated in *Xenopus* extract, they often aggregated, so that the *Xenopus* extract surrounded a number of different nuclei with a continuous nuclear envelope. All of the DNA within each of these 'multinuclear aggregates' started to replicate at about the same time despite originally being derived from a number of chicken nuclei.

With nuclei replicating asynchronously in the *Xenopus* extract, some nuclei may

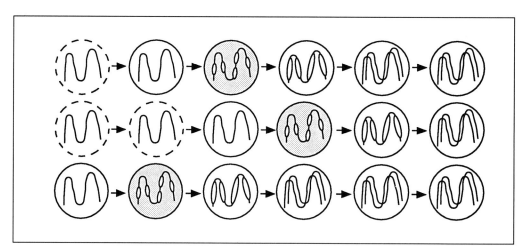

Fig. 5 Asynchronous replication of nuclei coincubated in *Xenopus* egg extracts. Summary of the flow cytometry data of the kinetics of nuclear replication in the *Xenopus* system (51). Three nuclei replicating together in the cell-free system are shown, each with a single chromosome for simplicity. Different nuclei complete nuclear assembly (as shown by a complete circle surrounding the chromosome) at different times. Once nuclei are assembled, they receive a signal to start replication (shading) which induces a burst of near synchronous initiation events within the nucleus. Despite the persistence of the initiation signal, no DNA replicates more than once.

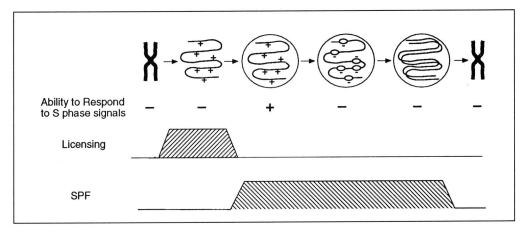

Fig. 6 Model to explain replication control in the *Xenopus* system. A single nucleus is shown as it passes through a complete cell cycle. During late mitosis, prior to nuclear envelope assembly, the DNA becomes 'licensed' to undergo DNA replication. Once assembled into a nucleus, the licensed DNA is capable of initiating DNA replication in response to the presence of the S phase inducer SPF. However, the license is destroyed as the DNA is replicated. Only following passage through mitosis does the nucleus once again become competent to undergo further DNA replication.

only be starting to replicate after others have replicated fully; despite this, no replicated nuclei undergo a further round of DNA replication (1, 51). This means that replicated and unreplicated nuclei must differ, as only the unreplicated nuclei have the potential to replicate in the current cell cycle. These results are consistent with the classic cell fusion studies of Rao and Johnson (56), where HeLa cells at different stages of the cell cycle were fused, and the replication timing of the different nuclei measured. In fusions of G1- or S-phase cells and G2-phase cells, the G1- and S-phase nuclei underwent normal DNA replication in the hybrid; however, the G2 nuclei did not replicate until after passage through mitosis.

The flow cytometry results provide the basis for understanding the way that DNA replication is controlled in the eukaryotic cell cycle, and is outlined in Fig. 6 (9, 51, 57). Replication control can be divided into two distinct components: the DNA template in the nucleus, and activities present in the cytoplasm which act on this nuclear substrate. The nucleus can be in one of two states: either capable or incapable of responding to S-phase inducers by undergoing DNA replication. As described below in Section 4, the ability of nuclei to respond to the S-phase inducers requires DNA to have been 'licensed' for DNA replication, a process which normally occurs during mitosis. The cytoplasm also provides an 'S phase promoting factor' (SPF) which acts on intact licensed nuclei to induce them to initiate DNA replication. Once a licensed nucleus has received the SPF signal it can progress through all the stages of replication (including the initiation, elongation, and termination of early- and late-firing origins) without any additional signals from the cytoplasm. As described below in Section 5, the SPF signal appears to be generated by cyclin-dependent protein kinases (CDKs).

4. Role of licensing factor in controlling DNA replication

4.1 The licensing factor model

The inability of replicated nuclei to respond to S-phase inducers present in *Xenopus* extracts can be demonstrated directly (9, 58, 59). Replicated 'G2' nuclei were generated by allowing DNA to replicate in *Xenopus* extract supplemented with the protein synthesis inhibitor cycloheximide to prevent them from entering mitosis. When these G2 nuclei were isolated from the extracts and transferred to fresh extract they remained incapable of further replication (9). However, if the G2 nuclei were allowed to progress into mitosis and undergo nuclear envelope breakdown and chromosome condensation, they could then undergo a further round of DNA replication when added back to fresh extract. Therefore some metaphase process had permitted G2 nuclei to revert to the responsive G1 state. To identify what this metaphase process is, G2 nuclei were subjected to various treatments to see whether they could induce G2 nuclei to rereplicate in fresh extract without entering metaphase (9). Agents that caused nuclear envelope permeabilization, such as lysolecithin or phospholipase, left the G2 nuclei capable of rereplicating in fresh extract. This nuclear envelope permeabilization presumably mimics the effect of nuclear envelope breakdown which normally occurs during mitosis in higher eukaryotes. Similar results have been obtained in cell-free extracts of *Drosophila* eggs (2).

Figure 7 outlines a model proposed to explain these results (9, 60). An essential replication factor called replication licensing factor (RLF) can bind to DNA during late mitosis or early interphase before nuclear assembly has occurred. RLF cannot cross the nuclear envelope, so that once nuclear assembly is complete, active RLF is only present in the nucleus where bound to DNA. On entry into S phase, each molecule of RLF bound to DNA supports a single initiation event, after which it is inactivated or destroyed. Thus in G2, no active RLF remains in the nucleus, and the nuclear envelope must be transiently permeabilized (as normally occurs during mitosis) to allow a further round of DNA replication to be licensed.

Results consistent with this model were also obtained on addition of nuclei from mammalian tissue culture cells into *Xenopus* eggs (61) or egg extracts (58). Although intact G1 nuclei replicated after transfer to *Xenopus* extract, G2 nuclei replicated only if they had been permeabilized prior to transfer. When permeabilized G2 nuclei were resealed using nuclear envelope precursors prior to exposure to *Xenopus* extract, they remained incapable of rereplicating in *Xenopus* extract (59). However, if they were exposed to *Xenopus* egg extract prior to resealing, replication could occur. Quiescent nuclei, however, may lose their license, as quiescent 3T3 fibroblasts required nuclear envelope permeabilization before they were competent to replicate in *Xenopus* extract (62).

4.2 Identification of licensing factor components

RLF has recently been subjected to biochemical fractionation (63). Activity resolved into two components, RLF-M and RLF-B, both of which were required for licensing. RLF-M was purified to apparent homogeneity, and shown to consist of a complex of at least three polypeptides, with molecular weights of 92, 106, and 115 kDa. The 106 kDa polypeptide is the *Xenopus* Mcm3 protein (see below), and the other two polypeptides seem likely to be other members of the MCM family. Consistent with the licensing factor model, both RLF-B and RLF-M were required for each successive round of DNA replication. *Xenopus* Mcm3 associated with chromatin in G1 but was removed during replication. The dissociation of Mcm3 from DNA during DNA replication is presumably the reason why the RLF-M complex must bind to chromatin prior to each round of DNA replication (63). Consistent results have also been obtained by immunodepletion of Mcm3-containing complexes from *Xenopus* extract (64, 65). Further, immunodepletion of Mcm3 from *Xenopus* extracts specifically inhibited it from replicating G2 nuclei, but not G1 nuclei (64).

MCM ('**m**inichromosome **m**aintenance') genes are members of a family of genes first identified in *S. cerevisiae* that are involved in the replication of specific replication origins (66–70) (see Chapter 7). The yeast Cdc46 gene product, which is equivalent to Mcm5 (70), had previously been suggested to be an RLF homologue, since the protein is present in nuclei during late mitosis and G1, but disappears from the nucleus at the start of S phase (69). Mcm2 and Mcm3 proteins also show a similar change in subcellular distribution (71, 72). Members of the MCM gene family are found throughout the eukaryotic kingdom, with homologues in fission yeast (73–75), *Xenopus* (73), and mammals (76–79). In mammalian cells, DNA replication was blocked following microinjection of antibodies to BM28, a human Mcm2 homologue (78), and to murine P1, an Mcm3 homologue (79). The licensing system therefore appears to be common to all eukaryotes.

4.3 Cell cycle control of licensing factor

Xenopus extracts treated with protein kinase inhibitors such as 6-dimethylamino-purine (6-DMAP) prior to exit from metaphase, were left functionally devoid of RLF activity, though able to efficiently replicate DNA that had already been licensed (60, 80, 81). Such 6-DMAP-treated extracts provide a simple assay for RLF activity which has been exploited to purify RLF-M from *Xenopus* extracts as described above (63) as well as to measure cytoplasmic RLF levels during the *Xenopus* cell cycle (60) (Fig. 7). Licensing factor levels were low during the early stages of mitosis. However, shortly after the metaphase–anaphase transition, levels rose very abruptly so that the decondensing telophase chromosomes could become licensed prior to complete nuclear envelope assembly. The appearance of RLF at this stage of the *Xenopus* cell cycle took place in the absence of new protein synthesis and was presumably due to posttranslational modification of a protein already present. Since 6-DMAP failed to inhibit RLF when added to extracts just a few minutes after

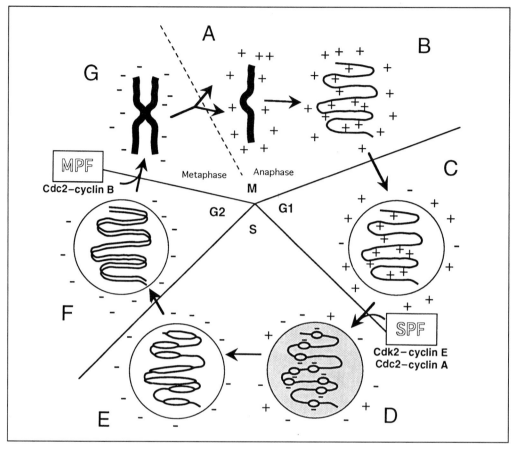

Fig. 7 The licensing factor model to explain why DNA is replicated only once in each cell cycle (9, 60). (a) On exit from metaphase, inactive RLF (−) is rapidly activated (+) and the chromosomal DNA decondenses. (b) RLF binds to DNA at potential sites of initiation. (c) DNA is assembled into a nucleus surrounded by a complete nuclear envelope. (d) A cytoplasmic activity ('SPF'), generates an intranuclear signal (stippled) that leads to initiation at licensed sites. Each molecule of RLF can support only a single initiation event, after which it is inactivated or destroyed (−). (e) The chromosome is replicated by forks initiated at licensed sites. (f) Fully replicated G2 nuclei cannot rereplicate due to exclusion of RLF from the DNA by the nuclear envelope. (g) Entry into mitosis is accompanied by nuclear envelope breakdown and chromosome condensation. (h) Sister chromatids separate and are partitioned to the daughter cells. During the earlier parts of the cycle (a–e) cytoplasmic RLF activity has slowly decayed. Reproduced from Blow (60).

progression into anaphase, this posttranslational modification is likely to be the real target of inhibition by 6-DMAP (60, 80, 81). Active RLF that failed to bind DNA and hence remained in the cytoplasm was unstable, and had disappeared by mid S phase (60). This provides a further mechanism to prevent the illegitimate re-replication of DNA, since even if the nuclear envelope became damaged in G2, no cytoplasmic RLF would be present at this time so no further initiation could take place.

Entry into mitosis is induced by a cyclin-dependent kinase called MPF ('mitosis promoting factor'; see below). The ability of protein kinase inhibitors to prevent RLF activation correlates closely with their ability to inhibit MPF (60, 80–82). This correlation is particularly notable with the purine analogue olomoucine, which is a highly specific inhibitor of cyclin-dependent kinases (81), and suggests that MPF plays an important role in RLF activation at the metaphase–anaphase transition. This would mean that MPF performs two functions to allow DNA to be licensed: firstly it causes nuclear envelope breakdown by inducing entry into mitosis, and secondly it leads to RLF activation. A central role for MPF in controlling RLF function is consistent with experiments which implicate the corresponding cdc2:cdc13 kinase (MPF equivalent) in preventing rereplication in fission yeast (83–85) (see Chapter 8).

5. SPF—the signal to initiate replication

As outlined above, virtually as soon as nuclear assembly is complete in the *Xenopus* system, each nucleus undergoes a coordinated burst of initiation events. The intranuclear signal that generates this burst of initiation events is unknown, but appears to be closely associated with cyclin-dependent kinases (CDKs). CDKs are small protein kinase subunits, activated by complexing with a cyclin partner, that play an important role in cell cycle regulation (see Chapter 7). In metazoans the cyclins are grouped into families A–F, and CDKs are grouped into families 1–4. A- and B-type cyclins can activate Cdc2 (Cdk1) on passage into mitosis to form the mitotic inducer (MPF), which as described above plays a role in licensing DNA. In addition, other cyclins and CDKs appear to be involved in the initiation of DNA replication, by providing SPF activity.

5.1 Requirement for CDKs in the initiation of replication

The cell cycle protein p13^{suc1} (see Chapter 8) binds specifically and tightly to CDKs, and can be used to affinity deplete *Xenopus* extracts of SPF activity. Such suc1-depleted extracts were specifically unable to support the initiation of DNA replication, although other processes such as nuclear assembly occurred normally (82). In extracts treated with the protein synthesis inhibitor cycloheximide, SPF activity appears to be dependent on the Cdk2 protein as immunodepletion of Cdk2 also caused a similar block to DNA replication (86, 87). Similarly, the CDK-inhibitor p21^{Cip1} inhibited replication in cycloheximide-treated extracts at concentrations comparable to the concentration of endogenous Cdk2 (88, 89). SPF function in *Xenopus* is specifically required for the initiation of DNA replication, since no effect on replication fork movement or complementary strand synthesis was seen in CDK-inhibited extracts (31, 82, 86, 88, 89). These results are consistent with micro-injection experiments in mammalian cells which also suggest a requirement for Cdk2 in progression through S phase (90).

In the presence of ongoing translation, Cdc2 also seems capable of providing SPF activity (91). DNA replication can be restored to suc1-depleted extracts by inducing translation of Cdc2 mRNA. In contrast, Cdk2 mRNA restored DNA replication much less efficiently. As described below, the differential ability of Cdk2 and Cdc2 to provide SPF activity depending on the translational activity of the extracts probably reflects the different cyclins present under these conditions.

CDKs seem likely to be responsible for generating the coordinated burst of initiation events that are seen as nuclei start to replicate. The substrates of SPF phosphorylation may be initiation proteins themselves, such as RLF or associated proteins. Although most cyclin-dependent kinases efficiently phosphorylate histone H1, there is only a low histone H1 kinase activity associated with SPF present in *Xenopus* extracts (88, 92). This may indicate that the SPF kinase has a much higher affinity for its physiological substrate than it does for histone H1. Indeed, high levels of H1 kinase induced by cyclin A (and cyclin B) were inhibitory to DNA replication since they drove extracts into mitosis, when the initiation of replication cannot occur (7, 88, 92).

5.2 Cyclins involved in the initiation of replication

Three different classes of cyclin, A, B, and E, have been identified in *Xenopus* eggs. Cyclins A and B are predominantly bound to Cdc2 in the early embryo, though cyclin A can also be found complexed with Cdk2 later in development (92–94). The cyclin B–Cdc2 complex forms the mitotic inducer MPF (95–98). In contrast, cyclin E is found exclusively complexed with Cdk2 (99). When Cdk2 activity was blocked by the inhibitor p21[Cip1], DNA replication could be rescued by recombinant A- or E-type cyclins but not B-type cyclins (88). Similarly, A- and E-type cyclins but not B-type cyclins restored DNA synthesis to extracts depleted of CDKs with p13[suc1] beads (92, 100). This suggests that SPF activity is capable of being provided by either A- or E-type cyclins, and is consistent with results implicating cyclin E (101, 102) as well as cyclin A (103–105) in DNA replication in other metazoans. Cyclin A also has MPF activity, and can induce mitotic events such as nuclear envelope breakdown and chromosome condensation. Low concentrations of cyclin A show SPF activity, whilst higher concentrations show MPF activity (92).

Cyclins A and B are abundantly translated in the early embryo, but are degraded on exit from metaphase. In contrast, cyclin E translation is much lower, but the protein remains stable throughout the cell cycle (94). Therefore cyclin E predominates in extracts treated with protein kinase inhibitors, whilst cyclins A and B become more important in actively translating extracts (Fig. 8). *Xenopus* initiates DNA replication almost immediately after template DNA has been assembled into interphase nuclei (1, 7, 21, 22, 51). This means that they lack an appreciable G1 phase, and that nuclear assembly, not SPF, is rate limiting for the initiation of DNA replication (Fig. 7, 'nuclear assembly'). In extracts treated with protein synthesis inhibitors, cyclin A is absent, and all SPF activity is therefore dependent on cyclin E (or any other mitotically stable cyclins). Since protein synthesis inhibitors do not affect progress through S phase in the *Xenopus* system (4, 9), this suggests that even in the absence of cyclin A, nuclear assembly, but not cyclin E–Cdk2 is still the rate-limiting activity for DNA

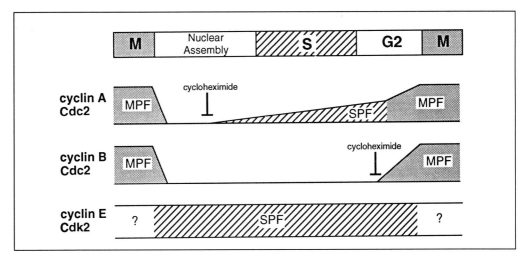

Fig. 8 Cartoon showing the proposed roles of cyclins A, B, and E in *Xenopus* egg extracts. The top panel shows the cell cycle events taking place in *Xenopus* extracts. Kinase activities of cyclin A–Cdc2, cyclin B–Cdc2 and cyclin E–Cdk2 at the different times are by the boxed regions below. In the presence of cycloheximide, levels of cyclins A and B remain low, whilst cyclin E is largely unaffected. MPF (stippled) and SPF (diagonal lines) activities associated with these kinases are also indicated; SPF activity of cyclin E–Cdk2 during mitosis is undetermined. Redrawn from Strausfeld *et al.* (92).

replication. However, under these conditions all SPF is dependent on Cdk2 and cyclin E, since DNA replication is inhibited if it is immunodepleted (86, 100).

5.3 The coordination of S and M phases in the early *Xenopus* embryo

During progress through a normal S phase in *Xenopus*, with ongoing protein synthesis, cyclin A–Cdc2 kinase builds up (93) and adds to the total SPF activity also provided by preexisting Cdk2–cyclin E (Fig. 8). The role of cyclin A may be to induce initiation at any replicons that have not already fired when the nucleus started to replicate. If translation is permitted following p13^{suc1} bead depletion, then cyclin A will be the only cyclin present capable of contributing to SPF activity. Since cyclin A preferentially associated with Cdc2 in the *Xenopus* embryo (92, 93), this would explain why translation of Cdc2 is more effective at restoring DNA replication than translation of Cdk2 (91). Cyclin A may also be required to delay mitosis if S phase is incomplete (106), and so therefore appears to have two different functions in mediating the progression from S phase to mitosis. Despite the continuing presence of SPF activity after DNA replication is complete, rereplication does not occur because RLF activity within the nucleus has been destroyed as a result of DNA replication.

As cyclin A and cyclin B kinases build up later in the cell cycle, kinase levels become sufficient to induce entry into mitosis. In order for DNA to become relicensed for another round of DNA replication, the nuclear envelope must be broken down, and cytoplasmic RLF must be reactivated. Both of these processes depend on activation of MPF (largely contributed by Cdc2–cyclin B) and passage through mitosis (9, 60, 81). The coordination of the different cyclin-dependent kinases is therefore responsible for the regulated replication of DNA during the cell cycle.

6. Prospects

At present, the *Xenopus* system has allowed us to build up a general picture of the way that DNA replication is controlled within the cell cycle. The basic elements are: (i) a requirement for nuclear assembly in DNA replication; (ii) the licensing of DNA early in the cell cycle prior to entry into S phase; and (iii) the action of SPF within nuclei to induce initiation on licensed DNA. Since each of these features is present throughout the eukaryotic kingdom (interphase nuclei, Mcm proteins, and cyclin-dependent kinases) it seems likely that this general scheme of regulation is common to all eukaryotes. The field is moving rapidly and it seems likely that further details will soon be known, such as the precise identification of RLF-B and SPF. This may also give us some insight into more complex problems such as the function of SPF in initiation, or the role of other DNA replication proteins such as RP-A in this process. However, many of these questions are likely to depend on a combination of experimental approaches, as described in the other chapters of this book.

Acknowledgements

J.J.B. is a Lister Institute Research Fellow.

References

1. Blow, J. J. and Laskey, R. A. (1986) Initiation of DNA replication in nuclei and purified DNA by a cell-free extract of *Xenopus* eggs. *Cell*, **47**, 577.
2. Crevel, G. and Cotterill, S. (1991) DNA replication in cell-free extracts from *Drosophila melanogaster*. *EMBO J.*, **10**, 4361.
3. Lohka, M. J. and Masui, Y. (1983) Formation in vitro of sperm pronuclei and mitotic chromosomes induced by amphibian ooplasmic components. *Science*, **220**, 719.
4. Harland, R. M. and Laskey, R. A. (1980) Regulated DNA replication of DNA microinjected into eggs of *Xenopus laevis*. *Cell*, **21**, 761.
5. Wangh, L. J. (1989) Injection of *Xenopus* eggs before activation, achieved by control of extracellular factors, improves plasmid DNA replication after activation. *J. Cell Sci.*, **93**, 1.
6. Sanchez, J. A., Marek, D., and Wangh, L. J. (1992) The efficiency and timing of plasmid DNA replication in *Xenopus* eggs: correlations to the extent of prior chromatin assembly. *J. Cell Sci.*, **103**, 907.
7. Blow, J. J. and Sleeman, A. M. (1990) Replication of purified DNA in *Xenopus* egg extracts is dependent on nuclear assembly. *J. Cell Sci.*, **95**, 383.
8. Lohka, M. J. and Masui, Y. (1984) Effects of Ca^{2+} ions on the formation of metaphase

chromosomes and sperm pronuclei in cell-free preparations from unactivated *Rana pipiens* eggs. *Dev. Biol.*, **103**, 434.

9. Blow, J. J. and Laskey, R. A. (1988) A role for the nuclear envelope in controlling DNA replication within the cell cycle. *Nature*, **332**, 546.

10. Kimelman, D., Kirschner, M., and Scherson, T. (1987) The events of the midblastula transition in *Xenopus* are regulated by changes in the cell cycle. *Cell*, **48**, 399.

11. Hara, K., Tydeman, P., and Kirschner, M. (1980) A cytoplasmic clock with the same period as the division cycle in *Xenopus* eggs. *Proc. Natl. Acad. Sci. USA*, **77**, 462.

12. Gerhart, J., Wu, M., and Kirschner, M. (1984) Cell cycle dynamics of an M-phase-specific cytoplasmic factor in *Xenopus laevis* oocytes and eggs. *J. Cell Biol.*, **98**, 1247.

13. Dasso, M. and Newport, J. W. (1990) Completion of DNA replication is monitored by a feedback system that controls the initiation of mitosis in vitro: studies in *Xenopus*. *Cell*, **61**, 811.

14. Kornbluth, S., Smythe, C., and Newport, J. W. (1992) In vitro cell cycle arrest induced by using artificial DNA templates. *Mol. Cell. Biol.*, **12**, 3216.

15. Smythe, C. and Newport, J. W. (1992) Coupling of mitosis to the completion of S phase in *Xenopus* occurs via modulation of the tyrosine kinase that phosphorylates p34^{cdc2}. *Cell*, **68**, 787.

16. Philpott, A., Leno, G. H., and Laskey, R. A. (1991) Sperm decondensation in *Xenopus* egg cytoplasm is mediated by nucleoplasmin. *Cell*, **65**, 569.

17. Philpott, A. and Leno, G. H. (1992) Nucleoplasmin remodels sperm chromatin in *Xenopus* egg extracts. *Cell*, **69**, 759.

18. Lohka, M. J. and Masui, Y. (1984) Roles of cytosol and cytoplasmic particles in nuclear envelope assembly and sperm pronuclear formation in cell-free preparations from amphibian eggs. *J. Cell Biol.*, **98**, 1222.

19. Vigers, G. P. and Lohka, M. J. (1992) Regulation of nuclear envelope precursor functions during cell division. *J. Cell Sci.*, **102**, 273.

20. Newmeyer, D. D., Lucocq, J. M., Burglin, T. R., and De Robertis, E. M. (1986) Assembly in vitro of nuclei active in nuclear protein transport: ATP is required for nucleoplasmin accumulation. *EMBO J.*, **5**, 501.

21. Newport, J. (1987) Nuclear reconstitution in vitro: stages of assembly around protein-free DNA. *Cell*, **48**, 205.

22. Sheehan, M. A., Mills, A. D., Sleeman, A. M., Laskey, R. A., and Blow, J. J. (1988) Steps in the assembly of replication-competent nuclei in a cell-free system from *Xenopus* eggs. *J. Cell Biol.*, **106**, 1.

23. Newmeyer, D. D., Finlay, D. R., and Forbes, D. J. (1986) In vitro transport of a fluorescent nuclear protein and exclusion of non-nuclear proteins. *J. Cell Biol.*, **103**, 2091.

24. Newport, J. W., Wilson, K. L., and Dunphy, W. G. (1990) A lamin-independent pathway for nuclear envelope assembly. *J. Cell Biol.*, **111**, 2247.

25. Meier, J., Campbell, K. H., Ford, C. C., Stick, R., and Hutchison, C. J. (1991) The role of lamin LIII in nuclear assembly and DNA replication, in cell-free extracts of *Xenopus* eggs. *J. Cell Sci.*, **98**, 271.

26. Jenkins, H., Holman, T., Lyon, C., Lane, B., Stick, R., and Hutchison, C. (1993) Nuclei that lack a lamina accumulate karyophilic proteins and assemble a nuclear matrix. *J. Cell Sci.*, **106**, 275.

27. Hola, M., Castleden, S., Howard, M., and Brooks, R. F. (1994) Initiation of DNA synthesis by nuclei from scrape-ruptured quiescent mammalian cells in high-speed supernatants of *Xenopus* egg extracts. *J. Cell Sci.*, **107**, 3045.

28. Forbes, D. J., Kirschner, M. W., and Newport, J. W (1983) Spontaneous formation of nucleus-like structures around bacteriophage DNA microinjected into *Xenopus* eggs. *Cell*, **34**, 13.

29. Méchali, M. and Harland, R. M. (1982) DNA synthesis in a cell-free system from *Xenopus* eggs: priming and elongation on single-stranded DNA in vitro. *Cell*, **30**, 93.

30. Cox, L. S. (1992) DNA replication in cell-free extracts from *Xenopus* eggs is prevented by disrupting nuclear envelope function. *J. Cell Sci.*, **101**, 43.

31. Shivji, M. K. K., Grey, S. J., Strausfeld, U. P., Wood, R. D., and Blow, J. J. (1994) Cip1 inhibits DNA replication but not PCNA-dependent nucleotide excision-repair. *Curr. Biol.*, **4**, 1062.

32. Finlay, D. R., Newmeyer, D. D., Price, T. M., and Forbes, D. J. (1987) Inhibition of in vitro nuclear transport by a lectin that binds to nuclear pores. *J. Cell Biol.*, **104**, 189.

33. Finaly, D. R. and Forbes, D. J. (1990) Reconstitution of biochemically altered nuclear pores: transport can be eliminated and restored. *Cell*, **60**, 17.

34. Mills, A. D., Blow, J. J., White, J. G., Amos, W. B., Wilcock, D., and Laskey, R. A. (1989) Replication occurs at discrete foci spaced throughout nuclei replicating in vitro. *J. Cell Sci.*, **94**, 471.

35. Moir, R. D., Montag Lowy, M., and Goldman, R. D. (1994) Dynamic properties of nuclear lamins: lamin B is associated with sites of DNA replication. *J. Cell Biol.*, **125**, 1201.

36. Hutchison, C. J., Bridger, J. M., Cox, L. S., and Kill, I. R. (1995) Weaving a pattern from disparate threads: lamin function in nuclear assembly and DNA replication. *J. Cell Sci.*, **107**, 3259.

37. Dasso, M., Nishitani, H., Kornbluth, S., Nishimoto, T., and Newport, J. W. (1992) RCC1, a regulator of mitosis, is essential for DNA replication. *Mol. Cell. Biol.*, **12**, 3337.

38. Kornbluth, S., Dasso, M., and Newport, J. (1994) Evidence for a dual role for TC4 protein in regulating nuclear structure and cell cycle progression. *J. Cell Biol.*, **125**, 705.

39. Adachi, Y. and Laemmli, U. K. (1992) Identification of nuclear pre-replication centers poised for DNA synthesis in *Xenopus* egg extracts: immunolocalization study of replication protein A. *J. Cell Biol.*, **119**, 1.

40. Adachi, Y. and Laemmli, U. K. (1994) Study of the cell cycle-dependent assembly of the DNA pre-replication centres in *Xenopus* egg extracts. *EMBO J.*, **13**, 4153.

41. Hutchison, C. and Kill, I. (1989) Changes in the nuclear distribution of DNA polymerase alpha and PCNA/cyclin during the progress of the cell cycle, in a cell-free extract of *Xenopus* eggs. *J. Cell Sci.*, **93**, 605.

42. Méchali, M. and Kearsey, S. (1984) Lack of specific sequence requirement for DNA replication in *Xenopus* eggs compared with high sequence specificity in yeast. *Cell*, **38**, 55.

43. Blow, J. J., Dilworth, S. M., Dingwall, C., Mills, A. D., and Laskey, R. A. (1987) Chromosome replication in cell-free systems from *Xenopus* eggs. *Phil. Trans. R. Soc. Lond. B*, **317**, 483.

44. Mahbubani, H. M., Paull, T., Elder, J. K., and Blow, J. J. (1992) DNA replication initiates at multiple sites on plasmid DNA in *Xenopus* egg extracts. *Nucleic Acids Res.*, **20**, 1457.

45. Hyrien, O. and Mechali, M. (1992) Plasmid replication in *Xenopus* eggs and egg extracts: a 2D gel electrophoretic analysis. *Nucleic Acids Res.*, **20**, 1463.

46. Hartl, P., Olson, E., Dang, T., and Forbes, D. J. (1994) Nuclear assembly with lambda DNA in fractionated *Xenopus* egg extracts: an unexpected role for glycogen in formation of a higher order chromatin intermediate. *J. Cell Biol.*, **124**, 235.

47. Cox, L. S. and Laskey, R. A. (1991) DNA replication occurs at discrete sites in pseudonuclei assembled from purified DNA in vitro. *Cell*, **66**, 271.

48. McTiernan, C. F. and Stambrook, P. J. (1984) Initiation of SV40 DNA replication after microinjection into *Xenopus* eggs. *Biochim. Biophys. Acta*, **782**, 295.

49. Hyrien, O. and Méchali, M. (1993) Chromosomal replication initiates and terminates at random sequences but at regular intervals in the ribosomal DNA of *Xenopus* early embryos. *EMBO J.*, **12**, 4511.

50. Callan, H. G. (1972) Replication of DNA in the chromosomes of eukaryotes. *Proc. R. Soc. Lond. B*, **181**, 19.

51. Blow, J. J. and Watson, J. V. (1987) Nuclei act as independent and integrated units of replication in a *Xenopus* cell-free system. *EMBO J.*, **6**, 1997.

52. Buongiorno-Nardelli, M., Micheli, G., Carri, M. T., and Marilley, M. (1982) A relationship between replicon size and supercoiled loop domains in the eukaryotic genome. *Nature*, **298**, 100.

53. Marilley, M. and Gassend Bonnet, G. (1989) Supercoiled loop organization of genomic DNA: a close relationship between loop domains, expression units, and replicon organization in rDNA from *Xenopus laevis*. *Exp. Cell Res.*, **180**, 475.

54. Gilbert, D. M., Miyazawa, H., Nalleseth, F. S., Ortega, J. M., Blow, J. J., and DePamphilis, M. L. (1993) Site-specific initiation of DNA replication in metazoan chromosomes and the role of nuclear organization. *Cold Spring Harbor Symp. Quant. Biol.*, **58**, 475.

55. Leno, G. H. and Laskey, R. A. (1991) The nuclear membrane determines the timing of DNA replication in *Xenopus* egg extracts. *J. Cell Biol.*, **112**, 557.

56. Rao, P. N. and Johnson, R. T. (1970) Mammalian cell fusion: studies on the regulation of DNA synthesis and mitosis. *Nature*, **225**, 159.

57. Blow, J. J. (1995) S phase and its regulation, in *Cell cycle control*. Hutchison, C. J. and Glover, D. M. (eds.). Oxford University Press, Oxford, p. 177.

58. Leno, G. H., Downes, C. S., and Laskey, R. A. (1992) The nuclear membrane prevents replication of human G2 nuclei but not G1 nuclei in *Xenopus* egg extract. *Cell*, **69**, 151.

59. Coverley, D., Downes, C. S., Romanowski, P., and Laskey, R. A. (1993) Reversible effects of nuclear membrane permeabilization on DNA replication: evidence for a positive licensing factor. *J. Cell Biol.*, **122**, 985.

60. Blow, J. J. (1993) Preventing re-replication of DNA in a single cell cycle: evidence for a Replication Licensing Factor. *J. Cell Biol.*, **122**, 993.

61. De Roeper, A., Smith, J. A., Watt, R. A., and Barry, J. M. (1977) Chromatin dispersal and DNA synthesis in G1 and G2 HeLa cell nuclei injected into *Xenopus* eggs. *Nature*, **265**, 469.

62. Leno, G. H. and Munshi, R. (1994) Initiation of DNA replication in nuclei from quiescent cells requires permeabilisation of the nuclear membrane. *J. Cell Biol.*, **127**, 5.

63. Chong, J. P. J., Mahbubani, M. H., Khoo, C.-Y., and Blow, J. J. (1995) Purification of an Mcm-containing complex as a component of the DNA replication licensing system. *Nature*, **375**, 418.

64. Madine, M. A., Khoo, C.-Y., Mills, A. D., and Laskey, R. A. (1995) MCM3 complex required for cell cycle regulation of DNA replication in vertebrate cells. *Nature*, **375**, 421.

65. Kubota, Y., Mimura, S., Nishimoto, S., Takisawa, H., and Nojima, H. (1995) Identification of the yeast MCM3-related protein as a component of *Xenopus* DNA Replication Licensing Factor. *Cell*, **81**, 601.

66. Maine, G. T., Sinha, P., and Tye, B. K. (1984) Mutants of *S. cerevisiae* defective in the maintenance of minichromosomes. *Genetics*, **106**, 365.

67. Gibson, S. I., Surosky, R. T., and Tye, B. K. (1990) The phenotype of the minichromosome maintenance mutant mcm3 is characteristic of mutants defective in DNA replication. *Mol. Cell. Biol.*, **10**, 5707.

68. Maiti, A. K. and Sinha, P. (1992) The mcm2 mutation of yeast affects replication, rather than segregation or amplification of the two micron plasmid. *J. Mol. Biol.*, **224**, 545.

69. Hennessy, K. M., Clark, C. D., and Botstein, D. (1990) Subcellular localization of yeast CDC46 varies with the cell cycle. *Genes Dev.*, **4**, 2252.

70. Chen, Y., Hennessy, K. M., Botstein, D., and Tye, B. K. (1992) CDC46/MCM5, a yeast protein whose subcellular localization is cell cycle-regulated, is involved in DNA replication at autonomously replicating sequences. *Proc. Natl. Acad. Sci. USA*, **89**, 10 459.

71. Yan, H., Gibson, S., and Tye, B. K. (1991) Mcm2 and Mcm3, two proteins important for ARS activity, are related in structure and function. *Genes Dev.*, **5**, 944.

72. Yan, H., Merchant, A. M., and Tye, B. K. (1993) Cell cycle-regulated nuclear localization of MCM2 and MCM3, which are required for the initiation of DNA synthesis at chromosomal replication origins in yeast. *Genes Dev.*, **7**, 2149.

73. Coxon, A., Maundrell, K., and Kearsey, S. E. (1992) Fission yeast cdc21+ belongs to a family of proteins involved in an early step of chromosome replication. *Nucleic Acids Res.*, **20**, 5571.

74. Miyake, S., Okishio, N., Samejima, I., Hiraoka, Y., Toda, T., Saitoh, I., and Yanagida, M. (1993) Fission yeast genes nda1+ and nda4+, mutations of which lead to S-phase block, chromatin alteration and Ca^{2+} suppression, are members of the CDC46/MCM2 family. *Mol. Biol. Cell*, **4**, 1003.

75. Forsburg, S. L. and Nurse, P. (1994) The fission yeast cdc19(+) gene encodes a member of the mcm family of replication proteins. *J. Cell Sci.*, **107**, 2779.

76. Thömmes, P., Fett, R., Schray, B., Burkhart, R., Barnes, M., Kennedy, C., Brown, N. C., and Knippers, R. (1992) Properties of the nuclear P1 protein, a mammalian homologue of the yeast Mcm3 replication protein. *Nucleic Acids Res.*, **20**, 1069.

77. Hu, B., Burkhart, R., Schulte, D., Musahl, C., and Knippers, R. (1993) The P1 family: a new class of nuclear mammalian proteins related to the yeast Mcm replication proteins. *Nucleic Acids Res.*, **21**, 5289.

78. Todorov, I. T., Pepperkok, R., Philipova, R. N., Kearsey, S. E., Ansorge, W., and Werner, D. (1994) A human nuclear protein with sequence homology to a family of early S phase proteins is required for entry into S phase and for cell division. *J. Cell Sci.*, **107**, 253.

79. Kimura, H., Nozaki, N., and Sugimoto, K. (1994) DNA polymerase α associated protein P1, a murine homolog of yeast MCM3, changes its intranuclear distribution during the DNA synthetic period. *EMBO J.*, **13**, 4311.

80. Kubota, Y. and Takisawa, H. (1993) Determination of initiation of DNA replication before and after nuclear formation in *Xenopus* egg cell free extracts. *J. Cell Biol.*, **123**, 1321.

81. Vesely, J., Havlicek, L., Strnad, M., Blow, J. J., Donnella-Deana, A., Pinna, L., Letham, D. S., Kato, J., Detivaud, L., Leclerc, S., and Meijer, L. (1994) Inhibition of cyclin-dependent kinases by purine analogues. *Eur. J. Biochem.*, **224**, 771.

82. Blow, J. J. and Nurse, P. (1990) A cdc2-like protein is involved in the initiation of DNA replication in *Xenopus* egg extracts. *Cell*, **62**, 855.

83. Broek, D., Bartlett, R., Crawford, K., and Nurse, P. (1991) Involvement of p34cdc2 in establishing the dependency of S phase on mitosis. *Nature*, **349**, 388.

84. Moreno, S. and Nurse, P. (1994) Regulation of progression through the G1 phase of the cell cycle by the rum1+ gene [see comments]. *Nature*, **367**, 236.

85. Hayles, J., Fisher, D., Woollard, A., and Nurse, P. (1994) Temporal order of S phase and mitosis in fission yeast is determined by the state of the p34cdc2-mitotic B cyclin complex. *Cell*, **78**, 813.

86. Fang, F. and Newport, J. W. (1991) Evidence that the G1-S and G2-M transitions are controlled by different cdc2 proteins in higher eukaryotes. *Cell*, **66**, 731.

87. Fang, F. and Newport, J. (1993) Distinct roles of cdk2 and cdc2 in RP-A phosphorylation during the cell cycle. *J. Cell Sci.*, **106**, 983.

88. Strausfeld, U. P., Howell, M., Rempel, R., Maller, J. L., Hunt, T., and Blow, J. J. (1994) Cip1 blocks the initiation of DNA replication in *Xenopus* extracts by inhibition of cyclin-dependent kinases. *Curr. Biol.*, **4**, 876.

89. Chen, J., Jackson, P. K., Kirschner, M. W., and Dutta, A. (1995) Separate domains of p21 involved in the inhibition of Cdk kinase and PCNA. *Nature*, **374**, 386.

90. Pagano, M., Pepperkok, R., Lukas, J., Baldin, V., Ansorge, W., Bartek, J., and Draeta, G. (1993) Regulation of the cell cycle by the cdk2 protein kinase in cultured human fibroblasts. *J. Cell Biol.*, **121**, 101.

91. Chevalier, S., Tassan, J.-P., Cox, R., Philippe, M., and Ford, C. (1995) Both cdc2 and cdk2 promote S phase initiation in *Xenopus* egg extracts. *J. Cell Sci.*, **108**, 1831.

92. Strausfeld, U. P., Howell, M., Descombes, P., Rempel, R., Maller, J. L., Hunt, T., and Blow, J. J. (1995) Both cyclin A and cyclin E have S phase Promoting Factor (SPF) activity in *Xenopus* egg extracts. *J. Cell Sci.*, in press.

93. Minshull, J., Golsteyn, R., Hill, C. S., and Hunt, T. (1990) The A- and B-type cyclin associated cdc2 kinases in *Xenopus* turn on and off at different times in the cell cycle. *EMBO J.*, **9**, 2865.

94. Howe, J. A., Howell, M., Hunt, T., and Newport, J. W. (1995) Identification of a developmental timer regulating the stability of embryonic cyclin A and a new somatic A-type cyclin at gastrulation. *Genes Dev.*, **9**, 1164.

95. Dunphy, W. G., Brizuela, L., Beach, D., and Newport, J. (1988) The *Xenopus* cdc2 protein is a component of MPF, a cytoplasmic regulator of mitosis. *Cell*, **54**, 423.

96. Gautier, J., Norbury, C., Lohka, M., Nurse, P., and Maller, J. (1988) Purified maturation-promoting factor contains the product of a *Xenopus* homolog of the fission yeast cell cycle control gene cdc2+. *Cell*, **54**, 433.

97. Murray, A. W. and Kirschner, M. W. (1989) Cyclin synthesis drives the early embryonic cell cycle. *Nature*, **339**, 275.

98. Minshull, J., Blow, J., and Hunt, T. (1989) Translation of cyclin mRNA is necessary for extracts of activated *Xenopus* eggs to enter mitosis. *Cell*, **56**, 947.

99. Rempel, R. E., Sleight, S. B., and Maller, J. L. (1995) Maternal *Xenopus* cdk2-cyclin E complexes function during meiotic and early embryonic cell cycles that lack a G1 phase. *J. Biol. Chem.*, **270**, 6843.

100. Jackson, P. K., Chevalier, S., Philippe, M., and Kirschner, M. (1995) Early events in DNA replication require cyclin E and are blocked by p21Cip1. *J. Cell Biol.*, **130**, 755.

101. Ohtsubo, M. and Roberts, J. M. (1993) Cyclin-dependent regulation of G1 in mammalian fibroblasts. *Science*, **259**, 1908.

102. Knoblich, J. A., Sauer, K., Jones, L., Richardson, H., Saint, R., and Lehner, C. F. (1994) Cyclin E controls S phase progression and its down-regulation during *Drosophila* embryogenesis is required for the arrest of cell proliferation. *Cell*, **77**, 107.

103. Girard, F., Strausfeld, U., Fernandez, A., and Lamb, N. J. (1991) Cyclin A is required for the onset of DNA replication in mammalian fibroblasts. *Cell*, **67**, 1169.

104. Pagano, M., Pepperkok, R., Verde, F., Ansorge, W., and Draetta, G. (1992) Cyclin A is required at two points in the human cell cycle. *EMBO J.*, **11**, 961.

105. Zindy, F., Lamas, E., Chenivesse, X., Sobczak, J., Wang, J., Fesquet, D., Henglein, B., and Brechot, C. (1992) Cyclin A is required in S phase in normal epithelial cells. *Biochem. Biophys. Res. Commun.*, **182,** 1144.

106. Walker, D. H. and Maller, J. L. (1991) Role for cyclin A in the dependence of mitosis on completion of DNA replication. *Nature*, **354,** 314.

7 | Cell cycle control of DNA replication in *Saccharomyces cerevisiae*

ETIENNE SCHWOB and KIM NASMYTH

1. Introduction

Living organisms grow and divide to produce genetically identical progeny. To divide continuously, a eukaryotic cell must duplicate all of its constituents including DNA, proteins, ribosomes, organelles, etc., and segregate them more or less equally to daughter cells. To perpetuate its genetic information, a cell must, however, duplicate and segregate its chromosomes with high fidelity (10^{-6} mutation/bp and 10^{-5} chromosome loss/cell division in yeast). An important question in cell biology is therefore how cells manage to replicate and segregate faithfully their chromosomes? Furthermore, how is cell division regulated by cell growth and external conditions? This formidable task is achieved following the instructions of a programme, the cell cycle, which integrates the many biochemical events needed for chromosome duplication, transmission, and cell separation over time and space. For example, this programme ensures that DNA is replicated before sister chromatids begin to separate, or that sisters move to opposite poles. When some events of the cell cycle fail, genome integrity is maintained by surveillance mechanisms, also named checkpoints, which stop the programme until the problem is fixed. Thus, checkpoint controls reinforce the temporal order and dependency of cell cycle events. It is thought that a cause for the rapid evolution of malignant cells might be the loss of mitotic fidelity due to perturbations of the cell cycle or its checkpoints (1). Of the many cell cycle events, DNA replication and mitosis are the most striking and perhaps the most highly regulated. In this chapter, we will focus on how DNA replication is integrated in the cell cycle of the budding yeast *Saccharomyces cerevisiae*. The first part opens with the problems of DNA replication and its interactions with the cell cycle. We will then describe the discovery of *CDC28* as the main cell cycle regulator and how this protein kinase is regulated. The core of the chapter will deal with how the Cdc28 kinase might initiate DNA replication. Finally, we will consider what might determine the strict order of DNA replication and mitosis in the cell cycle.

2. DNA replication and the cell cycle: what are the problems?

2.1 Cell cycle phases

In contrast to most cellular components which are present in large numbers and continuously synthesized, DNA is present in only one or two copies per cell and its replication is confined to a narrow window of interphase. It was the latter discovery (2) which led to the division of the cell cycle in four periods with respect to the chromosome cycle: a first gap period (G1) precedes DNA synthesis (S phase); S is followed by a second gap (G2) in which cells prepare for mitosis (M phase). After nuclear division and cytokinesis, the cycle is completed and cells are back in G1. When grown in rich medium, the doubling time of an *S. cerevisiae* culture is approximately 90 min. S phase and nuclear division occupy a constant and minor fraction of the yeast cell cycle (about 15–20 min each) (3, 4) suggesting that growth or other controls are limiting for cell division. Indeed, small daughter cells which need about 120 min for division, extend their G1 period but not their S or M phases. The sudden nature of chromosome duplication and segregation led people to wonder what could be the inducer of S and M.

2.2 DNA replication

The yeast genome (14 Mb) is carried by 16 chromosomes which vary in size between 240 kb and 1640 kb (5). DNA replication is initiated at multiple well-defined sites called replication origins which can also serve as autonomously replicating sequences (ARSs) for plasmids (see Chapter 3). DNA is replicated bi-directionally from 250–400 such origins at a fork rate of about 3–6 kb/min (6). Not all origins are activated synchronously during S phase. The timing of origin firing seems determined by chromosomal context since late origins are activated early when they are moved away from telomeres, and conversely (7). DNA replication is initiated at the vicinity of ARSs and needs the presence of a multi-protein complex bound to its core consensus sequence (8) (see Chapter 3). Binding of this origin recognition complex (ORC) cannot, however, determine the onset of DNA replication because it is bound to origins throughout the cell cycle (9). Significant progress is being made towards the identification of the cell cycle-dependent inducer of S phase but its targets in the DNA replication machinery are not yet identified.

2.3 Controls

Several controls ensure the precise timing and duration of DNA replication, and coordinate S phase with other cell cycle events. We shall operationally classify them in four types (Fig.1):

- dependency of S phase on a preceding mitosis;
- S phase induction;

Fig. 1 Order of cell cycle phases: gap 1 (G1), DNA synthesis (S), gap 2 (G2), mitosis (M). Cell cycle controls of DNA replication (from left to right): dependency of S on M; timing of initiation of DNA synthesis (iDS); block to reinitiation during S and rereplication during G2; dependency of M on completion of S.

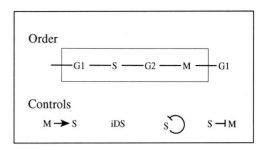

- block to reinitiation during S and rereplication in G2/M;
- dependency of mitosis on the completion of S phase.

To maintain constant ploidy, duplication of the genome (S phase) must be both preceded and followed by chromosome segregation (mitosis). The dependency of S on M ensures the alternation of DNA replication and mitosis. Only in very few cases do cells depart from this general rule, such as during the endoreplication cycles of *Drosophila* embryos. Several mutations and chemical treatments which allow rereplication in the absence of mitosis point towards a mechanism involving mitosis-promoting factor (MPF, a complex between $p34^{cdc2}/CDC28$ and B-type cyclins) and/or remodelling of the nuclear membrane (10) (see Chapter 6).

The size at which DNA synthesis commences (iDS) is precisely controlled. The observation that there is much less variation in the size of mouse fibroblasts at iDS than there is at their birth led to the notion that there exists a 'critical initiation mass' for the G1/S transition (11). Fusion of HeLa cells at various stages of the cell cycle showed that S phase cells contain a positive activator of DNA replication which is absent in G1 cells and declines in G2 cells (12). This cell cycle-dependent activator, called S phase promoting factor (SPF), can hasten the onset of S phase in the G1 nucleus. The identification of SPF, its regulation, and how it activates the replication machinery are major issues of current cell cycle research (see Chapter 6).

A remarkable feature of eukaryotic DNA replication is that no piece of DNA is left unreplicated nor is it replicated more than once within a cycle. Thus reinitiation from a given origin is prevented during S phase. What distinguishes a prereplicative origin from an origin that has already fired? The answer to this question might be relevant to what prevents G2 cells from rereplicating their DNA. The observation that S-phase cells cannot induce DNA replication in the G2 nucleus (12) indicates that the G2 nucleus or replicated DNA is refractory to SPF. The block to replication is only erased as cells pass mitosis, suggesting that the same mechanism could be responsible for the block to rereplication and the dependency of S on M.

The HeLa cell fusion experiments suggested the existence of another type of control by which S phase cells retard the onset of mitosis in G2 nuclei until DNA replication is finished (12). This indicated that S phase cells contain, in addition to SPF, substances capable of inhibiting mitosis. What these substances are is not known. The discovery of mutations which abolish cell cycle arrest in G2 due to the presence of unreplicated or damaged DNA, i.e., disrupt the dependency of M on S, have led to the notion of checkpoint controls (13).

Some of the controls described above are dispensable for basic cell cycles, as in early embryonic development, where S-phase follows mitosis rapidly without intervening periods. While most cells need to grow before they can divide, the cleavage divisions of fertilized eggs occur very rapidly in the absence of growth (14). In *Xenopus* eggs, the onset of mitosis is also largely independent of the completion of DNA synthesis (15, 16), which suggests that some controls operating in somatic cell cycles are overridden in early embryonic cycles (17) (see Chapter 6). From the comparison of early embryonic and somatic cell cycles, it is obvious that cells can divide much faster than they grow. The fact that S phase and mitosis normally do not restrain the pace of cell division emphasizes the predominant role of growth and other cell cycle controls in the regulation of cell proliferation.

3. The Cdc28 protein kinase

The picture that emerged from cell fusion studies is one in which the abundance of S and M phase inducers oscillate during the cycle with different phases. Candidates for the S phase inducer (SPF) were first identified by genetic studies in the yeast *S. cerevisiae* (18, 19).

3.1 *CDC28* and the concept of START

Budding yeast divides asymmetrically generating a large mother and a smaller daughter cell. The size of the bud, which first appears as cells enter S phase, is a good indication of the cell cycle position. Using this simple morphological marker, Hartwell and co-workers screened a collection of temperature-sensitive lethal mutants for those which would arrest uniformly at a given stage of the cell cycle (20). Many of these cell division cycle or cdc mutants, originally defining 32 genes, could be ordered in three independent pathways, one leading to bud emergence and cytokinesis (*CDC24*), one needed for spindle pole body (SPB) duplication (*CDC31*), and one leading to DNA replication and nuclear division (*CDC4* and *CDC7*) (19, 21, 22). Only one mutant, cdc28, could not undergo any event of the three pathways at the restrictive temperature and arrested as unbudded cells with a single unreplicated nucleus (Fig. 2). In addition to occupying the earliest detectable function in the sequence of dependent events in the cycle, the step controlled by *CDC28* was also the focus for cell cycle control by pheromones (21) and nutrients (23). The hypothetical event activated by *CDC28* was therefore termed the START of the cell cycle (19).

START leads to bud emergence, SPB duplication, DNA replication, but as we now know, these events which occur at approximately the same time during the cell cycle are probably triggered by different forms of the Cdc28 kinase. According to the original definition, START must therefore be referred to as 'window of time' in which these various events occur, or the collection of these events. Considering

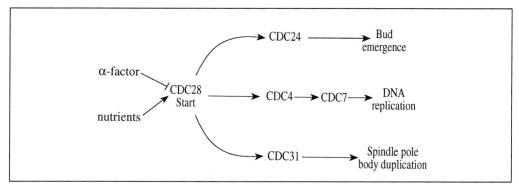

Fig. 2 The concept of START. The step controlled by Cdc28 is needed for all aspects of cell division, and is the focus of control by nutrients and pheromones. It was the earliest function in the sequence of dependent events, and therefore called the 'START' of the cell cycle. We now know that several distinct Cdc28 activities are needed for the completion of START. Spindle formation and nuclear division, which are not indicated, depend on both *CDC4* and *CDC31*.

that a cascade of Cdc28 kinase activities might be required for cell cycle progression, an alternative and more modern definition of START could be the first *CDC28*-dependent function of the cell cycle.

It is worth considering why cell cycle researchers cherished *CDC28* more than other *CDC* genes. There existed other mutants, like *cdc19* or *cdc35* (encoding pyruvate kinase and adenylate cyclase, respectively) that also arrest prior to START, but unlike *cdc28* they fail to increase in mass at the restrictive temperature. However, when new START mutants behaving like *cdc28* were discovered (24, 25), they did not get much attention either. The idea that *CDC28* might encode a key cell cycle regulator crystallized when it was found that the *cdc2*+ gene from fission yeast, in which there were alleles that advanced mitosis (26), and *CDC28* could complement each other (27) and encoded homologous protein kinases (28, 29) (see also Chapter 8). The *CDC28/cdc2*+ genes are now seen as playing a major role in cell cycle control and, quite ironically, one can say that their great destiny parallels the fortune with which they were discovered. Indeed, among the 150 *cdc* mutants, there was only one allele of *CDC28*; moreover, its detection was aided by an unlinked secondary mutation (30). Similarly, of the 25 *S. pombe* mutants that underwent mitosis at a reduced size, 24 were in *wee1* and only one in *wee2/cdc2* (3). A good lesson for people embarking on a genetic screen!

3.2 Cyclins

Despite the differences in their apparent function (*cdc28* mutants arrested in G1, *cdc2* mutants in G2), some facts hinted that *CDC28* and *cdc2*+ might have a role both in G1 and G2 (32, 33). Both gene products have kinase activity (34, 35), but it was not clear how the same kinase could trigger such different functions as DNA replication and mitosis. The answer came from the discovery that Cdc28 and cdc2 are only active when bound to a class of proteins called cyclins (36), which were originally discovered by virtue of oscillations in their abundance during sea urchin cleavage divisions (37, 38). All cyclins have a homologous region called the 'cyclin box' and some carry a 'destruction box' required for their cell cycle-regulated proteolysis. There exist different kinds of cyclins in yeast and metazoans, some being present in S, G2, and M (cyclins A and B) and others in G1 (cyclins Cln, D, and E). It is thought that their sequential activation of the Cdc28/cdc2 kinase in yeast and of various cyclin-dependent kinases (CDKs) in mammals, determines the timing of many key cell cycle events (39, 40).

3.2.1 Cln cyclins

S. cerevisiae contains three Cln cyclins (Cln1, 2, and 3) which activate the Cdc28 protein kinase and trigger the various events of START. The *CLN1* and -2 genes were discovered as multicopy suppressors of a temperature-sensitive *cdc28* mutation (41). The *WHI1-1* and *DAF1-1* dominant gain of function mutations identified another gene, now renamed *CLN3*, by virtue of reducing the critical cell size for START (42, 43) or conferring resistance to mating pheromone-induced cell cycle arrest (44). The products of the *CLN1* and *CLN2* genes are closely related to each other (57% identity) and distantly related to A and B cyclins. The Cln3 protein, which also contains a cyclin homology box, is only 20–25% identical to Cln1 and Cln2 (Fig. 3). Genetic analysis showed that any one of the three Cln cyclins is sufficient for viability but deletion of all three causes arrest before START (45, 46). This

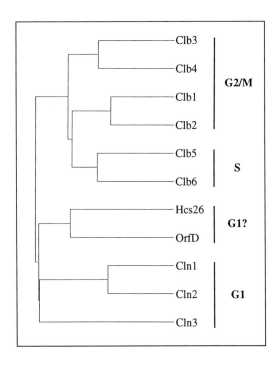

Fig. 3 Phylogenetic comparison of yeast cyclins. B type cyclins (Clbs) are required for DNA replication and nuclear division. Cln cyclins bind to Cdc28 and are required for START. Hcs26 and OrfD (renamed Plc1 and 2) bind to the Pho85 kinase and might contribute to G1 progression. Complete sequences were aligned using the Pileup subroutine of the UWGCG package; length of horizontal bars is a measure of evolutionary distance.

led to the proposal (erroneous as we shall see below) that Cln1, Cln2, and Cln3 are redundant cyclins which have equivalent roles in triggering cell cycle progression. *HCS26*, which was identified along with *CLN1,2* as a multicopy suppressor of *swi4* deletion (47), and *ORFD* (48) encode proteins with similarities to cyclins. These cyclin-like proteins activate the Pho85 protein kinase and seem to be involved in G1 progression, at least when normal passage through START is compromised (49, 50).

3.2.2 Clb cyclins

Four genes, *CLB1, CLB2, CLB3*, and *CLB4*, encode proteins homologous to B-type mitotic cyclins from other organisms. They were discovered owing to this homology (51) and as multi-copy suppressors of the *cdc28-1N* mutation which produces a mitotic arrest (52). These cyclins appear to be organized in pairs, Clb1 being more similar to Clb2 (62% identity) and Clb3 to Clb4 (50% identity) (Fig. 3). The Clb1,2 pair is more closely related to cdc13 from *S. pombe* or to B-type cyclins from animals. Of the four proteins, Clb2 seems to have the most important role in mitosis (53, 54). Inactivation of all four genes is lethal and causes cells to arrest in G2 with duplicated but unseparated spindle pole bodies (54). Thus, the Clb1−4 cyclins are required for the formation of a mitotic spindle but not for DNA replication (55). In a separate study, it was proposed that diploid cells lacking these cyclins are also unable to complete DNA replication (53).

CLB5 and *CLB6* encode another pair of related proteins (50% identity) which are also members of the B-type class of cyclins. Interestingly, *CLB5* and *CLB6* are adjacent to *CLB2* and *CLB1* respectively, which suggest that the two loci arose by interchromosomal duplication (56, 57). *CLB5* and *CLB6* are not essential genes. Deletion of the former prolongs S phase whereas deletion of both genes delays the onset of DNA replication relative to other post-START events such as budding (56, 58). Ectopic expression of *CLB5* suppresses the lethality of a *cln1,2,3* triple mutant, indicating that Cln cyclins might merely be needed for the expression and/or activation of Clb5,6. Taken together these two observations strongly suggest that the Clb5 and Clb6 cyclins are more directly involved in triggering DNA replication than Cln cyclins. Clb5 may have an additional role, at least when Clb3 and Clb4 are missing, in the formation of the mitotic spindle (56).

3.3 Transcriptional control of cyclin synthesis

From the 11 yeast cyclins or cyclin-like proteins enumerated above, all except one (Cln3) display cell cycle regulated transcription. *CLN1* and *-2, HCS26, ORFD, CLB5* and *-6* transcripts are absent or low in early G1 but appear abruptly around START. They peak at the G1/S boundary and later decline during S/G2. The *CLB3,4* and *CLB1,2* genes are expressed in late S and late G2, respectively (59). Since the kinase activity associated with these cyclins parallels their transcription (60–64), it is conceivable that the activation of the yeast cyclin-dependent kinases is at least partly dictated by regulated transcription of cyclins (Fig. 4). Understanding the regulation of cyclin transcription might therefore help to explain certain aspects of cell cycle dynamics.

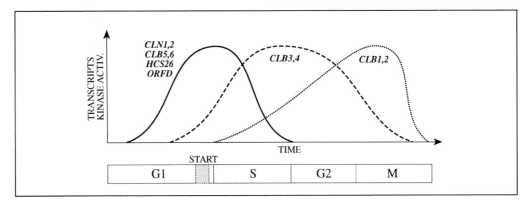

Fig. 4 Cell cycle-dependent transcription of yeast cyclins. The kinase activity associated with these cyclins parallels their transcription pattern.

While little is known about the regulation of G2 cyclin transcription, a quite detailed picture of late G1-specific transcription has been drawn. A large family of genes are transcribed exclusively at the G1/S transition, including cyclins and many DNA synthesis genes. In contrast to the latter, *CLN1* and *-2*, *CLB5* and *-6*, and perhaps also *CDC6* (157), encode unstable proteins which need to be resynthesized at each cell cycle. Their transcriptional activation might therefore be important for the G1/S transition (see Section 4.1.2). The late G1-specific genes can be subdivided in two groups according to the *cis*-acting sequences found within their promoters. The first group has a sequence motif called the SCB element which acts as a late G1-specific UAS (upstream activating sequence) element (65–67). The promoters of *CLN1*, *CLN2*, and *HCS26* contain such elements (47, 60, 68). The second group of genes, which includes those coding for DNA metabolism enzymes and the cyclin-encoding genes *CLB5* and *CLB6*, contain sequences similar to the MluI cell cycle box (MCB element) in their promoters (69, 70). Two related transcription factors bind to these elements and confer late G1-specific transcription. SBF and MBF (also called DSC1) share a common regulatory subunit, Swi6 (71, 72), but contain different site-specific DNA binding subunits: Swi4 for SBF (67, 73) and Mbp1 for MBF (74) (Fig. 5).

Following S phase, transcription of late G1-specific genes is down regulated. Repression of SCB-driven genes in G2 is dependent on the mitotic cyclins Clb 1–4 and might occur via the direct interaction of the Clb2/Cdc28 kinase with Swi4 (55). Interestingly, repression of MCB-driven genes still occurs in the absence of Clb 1–4 which indicates that the mechanisms by which SBF and MBF are down regulated are different.

3.4 Posttranslational regulation of the Cdc28 protein kinase

While *de novo* synthesis of cyclins is needed for cell cycle progression, their cell cycle-regulated transcription is not obligatory, as shown by the normal proliferation

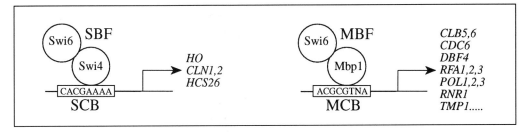

Fig. 5 The SBF and MBF transcription factors are responsible for late G1-specific expression of a number of genes involved in START and DNA replication. SBF, SCB-binding factor; SCB, Swi4,6 cell cycle box; HO, mating type-switching endonuclease; MBF, MCB-binding factor; MCB, *MluI* cycle box; Mbp1, *MluI*-binding protein; *RFA1,2, and 3* encode the three subunits of replication factor A (Rpa); *POL1,2,3*, DNA polymerases; *RNR1*, ribonucleotide reductase; *TMP1*, thymidylate synthase.

of yeast cells expressing cyclins constitutively. Indeed, in some organisms like *S. pombe*, the transcription of cyclin genes is little or not at all regulated. How then is the periodicity of CDK activation generated in these cases?

3.4.1 Cdc28 phosphorylation

Work on fission yeast has revealed that the phosphorylation state of p34^{cdc2} is crucial for the onset of mitosis (75, 76) (Chapter 8). It was found that *wee1* mutants which advance mitosis and *cdc25* mutants which delay mitosis have antagonistic effects on the phosphorylation of Tyr15 of p34^{cdc2}. *wee1$^+$* encodes a protein kinase and *cdc25$^+$* a protein phosphatase. Tyr15 phosphorylation is inhibitory and prevents the onset of mitosis when DNA replication is not completed (77). The equivalent Tyr19 in Cdc28 is also phosphorylated in a cell cycle-dependent manner but, unlike in *S. pombe*, mutation of Tyr19 has little or no effect on the length of G2 or on the inhibition of nuclear division by unreplicated or damaged DNA (62, 78). So, despite having a similar pattern of cdc2/Cdc28 phosphorylation, the two yeasts seem to make different use of it.

In addition to dephosphorylation of Tyr15/19, all cdc2/Cdc28 kinases need phosphorylation of Thr161/169 for their activity. The CDK activating kinase (CAK) which catalyses this phosphorylation is itself a CDK, composed of a catalytic (p40^{MO15}) (79, 80) and regulatory (cyclin H) subunit (81, 82). CAK seems to be constitutively active throughout the cell cycle (83) and therefore is unlikely to have a role in the periodic activation of p34^{cdc2}. The Kin28 kinase is the closest yeast homologue of MO15, but CAK activity is not decreased in extracts immunodepleted for Kin28 (84).

3.4.2 Cyclin proteolysis

Cyclins which are obligatory activators of cdc2/Cdc28 kinases are, by definition, periodical proteins. The abrupt destruction of B-type cyclins at anaphase could, by itself, account for the disappearance of cdc2/Cdc28 kinase activity as cells exit mitosis. The proteolysis of B-type cyclins is dependent on a conserved 9 amino acid sequence (RxxLxxxxN) within their N-termini, called the cyclin destruction box

(85). All yeast B-type cyclins, except perhaps Clb6, contain sequences similar to the destruction box. Mutation or deletion of this sequence in Clb2 causes stabilization (86); a similar, but modest effect is seen with Clb5 (S. Irniger, unpublished results). Destruction of mitotic cyclins is important (87), although not required (86), for exit from mitosis. Clb2 proteolysis is not confined to mitotic cells but continues during the subsequent G1 period until the activation of G1 cyclins, which explains why Clb cyclins cannot accumulate before Cln cyclins (86). Cln cyclins are not subject to the same kind of cell cycle-regulated proteolysis, but are unstable throughout the cell cycle due to PEST sequences in their C-termini (60, 88, 89). Thus accumulation of Cln cyclins is dictated mainly by their rate of synthesis.

3.4.3 CDK inhibitors

The Cdc28 protein kinase is not only regulated by phosphorylation and by periodic association to cyclins. Analysis of Cdc28-associated proteins (34) and studies of pheromone-induced cell cycle arrest (90) led to the discovery of another class of cell cycle regulators, the CDK inhibitors (CKI) (91, 92). The Far1 protein is induced by α-factor to bind and inactivate the Cln–Cdc28 kinase, which leads to cell cycle arrest before START (93–95) (see Section 4.1.3). In contrast, the p40^{SIC1} CKI inhibits Clb–Cdc28, but not Cln–Cdc28 complexes (64, 96). While Far1 causes cell cycle arrest in response to an external factor (pheromones), Sic1 seems to be an intrinsic regulator of the cell cycle which controls entry into S phase (see Section 4.2.2).

4. Control of the initiation of DNA replication

Now that we have introduced the basic notions of the cell cycle and its main regulator, the Cdc28 protein kinase, let us consider what we know about the cascade of events which lead a growing yeast cell to initiate DNA replication.

4.1 Starting the cell cycle

4.1.1. Nutrient and size control

In most proliferating cells, cell growth is tightly coordinated with cell division. The consequence (if not the cause) is size homeostasis. Without such a coupling and depending on their nutritional status, cells would become larger or smaller at each cell division, a behaviour incompatible with sustained proliferation. In fact, cell size is remarkably constant in a given organism or cell type. How this is achieved remains a mystery. START-dependent events (like budding or DNA replication) occur only after cells have reached a critical size (11, 22, 97) or rate of protein synthesis (98, 99). This is true for most cells, but particularly obvious for budding yeast which divides asymmetrically to produce large mothers and smaller daughters. Mothers need little or no growth and therefore have a short G1 period. In contrast, daughter cells are born much below the critical cell size needed for START and must grow before they can begin the cell division programme. Growth and division can also be coordinated at different stages of the cell cycle, as in fission yeast where the predominant size control is over mitosis and not START (see Chapter 8). Normally, cell growth is required for cell division, as starved cells which stop increasing mass arrest in G1 before START (22). The dependence of cell division on growth is not reciprocal, however, as cell cycle mutants continue to increase in mass without dividing. This property is actually taken as diagnostic for true cell cycle mutants.

How do cells measure growth and how does growth trigger the START events? Some evidence suggests that cells measure the rate of protein synthesis rather than size *per se*. Indeed, the critical size for START is not absolute: it can be increased by lowering the rate of protein synthesis (using low doses of cycloheximide) or by changing the nutritional conditions (100). The accumulation of an unstable protein, which would depend directly on its rate of synthesis, could be sufficient for assessing the general rate of protein synthesis in a cell. The Cln3 cyclin is a good candidate for such a protein. Cln3 is unstable, its overexpression or stabilization triggers START prematurely, and finally it is an activator of the Cdc28 kinase required for START (43–45). Among G1 cyclins, Cln3 is the best candidate because it is present in early G1 cells and seems to act upstream of Cln1,2 (158, 61). When nutrients are limiting, it is possible that the protein synthesis rates remain too low for Cln3 to accumulate above its threshold level needed for activation of the Cdc28 kinase.

According to this hypothesis, increasing the rate of Cln3 synthesis should be sufficient to lower the critical size for START. However, when cells are grown on rich carbon sources, this size is not smaller but larger than when cells are grown on poor medium (97). A similar delay of START is observed when cells are exposed to increased intracellular cAMP levels (101), which somewhat contradicts previous observations showing that cAMP is needed for protein synthesis, and so indirectly also for Cln synthesis and START (102). The paradox was resolved by showing that cAMP (whose levels are higher in rich medium) has a dual effect on the activation of Cdc28: on the one hand, it stimulates protein synthesis and thereby Cln3 accumulation and START: on the other hand, high cAMP levels inhibit START by specifically down regulating the transcription of *CLN1*, *CLN2*, and other SBF/MBF-regulated genes, but not that of *CLN3* (103, 104). Balanced levels of cAMP might therefore be needed for the coordination of cell growth and cell division in budding yeast.

4.1.2 G1 cyclins and START

Overexpression of Cln1, Cln2, or Cln3 triggers START prematurely (61; E. Schwob, unpublished results). More significantly, moderate expression of Cln2 in early G1 cells also causes premature S-phase entry (158), indicating that G1 cyclins are rate-limiting activators of START. *CLN1,2* transcripts are absent in early G1 but raise abruptly around START. It is therefore thought that this late G1 burst of transcription is important for START. So how is the late G1 transcriptional programme activated? Some clue came from the discovery that *CLN1,2* transcription depends on SBF (Swi4 + Swi6) and an active Cdc28 kinase. Since Cln1 and Cln2 are also activators of Cdc28, it was proposed that Cln1,2 (and perhaps also Cln3) are both needed for and dependent on the Cdc28 kinase. One way to solve this paradox was to invoke a positive feedback loop by which the Cln–Cdc28 kinase activates SBF,

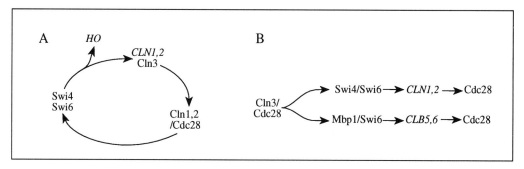

Fig. 6 Models for late G1 transcriptional activation. (a) Positive feedback loop; (b) linear model. The positive feedback hypothesis was invalidated by showing that Cln1,2 are not needed for their own expression; Cln3 is sufficient for full induction of *CLN1,2* and *CLB5,6* when cells reach the critical size for START.

which in turn activates *CLN1,2* transcription, thereby closing the loop by further activating Cdc28 (47, 68, 105, 106) (Fig. 6a). As predicted by this hypothesis, *CLN1,2* transcription is both dependent on an active Cdc28 and stimulated by G1 cyclin activity. The initial firing of the loop would be triggered by Cln3 accumulating above a certain threshold level. Such a positive feedback mechanism could explain the abruptness and irreversibility of the START transition. While it was shown that this loop can function, it was however not demonstrated that this is the way SBF is normally activated in small cells undergoing START. In fact, recent experiments show that Cln3 is sufficient for full activation of SBF and MBF at the critical cell size for START (158). Cln1 and Cln2 are therefore not needed for their own expression. Thus, the SBF/MBF transcription factors are activated according to a linear pathway, with Cln3 as the main actor (Fig. 6b). In *cln3* null mutants, transcription of SBF-, and MBF-regulated genes is delayed until cells reach a size several times larger than normal. It is worth considering why people favoured, at first, the positive feedback loop model. One reason is that genetic evidence (anyone of the Clns was sufficient for life) suggested that all three Cln cyclins had an equivalent function. Even though *CLN3* transcription was not dependent upon SBF, Cln3 function was put downstream of Swi4,6 because Swi6 was somehow required for the viability of *cln1 cln2* mutants (68).

Elevating Cln3 to the rank of prime activator of START leaves us with the unanswered question of how Cln3 activates SBF and MBF. It is probably not due to changes in the abundance of the transcription factors because SBF and MBF have been detected at all stages of the cell cycle (71, 107) and because their activation can occur in the absence of protein synthesis (108). The activity of these transcription factors must therefore be regulated posttranslationally, so that they are active in late G1 and inactive in G2. Could SBF and MBF be activated by the Cln3–Cdc28 protein kinase, either directly or indirectly, and what determines the timing of activation? Cdc28-dependent phosphorylation sites have been found in Swi4, but their mutation does not alter gene regulation (M. Neuberg, unpublished results). Swi6 could be a better candidate because it is part of both SBF and MBF. It is puzzling that

SBF/MBF activation is critically dependent upon Cln3 levels, although these do not appear to change during the cell cycle (61). In fact, if they do not change relative to total cell mass while the cell is growing, they will increase exponentially relative to some stable component like DNA. The Cln3 to DNA ratio might be a means of setting a threshold for Cdc28 activation, consistent with the observation that the critical cell size for START in diploids is twice that of haploids. Alternatively, an inhibitor of Cln3 kinase could be present in small cells and destroyed or titrated when cells reach the proper size for START. A problem is that Cln3-associated H1 kinase is so weak that accurate measurements of its activity during the cell cycle have not so far been possible. Nevertheless, if such an inhibitor were to exist, it should be possible to isolate it as a recessive wee (small size) mutant.

4.1.3 Regulation of START by pheromones

When haploid yeast cells encounter cells from the opposite mating type, diffusible peptide pheromones (**a**- and α-factor) bind to surface receptors, stimulate a hetero-trimeric G protein leading to the activation of protein kinases causing three main responses: cell cycle arrest, morphological changes (shmoo formation), and transcriptional induction of several genes needed for conjugation (109). These responses are conveyed by a MAP (mitogen-activated protein) kinase cascade pathway which later diverges to mediate the specific responses (110, 111). Like *cln1,2,3* triple mutants, pheromone-treated cells arrest before START, suggesting that G1 cyclin function might be inhibited. Conversely, Cln3 overexpression or gain of function mutations, as for the hyperstable Cln3-1 truncation, causes resistance to α-factor-induced G1 arrest (44, 88).

How does α-factor arrest the cell cycle? Transcripts from the *CLN1, CLN2* (44, 45, 60) and *CLB5, CLB6* genes (57, 58) are absent in pheromone-treated cells. If this were a direct effect of α-factor, as opposed to a consequence of pre-START arrest, it would result in the rapid loss of these unstable proteins. In contrast, *CLN3* transcripts and protein are slightly induced by α-factor (61). In fact, the effect of α-factor on Cln1,2 cannot be only transcriptional because cells expressing *CLN2* from an unregulated promoter still exhibit cell cycle arrest in response to α-factor, despite the presence of presence of normal levels of Cln2 protein (95). The discovery of *far1* mutants, which do not arrest in response to α-factor (90), provided the answer to how pheromones arrest the cell cycle. The Far1 protein binds all three Clns (94) and inhibits the Cdc28 kinase bound to Cln1 and Cln2 (95). It is not yet shown but likely that Far1 also inhibits Cln3–Cdc28. *FAR1* transcripts are induced by α-factor, but expression of *FAR1* from the *GAL* promoter does not cause G1 arrest, suggesting that an additional event is required. Far1, which is present in haploid G1 cells even in the absence of α-factor (112), needs to be activated by phosphorylation (93). Thus, pheromones arrest the cell cycle mainly by direct inhibition of the Cln–Cdc28 kinase although a decrease in *CLN1,2* transcription could be important for sustained arrest (95). An active proteolysis of B-type cyclins, which can trigger S in the absence of Clns, also seems important for sustained G1 arrest in α-factor (S. Irniger, unpublished results).

4.2   B-type cyclins trigger S phase

4.2.1   Post-START functions needed for S-phase entry

The activation of Cdc28 by the G1 cyclins Cln1,2,3 is necessary for START and DNA replication. The existence of the *cdc4, cdc34*, and *cdc53* class of mutants which arrest after START but before the initiation of DNA replication (22) suggests that the Cln–Cdc28 kinase is not sufficient for S-phase entry. Mutants of this class arrest in late G1 with multiple elongated buds, duplicated spindle pole bodies (but no spindle) (113) and high Cln–Cdc28 kinase activity (88). *CDC34* encodes an ubiquitin-conjugating enzyme thought to promote the degradation of proteins (114). Thus, S phase not only requires the synthesis but also the destruction of specific proteins. The Cdc7 protein kinase is needed for one of the last steps before DNA replication, after Cdc34, as *cdc7* mutants no longer require protein synthesis for replication and arrest with a short bipolar spindle. In contrast to Cdc7, whose function might not be required for premeiotic S phase (115), Cdc4 and Cdc34 are required for DNA replication in both the meiotic and mitotic cycle.

What is the role of Cdc34 and Cdc7 in the initiation of DNA replication? An important finding is that the Cdc28 kinase associated with B-type cyclins is inactive in *cdc34* mutants (62, 64). Could the Clb–Cdc28 kinase, rather than the Cln–Cdc28 form, be required for DNA replication? This possibility was examined by measuring the kinetics of S phase in mutants lacking different sets of Clb cyclins. Inactivation of the four mitotic Clb 1–4 cyclins causes cell cycle arrest before mitosis but has no effect on S phase (55). In contrast, cells lacking Clb5 and Clb6 are viable but S phase is delayed by 30 min relative to budding, suggesting a specific role of Clb5,6 in the initiation of DNA replication (56). Clb5 seems more important than Clb6 for S phase. In contrast to *clb6* mutants which apart from a delayed S phase have little or no phenotype, the inactivation of *CLB5* greatly extends S phase. Interestingly, *CLB6* deletion rescues the slow S phase of *clb5* mutants, as *clb5 clb6* double mutants progress rapidly through S phase after the initial delay. If Clb5 and Clb6 are the rate-limiting activators of DNA replication, why are they not essential? The answer is that other B-type cyclins trigger DNA replication in the absence of Clb5,6. Hence, early G1 cells lacking all known B-type cyclins (Clb1–6) form multiple elongated buds but fail altogether to enter S phase and form spindles. The phenotype is very similar to that of the *cdc34* class of mutants and consistent with their lack of Clb–Cdc28 kinase activity. A 15 min pulse of *CLB5* in late G1 *clb1–6* mutant cells is sufficient to induce a rapid and apparently full round of DNA replication (64). This demonstrates that activation of Cdc28 by B-type cyclins is required for DNA replication. Thus, Clns are not sufficient. Due to their pattern of expression, Clb5 and Clb6 are most probably the physiological inducers of DNA replication. In the absence of Clb5,6, S phase is delayed until Clb3,4 (or eventually Clb1,2) can substitute.

What is the Cln–Cdc28 kinase then needed for? The dependence of *CLB5,6* expression on Cln cyclins could alone account for the failure of *cln1,2,3* mutants to enter S phase because ectopic expression of *CLB5* is sufficient to rescue their lethality (58).

CLB5 is however not sufficient to trigger DNA replication as *CLB5* expression in early G1 cells does not cause premature S-phase entry (56).

4.2.2 Activation of the Clb5–Cdc28 kinase

Why does premature *CLB5* expression not advance S phase? Is the Clb5–Cdc28 kinase not the actual S phase inducer or is the kinase not active in early G1 cells? The latter explanation is correct. The Clb5–Cdc28 complex is formed in early G1 cells but no associated H1 kinase can be detected. What is the mechanism preventing activation of the kinase? G1 cells obtained either by centrifugal elutriation, α-factor treatment or *cdc28* arrest, contain an activity capable of inhibiting exogenously added active Clb5 kinase. Inhibition depends on the presence in the extract of the p40^{SIC1} protein, which is a known substrate and inhibitor of the Cdc28 kinase (34, 64, 96, 116). Sic1 is present in early G1 cells, disappears just before S phase, and then reaccumulates in the following late M or early G1 period. Expression of *CLB5* from the *GAL* promoter in early G1 cells lacking Sic1 leads to the immediate accumulation of active kinase and premature DNA replication, suggesting that Sic1 might negatively regulate S-phase entry (64). It is unclear whether this is also true or important in wild-type cells, where *CLB5* expression is restricted to late G1.

4.2.3 Proteolysis of the Sic1 CDK inhibitor is required for S-phase entry

Sic1 is present during the G1 period of the cell cycle and might prevent S phase by inhibiting the Clb–Cdc28 kinases. Do cells need to get rid of Sic1 to enter S phase? The importance of Sic1 degradation for S-phase entry can be assessed in *cdc34* mutants which do not degrade Sic1 and do not enter S phase. The key finding is that *cdc34 sic1* double mutants undergo DNA replication which demonstrates that Sic1 alone is responsible for the G1 arrest of *cdc34* mutants (64). Thus, the levels of Sic1 existing in *cdc34* mutant cells are sufficient to block S-phase entry. The observation that wild-type G1 cells contain the same amount of Sic1 as do *cdc34* mutants is of fundamental importance. It means that normal G1 cells contain a high potential of Clb–Cdc28 inhibitory activity, and underscores the absolute need to destroy Sic1 before cells can enter S phase.

How wild-type cells get rid of Sic1 in late G1 is an important but unanswered question. Some evidence points to regulated proteolysis. Sic1 is not degraded in *cdc34* mutants suggesting that Cdc34, which is an ubiquitin-conjugating enzyme, might be responsible for Sic1 proteolysis in late G1. The failure to degrade Sic1 is not due to stage-specific arrest because *clb1–6* sextuple mutants which arrest at a similar stage lose the *SIC1*-dependent inhibitory activity (T. Böhm, unpublished results). This also indicates that Clb–Cdc28 kinase activity is not required for Sic1 degradation. In contrast, phosphorylation of Sic1 by the Cln–Cdc28 kinase might be needed for Sic1 proteolysis. The lower electrophoretic mobility of Sic1 in *cdc34* mutants (high Cln–Cdc28 activity) than in *cdc28*-arrested cells (no Cln–Cdc28 activity) and the observation that Sic1 does not inhibit the Cln2–Cdc28 kinase (64) are consistent with such a hypothesis. A working model is therefore that the Cln1,2–Cdc28

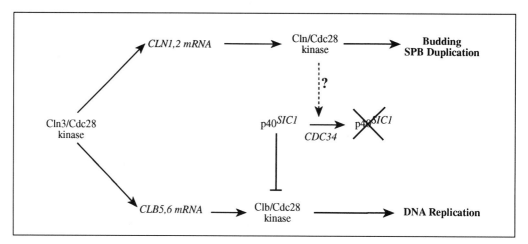

Fig. 7 Destruction of Sic1 is needed for S-phase entry. DNA replication depends on the Clb5,6–Cdc28 kinase but is controlled by the Cln–Cdc28 kinase. Transcription of *CLB5,6* requires Cln3 whereas Sic1 proteolysis, which is needed for activation of the Clb kinases, might require the Cln1,2–Cdc28 kinase.

kinase phosphorylates Sic1 in late G1 and targets it for proteolysis by the Cdc34 ubiquitin-mediated pathway. Cdc4 which encodes a protein with homology to β-transducins (117), and Cdc53 could help Cdc34 to perform its function. The dependence of Sic1 degradation on the Cln–Cdc28 kinase would ensure that the S phase kinase (Clb5,6–Cdc28) does not become active before the G1 Cln kinase (Fig. 7).

4.3 What links the cell cycle to the replication machinery?

The Clb5–Cdc28 kinase has properties of S-phase-promoting factor (SPF). First, Clb5 is an unstable protein which needs to be resynthesized at each cell cycle. The Clb5-associated H1 kinase is absent in G1, peaks during S phase and later declines. Second, Clb–Cdc28 kinases are required for DNA replication and in the absence of other B-type cyclins, Clb5 is an essential and very potent S phase inducer. Third, when Sic1 is absent precocious Clb5 expression advances the onset of S phase. The Cln–Cdc28 kinase has equivalent properties: it is also required and rate-limiting for DNA replication, but in contrast to the Clb–Cdc28 kinase the Cln kinase is also required for other START events like budding and spindle pole body duplication. Furthermore, overexpression of Clb5 (or Clb2) triggers S phase in the absence of Cln cyclins, but Clns cannot trigger S phase in the absence of Clbs. Thus, the Clb5,6–Cdc28 kinase could be the true S-phase inducer. Yet, the accumulation of an active Clb5–Cdc28 kinase is tightly controlled by the Cln–Cdc28 kinase, both at the transcriptional and posttranslational levels. Cln3 activates *CLB5,6* transcription and Cln1,2 might be needed for the stabilization of Clb cyclins and the destruction of the Sic1 inhibitor of Clb–Cdc28 kinases. Thus, it appears that a cascade of Cdc28 kinase activities (Cln3 → Cln1,2 → Clb5,6) is needed for DNA replication. While the successive events leading to the activation of the S-phase kinase begin to be clarified, very little is known about the downstream events. Does the Clb5–Cdc28 kinase activate the DNA replication machinery directly or are additional steps needed? Of course, many other gene products are required for DNA replication, but do they have a regulatory role for the onset of S phase? Are they targets of the Cdc28 kinase? In this section, we will consider some of the better candidates.

4.3.1 The Dbf4–Cdc7 protein kinase

CDC7 encodes a protein kinase whose function is needed just prior to the initiation of DNA synthesis (118). After release from *cdc7* arrest, DNA replication no longer requires protein synthesis (21). Cdc7 is downstream or on a parallel pathway to Clb5–Cdc28, because the latter is active in *cdc7*-arrested cells (64). Cdc7 is also involved in other aspects of DNA metabolism like replication-dependent DNA repair, transcriptional silencing and meiotic DNA recombination. How is the Cdc7 kinase regulated? It peaks in late G1, but Cdc7 protein levels are constant and in excess during the cell cycle (119–121). By analogy to CDKs, it was proposed that Cdc7 activity is regulated by periodic association with the Dbf4 protein. Dbf4 is also needed for the initiation of DNA replication (122) and was shown to interact with Cdc7 (119, 123). The *DBF4* promoter contains a MCB element and transcription is late G1 specific. It is postulated but not yet shown, that Dbf4 oscillates during the cell cycle and that this periodic accumulation drives Cdc7 kinase activity. An argument against a rate-limiting role of Dbf4 in Cdc7 kinase activation is that spores inheriting a *dbf4* null allele can divide two to four times (123). The Cdc7 kinase also needs phosphorylation. Cdc7 is inactive in *cdc28* mutants, and its activity is low in *cdc4* mutants (but see 120 for contradictory results), indicating that Cdc7 or Dbf4 might be activated by Clb–Cdc28 kinases (119, 120). Site-directed mutagenesis of all four Cdc28 consensus phosphorylation sites does not however alter Cdc7 function *in vivo* (118). Dbf4 also contains Cdc28 consensus sites which are perhaps important for the association and/or localization of Dbf4 and Cdc7. Genetic evidence supports a direct role for the Cdc7 kinase in DNA replication. Fusions of Dbf4 to the Gal4 activation domain *trans*-activate transcription through ARS elements, which suggests that Dbf4 binds to replication origins, probably via the origin-recognition complex (ORC) (124). Thus, the role of Dbf4 might be to recruit active Cdc7 to origins where it might phosphorylate some component of the DNA replication machinery. In agreement with this, preliminary evidence suggests that Cdc7 binds and phosphorylates Orc2 (125). The existence of a mutation (*bob1*) which bypasses Cdc7 and Dbf4 function (119) indicates that the Dbf4–Cdc7 kinase might not be absolutely required for DNA replication. The *bob1* mutation is recessive, which suggests that the wild-type protein might normally prevent initiation of DNA replication. This inhibition would be relieved when the Dbf4–Cdc7 kinase becomes active (Fig. 8). In this model, the Cdc7 kinase is not needed for initiation *per se*, but rather to remove a block to DNA replication. Interestingly, *orc2-1* mutants first arrest before S, but then enter an abortive round of DNA replication, suggesting that Orc2 also has a role in preventing aberrant DNA replication (126). It will be interesting to see what the role of Bob1 is and whether it is part of ORC.

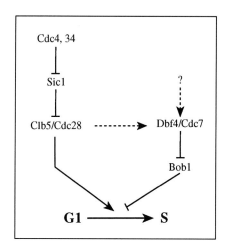

Fig. 8 The Cdc28 and Cdc7 kinases are both required for initiation of DNA replication. Recessive mutations in *BOB1* bypass the requirement of Cdc7 and Dbf4 for DNA replication, indicating that Bob1 might normally inhibit S-phase entry. Activation of the Db4–Cdc7 kinase might depend on the Clb5 kinase or on the destruction of Sic1.

4.3.2 Other genes required for initiation

ORC. The multiprotein origin-recognition complex (ORC) was purified from yeast by virtue of its ATP-dependent binding to ARSs (8; see Chapter 3 for a more complete description). ORC binds *in vitro* and *in vivo* to the conserved domain of ARS. Each of the six subunits is essential for viability and the phenotype of conditional *orc2* mutants is consistent with a role in initiation of DNA replication. Orc2 and Orc6 contain potential Cdc28 phosphorylation sites, so they could be the target for cell cycle-regulated protein kinases. Because ORC binds to origins throughout the cell cycle (9), initiation is however not likely to be regulated at the level of binding. An extended footprint is observed on ARS from the end of mitosis until the beginning of S phase (127), suggesting a conformational change of ORC or the binding of additional proteins to origins. This extended footprint is present in *cdc7* but absent in *cdc6* mutants (127), which both arrest with unreplicated DNA, suggesting that Cdc6 might be part of or needed for the formation of the so-called prereplicative complex.

CDC6. Like Cdc7 and Dbf4, Cdc6 is needed for DNA replication (128) and is a potential target of the Clb5–Cdc28 kinase. Whether Cdc6 functions before Cdc7, or vice versa, is not known. *cdc6* mutants show a high rate of plasmid loss which can be suppressed by the addition of more ARS elements on the plasmid (129). This indicates that plasmid loss is due to a replication defect, and that Cdc6 might be needed for ARS function, i.e., for origin firing. The observation that Orc6 over-expression is toxic in *cdc6* mutants (130) suggests that the two proteins might interact. After shift to the restrictive temperature, *cdc6-1* mutants first stop replication (131) and later undergo a lethal haploid mitosis (157). This haploid mitosis depends on mitotic cyclins, further indicating that Clb–Cdc28 kinases are active in *cdc6* mutants. Thus, Cdc6 is required both for DNA replication and for preventing mitosis. Cdc18, the *S. pombe* homologue of Cdc6, has similar properties (132). Clearly, Cdc6 function is needed for DNA replication, but is *de novo* synthesis of Cdc6 important for the onset of S phase? The transcriptional regulation of *CDC6* is complex: normally, the gene is expressed in late M (133), but upon release from α-factor arrest or following elutriation, there is a peak of transcription in late G1, consistent with the presence of MCB elements in the *CDC6* promoter (134). This late G1-specific expression is important for DNA replication, as turning off *CDC6* transcription in early G1 cells affects the kinetics of S phase (S. Piatti, unpublished results). It also indicates that maternal Cdc6 becomes limiting during an extended G1 period, suggesting that Cdc6 is an unstable protein. What could be the function of Cdc6 for the initiation of DNA replication? *CDC6* encodes a 58 kDa protein which can bind and hydrolyse ATP, but does not seem to bind to DNA or replication origins directly (133). It contains potential nuclear localization signals and PEST sequences found in other unstable proteins. Cdc6 can be detected only in late M and G1 cells (157), in which the origins are in the prereplicative state (127). It is possible that Cdc6 binds, perhaps via ORC, to replication origins after mitosis and remains there until DNA replication commences. Cdc6, which also contains potential Cdc28 phosphorylation sites, could be activated by the Clb5 kinase to initiate DNA replication.

CDC46 (MCM5), MCM2, MCM3. Two genetic approaches provided additional candidates for proteins that could be involved in early steps of DNA replication (135). *cdc46* and *cdc47* mutants were isolated as allele-specific suppressors of the *cdc45* and *cdc54* cold-sensitive mutants which arrest at the G1/S transition (136). Mutants in *CDC46* are also temperature-sensitive for division, with some alleles arresting before S phase and others late in S (137). ARS-specific defects of minichromosome maintenance was the basis for the isolation of the *mcm2*, *mcm3* and *mcm5* mutants (138). *MCM5* is identical to *CDC46*. Consistent with a role in DNA replication, these mutants show increased rates of recombination and chromosome loss, but conditional alleles arrest in late S phase rather than at the G1/S transition. This property can be explained by the selective defect in the use of particular replication origins (139). The Mcm family of proteins is conserved throughout evolution and their members share significant homology. These proteins are not redundant, however, as inactivation of each is lethal (137, 139). Much interest in the Mcm proteins originates from their cell cycle-dependent cellular localization. In yeast but not in all organisms, they are present in the nucleus exclusively from late M to early S, meeting the criteria for a 'licensing factor' postulated to explain the rereplication observed in *Xenopus* extracts after permeabilization of the nuclear membrane (140; see Chapter 6). Mcm2 and Mcm3 bind tightly to chromatin (139), but it is not known whether they bind specifically to ARS. The function of Mcm proteins might be regulated by cell cycle-dependent nuclear import. Mcm3 has a nuclear localization signal flanked by potential Cdc28 phosphorylation sites and its import into the nucleus might be triggered, analogous to Swi5, by the inactivation of the mitotic kinase (141). Cdc46 and Mcm2 do not have nuclear localization signals, but could be transported along with Mcm3. What is the function of the Mcm proteins? P1, the human homologue of Mcm3, copurifies with the DNA polymerase α–primase complex (142) which is required for the initiation of both leading and lagging strand replication. Mcm proteins contain a central domain with similarity to the ATPase motif found in helicases, but it is not known whether they have a role in unwinding DNA at origins.

RPA. RPA, also known as RF-A or ssb, is an heterotrimeric protein complex which binds tightly to single-stranded (ss) DNA and is required for an early step of SV40 DNA replication *in vitro* (see Chapter 2). It is also required for replication in intact nuclei in *Xenopus* egg extracts (143). Each of three subunits (70, 32, 14 kDa) of yeast RPA is essential for viability but, due to the lack of conditional mutants, it is not known whether they are also required for the initiation of DNA replication *in vivo* (144). The large subunit possesses ssDNA binding activity, but much of the attention has concentrated on the 32 kDa subunit because it displays cell cycle-dependent phosphorylation. It is phosphorylated during S phase, both in yeast and human extracts, and becomes dephosphorylated during mitosis (145). Although Rpa32 is phosphorylated *in vitro* by the mitotic p34^{cdc2} or S phase p33^{cdk2} kinases on a subset of sites also labelled *in vivo* (146), it is not clear whether this phosphorylation is a cause or a consequence of DNA replication. In fact, mutations in p34^{cdc2} sites do not impair RPA's ability to support SV40 T antigen-dependent DNA replication *in vitro* (147). In *Xenopus* egg extracts, RPA colocalizes with prereplicative centres already after mitosis when chromosomes decondense, and this assembly requires inactivation of the cyclin B–cdc2 kinase (143). Thus, RPA might be needed at the end of mitosis and not only when it gets phosphorylated during S phase.

5. Temporal order of DNA replication and mitosis

What restricts DNA from replicating more than once per cell cycle and what ensures the correct order of S phase and mitosis? Insights into this question first stemmed from the observation that nuclear synchrony is achieved rapidly after fusion of cells arrested at different stages of the cell cycle (12). S-phase cells can induce DNA replication in a G1 but not in a G2 nucleus, implying that the two nuclei are differently receptive to SPF. G2 cells do not contain a diffusible inhibitor of DNA replication because they do not delay replication in S-phase cells, suggesting that the state of the G2 DNA itself could be different. In contrast, S-phase cells contain a diffusible inhibitor of mitosis, explaining why mitosis awaits completion of DNA replication. Thus, two types of dependency controls, one ensuring that DNA replication is always preceded by mitosis and the second ensuring that mitosis does not begin before DNA replication is finished, are responsible for maintaining genome ploidy and integrity during cell division.

5.1 Dependency of S on M

Only in rare cases do cells depart from the strict order of DNA replication and mitosis, as in the endoreplication cycles of *Drosophila* embryonic development when several rounds of DNA replication occur without intervening mitoses (148). Certain chemical treatments and mutations have a similar effect. G2 nuclei can be induced to rereplicate in *Xenopus* egg extracts when the nuclear membrane is permeabilized and new extract is added (140). This was the basis for the DNA 'licensing factor' hypothesis which proposes that disassembly of the nuclear membrane during mitosis allows access to DNA of a factor which is essential for replication (see Chapter 6). Such a simple model is however not sufficient to explain DNA replication following closed mitoses (as in fungi) or when rereplication occurs in the complete absence of G2 and M (as in the *ts41* mutation of CHO cells) (149). Thus nuclear membrane breakdown is not necessary for DNA replication. Alternatively, access to DNA of licensing factor could be controlled by cell cycle-regulated nuclear import. What happens during mitosis that allows DNA replication in the subsequent cycle? *In vivo* footprinting experiments show that yeast replication origins exist in two states during the cell cycle (127). The prereplicative state, thought to be composed of ORC and additional proteins, is seen from late mitosis up to the initiation of DNA replication. The postreplicative state does not appear until the beginning of DNA replication.

What could be the signal for assembly of the prereplicative complex on origins? It seems to occur coincidentally with the destruction of the mitotic kinase, and the isolation of yeast mutants that rereplicate further points to a crucial role of the cdc2/Cdc28 kinase. Indeed in *S. pombe*, mutations in both components of MPF (the cdc2 kinase and the cdc13 B-type cyclin) lead to polyploidization by rereplication in the absence of mitosis (150, 151). Is destruction of the mitotic kinase sufficient for eliminating the block to rereplication? This does not seem to be the case in *S. cerevisiae* where inactivation of the Clb1,2,3,4 cyclins causes a G2 arrest, but not rereplication. This is perhaps due to the persistence in G2 of the Clb5,6–Cdc28 kinase needed for DNA replication. It is conceivable that the presence of any Clb–Cdc28 kinase activity would prevent assembly of the prereplicative complex and rereplication in G2. According to this model, inactivation of all Clb/Cdc28 kinases should allow reformation of prereplicative complexes on origins and rereplication in G2. This is indeed what happens when high levels of the p40^{SIC1} inhibitor of Clb–Cdc28 complexes are first induced, and then repressed, in nocodazole-arrested cells (159). It is not known whether inactivation of the Clb–Cdc28 kinases leads to nuclear import of Mcm/Cdc46 type proteins and whether this would be needed for rereplication. This treatment also causes reactivation of the late G1 transcriptional programme and rebudding, suggesting that destruction of the Cdc28 kinase bound to B-type cyclins resets the cycle into a G1 state. In this simple model, Clb–Cdc28 kinase activity is both needed for DNA replication and sufficient to prevent the reassembly of preinitiation complexes on origins. If these complexes were destroyed by the act of initiation or elongation, it would also explain why origins are not fired twice during S phase.

In *S. pombe*, deletion of *cdc13* leads to rereplication, suggesting that the cdc2 activity needed for DNA replication either does not prevent rereplication or disappears with time in cells arrested before mitosis (see Chapter 8). The long delay needed for rereplication in *cdc13* null mutants (151) indicates that the latter might be the case. The B-type cyclins Cig1 and Cig2, which are expressed at the G1/S boundary in *S. pombe*, could be components of the S-phase inducer (152). The *S. pombe* Rum1 protein, which allows rereplication when overexpressed (like Sic1) and leads to premature mitosis when deleted (153), could perform its function by inhibiting the cdc13–cdc2 kinase. Despite these similarities, there are some differences in how the two yeast regulate S and M. In cells arrested before START, overexpression of the mitotic kinase triggers S phase in budding yeast (86), but triggers mitosis in fission yeast (151). The reason could be either that the fission yeast cdc13–cdc2 kinase has no SPF activity or that some proteins needed for DNA replication are missing until after START. The fact that *S. pombe cdc18⁺* (the homologue of *CDC6*) is a critical target of the START-specific transcriptional programme (132) is consistent with the latter proposal. Thus, *S. pombe cdc10* mutants might arrest before S mainly because origins are not in a replication-competent state, whereas *S. cerevisiae cln1,2,3* mutants do not replicate due to the lack of Clb/Cdc28 activity.

Aided by the results described above, one can draw the following model for the control of DNA replication in yeast (Fig. 9). We believe that this model could apply for all eukaryotes. At the end of mitosis, when the activity of the Clb–Cdc28 kinases collapses, a factor binds to the ORC/origin complex forming the prereplicative state of origins. Cdc6 and Mcm2,3,5 may be part of or promote its formation. Only such origins are competent for DNA replication. At the G1/S transition, the Clb5,6–Cdc28 kinase becomes active and leads, maybe through phosphorylation of Cdc6 and Dbf4–Cdc7, to initiation of replication. The replicative fork then converts the origin to the postreplicative state by displacing or destroying the initiator proteins. Reassembly of the prereplicative complex is prevented by Clb–Cdc28 kinase activity which persists until the end of mitosis when cyclin proteolysis takes place. Destruction of the Clb kinase, which occurs only when mitosis is completed, is a prerequisite for formation of new prereplicative complexes on origins, thus ensuring the dependency of DNA replication on mitosis. Sic1 might contribute to the inactivation of the residual mitotic kinase or prevent the precocious activation of the Clb5,6 kinase, thus allowing prereplicative complexes to form on origins. Destruction of the prereplicative complex upon origin firing along with the continuous presence of Clb-Cdc28 kinase would also explain why origins can fire twice during S phase.

5.2 Dependency of M on the completion of S phase

To prevent segregation of partially duplicated chromosomes, surveillance mechanisms exist that detect incompletely replicated or damaged DNA and send a signal to inhibit mitosis (154). As a consequence, cells do not undergo nuclear division until replication is completed and DNA repaired. It is not clear what prevents mitosis because, in many cases, cells arrest with significant amounts of Clb2–Cdc28 kinase activity (62). What could be detected is a particular structure of replicating or damaged DNA, or some activity associated with it. Interestingly, a limited number of genes are required both for DNA replication and preventing mitosis, and could therefore

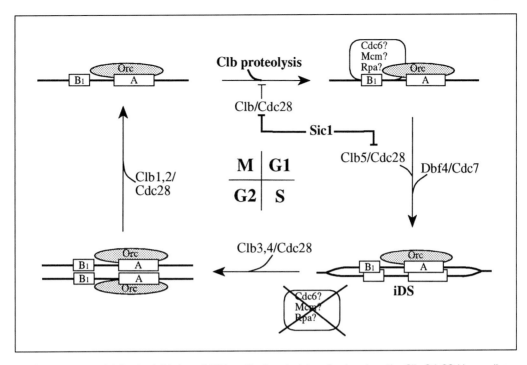

Fig. 9 Two-step model for the initiation of DNA replication. In late mitosis, when the Clb–Cdc28 kinase disappears due to cyclin proteolysis and the presence of Sic1, origins become competent for replication perhaps by binding of additional proteins to ORC. In late G1 the Clb5 kinase activates these origins, DNA replication begins and origins switch to the postreplicative state. Clb–Cdc28 kinases which are present in S, G2, and M may trigger the degradation or inhibit the reassembly of the prereplicative complex, thus preventing rereplication during the same cycle.

be involved in generating the signal that prevents mitosis. These are the *S. pombe* *cut5⁺/rad4⁺* and *cdc18⁺* genes, and the *CDC6* gene from *S. cerevisiae*. Mutants do not replicate, but undergo a lethal haploid mitosis (132, 155, 157). *cdc7* mutants, which do not replicate and contain an active mitotic kinase, do not, however, undergo haploid mitosis. Thus, *cdc7*-arrested cells might already signal that they have begun DNA replication, though this could be simply due to leakiness of the allele.

6. Conclusion and prospects

Last year, major strides were made towards the understanding of how DNA replication is regulated. It was shown that B-type cyclin kinases, and not just the Cln–Cdc28 kinase, trigger S phase in *S. cerevisiae*. At the same time, the isolation of rereplicating mutants in fission yeast indicated that MPF has a key role in ensuring the dependency of S phase on mitosis. It was shown that the state of replication origins changes after mitosis, suggesting that the loss of MPF (and not its increase as proposed by the licensing factor model) could be important for resetting origins from a post- to a prereplicative state. With these observations, a simplified model could be proposed to explain the block to rereplication and the dependency of S on M. Showing that a protein (Dbf4) required for the initiation of DNA replication could bind to origins provided the first link between the DNA replication machinery and a cell division cycle gene (*CDC7*). We have no doubt that rapid progress will follow and that the molecular details of the initiation of DNA replication will soon be at hand. Important questions remain however to be solved. For example, why are genes important for START turned on at a critical cell size? What is the signal that prevents mitosis when DNA replication is blocked? We know that Cdc28 bound to B-type cyclins can trigger both DNA replication and mitosis, so what maintains the order of these two events and what is the role of the Cln–Cdc28 kinase? Is it only needed for the proper timing of Clb kinases activation, or does it have specialized functions that Clbs cannot perform? Budding could be such a function: it depends on Clns (156), it is not triggered by Clbs (86), and it is special to *S. cerevisiae*. This might explain why G1 cyclins are less well conserved in evolution than B-type cyclins. Finally, we have gained insight into the cyclical nature of the cell cycle. The G1 period could be nothing else than a state with no Clb–Cdc28 kinase activity. Two reciprocal proteolytic switches seem to define the borders of G1. At the end of mitosis, cyclin proteolysis and the accumulation of Sic1 might be essential for inactivating Clb kinases, establishing the G1 state and prepare origins for replication. Towards the end of G1, accumulation of Clbs and destruction of Sic1, both triggered by the Cln–Cdc28 kinase, leads to activation of the Clb kinases, DNA replication and mitosis. If the accumulation of the Cln kinase were to depend on the prior inactivation of the mitotic kinase, M would directly lead to G1, and then G1 automatically to S and M. Thus, the inverse patterns of Clb and Sic1 accumulation and proteolysis might be the heart of the cell cycle oscillator.

Acknowledgements

We would like to thank all members of the Nasmyth and Ammerer labs for the intense scientific and intellectual exchanges that make yeast cell cycle research so stimulating and enjoyable in Vienna. We are particularly indebted to Christian Dahmann and Simonetta Piatti for their essential contribution in elaborating the new model of replication control presented here. We also thank John Diffley, Léon Dirick, Thomas Böhm, Stefan Irniger, and Manfred Neuberg for allowing us to cite unpublished results, and Chris Koch for discussions and critical reading of the manuscript.

References

1. Hartwell, L. (1992) Defects in a cell cycle checkpoint may be responsible for the genomic instability of cancer cells. *Cell*, **71**, 543.
2. Howard, A. and Pelc, S. R. (1951) Synthesis of nucleoprotein in bean root cells. *Nature*, **167**, 599.
3. Williamson, D. H. and Scopes, A. W. (1960) The behaviour of nucleic acids in synchronously dividing cultures of *Saccharomyces cerevisiae*. *Exp. Cell Res.*, **20**, 338.

4. Williamson, D. H. (1966) Nuclear events in synchronously dividing yeast cultures. In *Cell synchrony. Studies in biosynthetic regulation*. Cameron, I. L. and Padilla, G. L. (eds.). Academic Press, New York, p. 88.

5. Olson, M. V. (1991) Genome structure and organization in *Saccharomyces cerevisiae*. In *The molecular and cellular biology of the yeast Saccharomyces: Genome dynamics, protein synthesis and energetics*. Broach, J. R., Pringle, J. R., and Jones, E. R. (eds.). Cold Spring Harbor Laboratory, Cold Spring Harbor, NY, p. 1.

6. Campbell, J. L. and Newlon, C. S. (1991) Chromosomal DNA replication. In *The molecular and cellular biology of the yeast Saccharomyces: Genome dynamics protein synthesis and energetics*. Broach, J. R., Pringle, J. R., and Jones, E. R. (eds.). Cold Spring Harbor Laboratory, Cold Spring Harbor, NY, p. 41.

7. Ferguson, B. M. and Fangman, W. L. (1992) A position effect on the time of replication origin activation in yeast. *Cell*, **68**, 333.

8. Bell, S. P. and Stillman, B. (1992) ATP-dependent recognition of eukaryotic origins of DNA replication by a multiprotein complex. *Nature*, **357**, 128.

9. Diffley, J. F. X. and Cocker, J. H. (1992) Protein–DNA interactions at a yeast replication origin. *Nature*, **357**, 169.

10. Nurse, P. (1994) Ordering S phase and M phase in the cell cycle. *Cell*, **79**, 547.

11. Killander, D. and Zetterberg, A. (1965) A quantitative cytochemical investigation of the relationship between cell mass and the initiation of DNA synthesis in mouse fibroblasts *in vitro*. *Exp. Cell Res.*, **40**, 12.

12. Rao, P. N. and Johnson, R. T. (1970) Mammalian cell fusion: Studies on the regulation of DNA synthesis and mitosis. *Nature*, **225**, 159.

13. Hartwell, L. H. and Weinert, T. A. (1989) Checkpoints: Controls that ensure the order of cell cycle events. *Science*, **246**, 629.

14. Kirschner, M., Newport, J., and Gerhardt, J. (1985) The timing of early development events in *Xenopus*. *Trends Genet.*, **1**, 41.

15. Kimmelman, D., Kirschner, M., and Scherson, T. (1987) The events of the midblastula transition in *Xenopus* are regulated by changes in the cell cycle. *Cell*, **48**, 399.

16. Dasso, M. and Newport, J. W. (1990) Completion of DNA replication is monitored by a feedback system that controls the initiation of mitosis *in vitro*: Studies in *Xenopus*. *Cell*, **61**, 811.

17. Murray, A. W. and Kirschner, M. W. (1989) Dominoes and clocks: The union of two views of the cell cycle. *Science*, **246**, 614.

18. Hartwell, L. H. (1971) Genetic control of the cell division cycle in yeast. II. Genes controlling DNA replication and its initiation. *J. Mol. Biol.*, **59**, 183.

19. Hartwell, L. H., Culotti, J., Pringle, J. R., and Reid, B. J. (1976) Genetic control of the cell division cycle in yeast. *Science*, **183**, 46.

20. Hartwell, L. H., Culotti, J., and Reid, B. (1970) Genetic control of the cell division cycle in yeast. I. Detection of mutants. *Proc. Natl. Acad. Sci. USA*, **66**, 352.

21. Hereford, L. M. and Hartwell, L. H. (1974) Sequential gene function in the initiation of *Saccharomyces cerevisiae* DNA synthesis. *J. Mol. Biol.*, **84**, 445.

22. Pringle, J. R. and Hartwell, L. H. (1981) The *Saccharomyces cerevisiae* cell cycle. In *The molecular biology of the yeast Saccharomyces*. Strathern, J. N., Jones, E. W., and Broach, J. R. (eds.). Cold Spring Harbor Laboratory, Cold Spring Harbor, NY, pp. 97–142.

23. Johnston, G. C., Pringle, J. R., and Hartwell, L. H. (1977) Coordination of growth with cell division in the yeast *Saccharomyces cerevisiae*. *Exp. Cell Res.*, **105**, 79.

24. Reed, S. I. (1980) The selection of *S. cerevisiae* mutants defective in the start event of cell division. *Genetics*, **95**, 561.

25. Sutton, A., Immanuel, D., and Arndt, K. T. (1991) The SIT4 protein phosphatase functions in late G1 for progression into S phase. *Mol. Cell. Biol.*, **11**, 2133.

26. Nurse, P. and Thuriaux, P. (1980) Regulatory genes controlling mitosis in the fission yeast *Schizosaccharomyces pombe. Genetics*, **96**, 627.

27. Beach, D., Durkacz, B., and Nurse, P. (1982) Functionally homologous cell cycle control genes in budding and fission yeast. *Nature*, **300**, 706.

28. Hindley, J. and Phear, G. A. (1984) Sequence of the cell division cycle gene *cdc2* from *Schizosaccharomyces pombe*. Pattern of splicing and homology to protein kinases. *Gene*, **31**, 129.

29. Lörincz, A. T. and Reed, S. I. (1984) Primary structure homology between the product of yeast division control gene *CDC28* and vertebrate oncogenes. *Nature*, **307**, 183.

30. Hartwell, L. H. (1993) Getting started in the cell cycle. In *The early days of yeast genetics*. Hall, M. N. and Linder, P. (eds.). Cold Spring Harbor Laboratory, Cold Spring Harbor, NY, pp. 307–314.

31. Thuriaux, P., Nurse, P., and Carter, B. (1978) Mutants altered in the control coordinating cell division with cell growth in the fission yeast *Schizosaccharomyces pombe*. *Mol. Gen. Genet.*, **161**, 215.

32. Nurse, P. and Bisset, Y. (1981) Cell cycle gene required in G1 for commitment to cell division and in G2 for control of mitosis in fission yeast. *Nature*, **292**, 558.

33. Piggot, J. R., Rai, R., and Carter, B. L. A. (1982) A bifunctional gene product involved in two phases of the yeast cell cycle. *Nature*, **298**, 391.

34. Reed, S. I., Hadwiger, J. A., and Lörincz, A. (1985) Protein kinase activity associated with the product of the yeast cell division cycle gene *CDC28*. *Proc. Natl. Acad. Sci. USA*, **82**, 4055.

35. Simanis, V. and Nurse, P. (1986) The cell cycle control gene *cdc2+* of fission yeast encodes a protein kinase potentially regulated by phosphorylation. *Cell*, **45**, 261.

36. Draetta, G., Luca, F., Westendorf, J., Brizuela, L., Ruderman, J., and Beach, D. (1989) cdc2 protein kinase is complexed with both cyclin A and B: Evidence for proteolytic inactivation of MPF. *Cell*, **56**, 829.

37. Evans, T., Rosenthal, E. T., Youngblom, J., Distel, D., and Hunt, T. (1983) Cyclin: a protein specified by maternal mRNA in sea urchin eggs that is destroyed at each cleavage division. *Cell*, **33**, 389.

38. Hunt, T. (1989) Maturation promoting factor, cyclin and the control of M-phase. *Curr. Opin. Cell Biol.*, **1**, 268.

39. Nasmyth, K. (1993) Control of the yeast cell cycle by the Cdc28 protein kinase. *Curr. Opin. Cell Biol.*, **5**, 166.

40. Sherr, C. J. (1993) Mammalian G1 cyclins. *Cell*, **73**, 1059.

41. Hadwiger, J. A., Wittenberg, C., Richardson, H. E., de Barros Lopes, M., and Reed, S. I. (1989) A novel family of cyclin homologs that control G1 in yeast. *Proc. Natl. Acad. Sci. USA*, **86**, 6255.

42. Sudbery, P. E., Goodey, A. R., and Carter, B. L. A. (1980) Genes which control cell proliferation in the yeast *Saccharomyces cerevisiae*. *Nature*, **288**, 401.

43. Nash, R., Tokiwa, G., Anand, S., Erickson, K., and Futcher, A. B. (1988) The *WHI1+* gene of *Saccharomyces cerevisiae* tethers cell division to cell size and is a cyclin homolog. *EMBO J.*, **7**, 4335.

44. Cross, F. R. (1988) DAF1, a mutant gene affecting size control, pheromone arrest, and cell cycle kinetics of *Saccharomyces cerevisiae*. *Mol. Cell. Biol.*, **8**, 4675.

45. Richardson, H. E., Wittenberg, C., Cross, F., and Reed, S. I. (1989) An essential G1 function for cyclin-like proteins in yeast. *Cell*, **59**, 1127.

46. Cross, F. R. (1990) Cell cycle arrest caused by *CLN* gene deficiency in *Saccharomyces cerevisiae* resembles START-I arrest and is independent of the mating pheromone signalling pathway. *Mol. Cell. Biol.*, **10**, 6482.

47. Ogas, J., Andrews, B. J., and Herskowitz, I. (1991) Transcriptional activation of *CLN1*, *CLN2*, and a putative new G1 cyclin (*HCS26*) by *SWI4*, a positive regulator of G1-specific transcription. *Cell*, **66**, 1015.

48. Frohlich, K. U., Fries, H. W., Rudiger, M., Erdmann, R., Botstein, D., and Merke, D. (1991) Yeast cell cycle protein CDC48p shows full-length homology to the mammalian protein VCP and is a member of a protein family involved in secretion, peroxisome formation, and gene expression. *J. Cell Biol.*, **114**, 443.

49. Espinoza, F. H., Ogas, J., Herskowitz, I., and Morgan, D. O. (1994) Cell cycle control by a complex of the cyclin HCS26 (PLC1) and the kinase PHO85. *Science*, **266**, 1388.

50. Measday, V., Moore, L., Ogas, J., Tyers, M., and Andrews, B. (1994) The PLC2 (ORFD)-PHO85 cyclin-dependent kinase complex: a cell cycle regulator in yeast. *Science*, **266**, 1391.

51. Ghiara, J. B., Richardson, H. E., Sugimoto, K., Henze, M., Lew, D. J., Wittenberg, C., and Reed, S. I. (1991) A cyclin B homolog in *S. cerevisiae*: Chronic activation of the Cdc28 protein kinase by cyclin prevents exit from mitosis. *Cell*, **65**, 163.

52. Surana, U., Robitsch, H., Price, C., Schuster, T., Fitch, I., Futcher, A. B., and Nasmyth, K. (1991). The role of CDC28 and cyclins during mitosis in the budding yeast *S. cerevisiae*. *Cell*, **65**, 145.

53. Richardson, H. E., Lew, D. J., Henze, M., Sugimoto, K., and Reed, S. I. (1992) Cyclin B homologs in *Saccharomyces cerevisiae* function in S phase and in G2. *Genes Dev.*, **6**, 2021.

54. Fitch, I., Dahmann, C., Surana, U., Amon, A., Nasmyth, K., Goetsch, L., Byers, B., and Futcher, B. (1992) Characterization of four B-type cyclin genes of the budding yeast *Saccharomyces cerevisiae*. *Mol. Biol. Cell*, **3**, 805.

55. Amon, A., Tyers, M., Futcher, B., and Nasmyth, K. (1993) Mechanisms that help the yeast cell cycle clock tick: G2 cyclins transcriptionally activate their own synthesis and repress G1 cyclins. *Cell*, **74**, 993.

56. Schwob, E. and Nasmyth, K. (1993) *CLB5* and *CLB6*, a new pair of B cyclins involved in DNA replication in *Saccharomyces cerevisiae*. *Genes Dev.*, **7**, 1160.

57. Kühne, C. and Linder, P. (1993) A new pair of B-type cyclins from *Saccharomyces cerevisiae* that function early in the cell cycle. *EMBO J.*, **12**, 3437.

58. Epstein, C. B. and Cross, F. R. (1992) CLB5: a novel B cyclin from budding yeast with a role in S phase. *Genes Dev.*, **6**, 1695.

59. Koch, C. and Nasmyth, K. (1994) Cell cycle regulated transcription in yeast. *Curr. Opin., Cell Biol.*, **6**, 451.

60. Wittenberg, C., Sugimoto, K., and Reed, S. I. (1990) G1-specific cyclins of *S. cerevisiae*: Cell cycle periodicity, regulation by mating pheromone, and association with the p34^{CDC28} protein kinase. *Cell*, **62**, 225.

61. Tyers, M., Tokiwa, G., and Futcher, B. (1993) Comparison of the *S. cerevisiae* G1 cyclins: Cln3 may be an upstream activator of Cln1, Cln2 and other cyclins. *EMBO J.*, **12**, 1995.

62. Amon, A., Surana, U., Muroff, I., and Nasmyth, K. (1992) Regulation of p34^{CDC28} tyrosine phosphorylation is not required for entry into mitosis in *S. cerevisiae*. *Nature*, **355**, 368.

63. Grandin, N. and Reed, S. I. (1993) Differential function and expression of *Saccharomyces cerevisiae* B-type cyclins in mitosis and meiosis. *Mol. Cell Biol.*, **13**, 2113.

64. Schwob, E., Böhm, T., Mendenhall, M. D., and Nasmyth, K. (1994). The B-type cyclin kinase inhibitor p40^{SIC1} controls the G1 to S transition in *S. cerevisiae*. *Cell*, **79**, 233.

65. Nasmyth, K. (1985) A repetitive DNA sequence that confers cell-cycle START (*CDC28*)-dependent transcription on the *HO* gene in yeast. *Cell*, **42**, 225.

66. Breeden, L. and Nasmyth, K. (1987) Cell cycle control of the yeast *HO* gene: *cis*- and *trans*-acting regulators. *Cell*, **48**, 389.

67. Andrews, B. J. and Herskowitz, I. (1989) The yeast Swi4 protein contains a motif present in developmental regulators and is part of a complex involved in cell-cycle-dependent transcription. *Nature*, **342**, 803.

68. Nasmyth, K. and Dirick, L. (1991) The role of *SWI4* and *SWI6* in the activity of G1 cyclins in yeast. *Cell*, **66**, 995.

69. McIntosh, E. M. (1993) MCB elements and the regulation of DNA replication in yeast. *Curr. Genet.*, **24**, 185.

70. Lowndes, N. F., Johnson, A. L., and Johnston, L. H. (1991) Coordination of expression of DNA synthesis genes in budding yeast by cell cycle regulated trans factor. *Nature*, **350**, 247.

71. Dirick, L., Moll, T., Auer, H., and Nasmyth, K. (1992) A central role for *SWI6* in modulating cell cycle Start-specific transcription in yeast. *Nature*, **357**, 508.

72. Lowndes, F. L., Johnson, A. L., Breeden, L., and Johnston, L. H. (1992) Swi6 protein is required for transcription of the periodically expressed DNA synthesis genes in budding yeast. *Nature*, **357**, 505.

73. Primig, M., Sockanathan, S., Auer, H., and Nasmyth, K. (1992) Anatomy of a transcription factor important for the cell cycle in *Saccharomyces cerevisiae*. *Nature*, **358**, 593.

74. Koch, C., Moll, T., Neuberg, M., Ahorn, H., and Nasmyth, K. (1993) A role for the transcription factors Mbp1 and Swi4 in progression from G1 to S phase. *Science*, **261**, 1551.

75. Nurse, P. (1990) Universal control mechanism regulating onset of M-phase. *Nature*, **344**, 503.

76. Murray, A. W. (1993) Turning on mitosis. *Curr. Biol.*, **3**, 291.

77. Enoch, T. and Nurse, P. (1991) Coupling M phase and S phase: controls maintaining the dependence of mitosis on chromosome replication. *Cell*, **65**, 921.

78. Sorger, P. K. and Murray, A. W. (1992) S-phase feedback control in budding yeast independent of tyrosine phosphorylation of p34^{cdc28}. *Nature*, **355**, 365.

79. Fesquet, D., Labbé, J., Derancourt, J., Capony, J., Galas, S., Girard, F., Lorca, T., Shuttleworth, J., Dorée, M., and Cavadore, J. (1993) The *MO15* gene encodes the catalytic subunit of a protein kinase that activates cdc2 and other cyclin-dependent kinases (CDKs) through phosphorylation of Thr161 and its homologues. *EMBO J.*, **12**, 3111.

80. Solomon, M. J., Harper, J. W., and Shuttleworth, J. (1993) CAK, the p34^{cdc2} activating kinase, contains a protein identical or closely related to p40^{MO15}. *EMBO J.*, **12**, 3133.

81. Fisher, R. P. and Morgan, D. O. (1994) A novel cyclin associates with MO15/CDK7 to form the CDK-activating kinase. *Cell*, **78**, 713.

82. Makela, T., Tassan, J. P., Nigg, E. A., Frutiger, S., Hughes, G. J., and Weinberg, R. A. (1994) A cyclin associated with the CDK-activating kinase MO15. *Nature*, **371**, 254.

83. Poon, R. Y. C., Yamashita, K., Howell, M., Ershler, M. A., Belyavsky, A., and Hunt, T. (1994) Cell cycle regulation of the p34^{cdc2}/p33^{cdk2}-activating kinase p40^{MO15}. *J. Cell Sci.*, **107**, 2789.

84. Solomon, M. J. (1994) The function(s) of CAK, the p34^{cdc2}-activating kinase. *Trends Biol. Sci.*, **19**, 496.

85. Glotzer, M., Murray, A. W., and Kirschner, M. W. (1991) Cyclin is degraded by the ubiquitin pathway. *Nature*, **349**, 132.

86. Amon, A., Irniger, S., and Nasmyth, K. (1994) Closing the cell cycle circle in yeast: G2 cyclin proteolysis initiated at mitosis persists until activation of G1 cyclins in the next cycle. *Cell*, **77**, 1037.

87. Surana, U., Amon, A., Dowzer, C., McGrew, J., Byers, B., and Nasmyth, K. (1993) Destruction of the Cdc28–Clb kinase is not required for metaphase/anaphase transition in yeast. *EMBO J.*, **12**, 1969.

88. Tyers, M., Tokiwa, G., Nash, R., and Futcher, B. (1992) The Cln3–Cdc28 kinase complex of *S. cerevisiae* is regulated by proteolysis and phosphorylation. *EMBO J.*, **11**, 1773.

89. Cross, F. R. and Blake, C. M. (1993) The yeast Cln3 protein is an unstable activator of Cdc28. *Mol. Cell. Biol.*, **13**, 3266.

90. Chang, F. and Herskowitz, I (1990) Identification of a gene necessary for cell cycle arrest by negative growth factor of yeast: *FAR1* is an inhibitor of a G1 cyclin, *CLN2*. *Cell*, **63**, 999.

91. Nasmyth, K. and Hunt, T. (1993) Dams and sluices. *Nature*, **366**, 634.

92. Peter, M. and Herskowitz, I. (1994) Joining the complex: cyclin-dependent kinase inhibitory proteins and the cell cycle. *Cell*, **71**, 181.

93. Peter, M., Gartner, A., Horecka, J., Ammerer, G., and Herskowitz, I. (1993) FAR1 links the signal transduction pathway to the cell cycle machinery in yeast. *Cell*, **73**, 747.

94. Tyers, M. and Futcher, B. (1993). Far1 and Fus3 link the mating pheromone signal transduction pathway to three G1-phase Cdc28 kinase complexes. *Mol. Cell. Biol.*, **13**, 5659.

95. Peter, M. and Herskowitz, I. (1994) Direct inhibition of the yeast cyclin-dependent kinase Cdc28–Cln by Far1. *Science*, **265**, 1228.

96. Mendenhall, M. D. M. (1993) An inhibitor of p34^{CDC28} protein kinase activity from *Saccharomyces cerevisiae*. *Science*, **259**, 216.

97. Lörincz, A. and Carter, B. L. A. (1979) Control of cell size at bud initiation in *Saccharomyces cerevisiae*. *J. Gen. Microbiol.*, **113**, 287.

98. Popolo, L., Vanoni, M., and Alberghina, L. (1982) Control of the yeast cell cycle by protein synthesis. *Exp. Cell Res.*, **142**, 69.

99. Moore, S. A. (1988) Kinetic evidence for a critical rate of protein synthesis in the *Saccharomyces cerevisiae* yeast cell cycle. *J. Biol. Chem.*, **263**, 9674.

100. Cross, F., Roberts, J., and Weintraub, H. (1989) Simple and complex cell cycles. *Annu. Rev. Cell Biol.*, **5**, 341.

101. Baroni, M. D., Monti, P., Marconi, G., and Alberghina, L. (1992) cAMP-mediated increase in the critical cell size required for the G1 to S transition in *Saccharomyces cerevisiae*. *Exp. Cell Res.*, **201**, 299.

102. Futcher, A. B. (1990) Yeast cell cycle. *Curr. Opin. Cell Biol.*, **2**, 246.

103. Baroni, M. D., Monti, P., and Alberghina, L. (1994) Repression of growth-regulated G1 cyclin expression by cAMP in budding yeast. *Nature*, **371**, 339.

104. Tokiwa, G., Tyers, M., Volpe, T., and Futcher, B. (1994) Inhibition of G1 cyclin activity by the Ras/cAMP pathway in yeast. *Nature*, **371**, 342.

105. Cross, F. and Tinkelenberg, A. H. (1991) A potential feedback loop controlling *CLN1* and *CLN2* gene expression at the Start of the yeast cell cycle. *Cell*, **65**, 875.

106. Dirick, L. and Nasmyth, K. (1991) Positive feedback in the activation of G1 cyclins in yeast. *Nature*, **351**, 754.

107. Taba, M. R. M., Muroff, I., Lydall, D., Tebb, G., and Nasmyth, K. (1991) Changes in a SWI4,6-DNA-binding complex occur at the time of *HO* gene activation in yeast. *Genes Dev.*, **5**, 2000.

108. Marini, N. J. and Reed, S. I. (1992) Direct induction of G1-specific transcripts following reactivation of the Cdc28 kinase in the absence of *de novo* protein synthesis. *Genes Dev.*, **6**, 557.

109. Kurjan, J. (1993) The pheromone response pathway in *Saccharomyces cerevisiae*. *Annu. Rev. Genet.*, **27**, 147.

110. Neiman, A. M. (1993) Conservation and reiteration of a kinase cascade. *Trends Genet.*, **9**, 390.

111. Ammerer, G. (1994) Sex, stress and integrity: the importance of MAP kinases in yeast. *Curr. Opin. Genet. Dev.*, **4**, 90.

112. McKinney, J. D., Chang, F., Heintz, N., and Cross, F. (1993) Negative regulation of FAR1 at the Start of the yeast cell cycle. *Genes Dev.*, **7**, 833.

113. Byers, B. (1981) Cytology of the yeast life cycle. In *The molecular biology of the yeast Saccharomyces*. Strathern, J. N., Jones, E. W., and Broach, J. R. (eds.). Cold Spring Harbor Laboratory Press, Cold Spring Harbor, NY, pp. 59–96.

114. Goebl, M. G., Yochem, J., Jentsch, S., McGrath, J. P., Varshavsky, A., and Byers, B. (1988) The yeast cell cycle gene *CDC34* encodes a ubiquitin-conjugating enzyme. *Science*, **241**, 1331.

115. Schild, D. and Byers, B. (1978) Meiotic effects of DNA-defective cell division cycle mutations of *Saccharomyces cerevisiae*. *Chromosoma*, **70**, 109.

116. Nugroho, T. T. and Mendenhall, M. D. (1994) An inhibitor of yeast cyclin-dependent protein kinase plays an important role ensuring the genomic integrity of daughter cells. *Mol. Cell Biol.*, **14**, 3320.

117. Yochem, J. and Byers, B. (1987) Structural comparison of the yeast cell division cycle gene *CDC4* and a related pseudogene. *J. Mol. Biol.*, **195**, 223.

118. Sclafani, R. A. and Jackson, A. L. (1994) Cdc7 protein kinase for DNA metabolism comes of age. *Mol. Microbiol.*, **11**, 805.

119. Jackson, A. L., Pahl, P. M. B., Harrison, K., Rosamond, J., and Sclafani, R. A. (1993) Cell cycle regulation of the yeast Cdc7 protein kinase by association with the Dbf4 protein. *Mol. Cell. Biol.*, **13**, 2899.

120. Yoon, H. J., Loo, S., and Campbell, J. (1993) Regulation of *Saccharomyces cerevisiae CDC7* function during the cell cycle. *Mol. Biol. Cell*, **4**, 195.

121. Sclafani, R. A., Patterson, M., Rosamond, J., and Fangman, W. L. (1988) Differential regulation of the yeast *CDC7* gene during mitosis and meiosis. *Mol. Cell. Biol.*, **8**, 293.

122. Johnston, L. H. and Thomas, A. P. (1982) The isolation of new DNA synthesis mutants in the yeast *Saccharomyces cerevisiae*. *Mol. Gen. Genet.*, **186**, 439.

123. Kitada, K., Johnston, L. H., Sugino, T., and Sugino, A. (1992) Temperature-sensitive *cdc7* mutations of *Saccharomyces cerevisiae* are suppressed by the *DBF4* gene, which is required for the G1/S cell cycle transition. *Genetics*, **131**, 21.

124. Dowell, S. J., Romanowski, P., and Diffley, J. F. X. (1994) Interaction of Dbf4, the Cdc7 protein kinase regulatory subunit, with yeast replication origins *in vivo*. *Science*, **265**, 1243.

125. Barinaga, M. (1994) Yeast enzyme finds fame in link to DNA replication. *Science*, **265**, 1175.

126. Bell, S. P., Kobayashi, R., and Stillman, B. (1993) Yeast origin recognition complex functions in transcription silencing and DNA replication. *Science*, **262**, 1844.

127. Diffley, J. F. X., Cocker, J. H., Dowell, S., and Rowley, A. (1994) Two steps in the assembly of complexes at yeast replication origins *in vivo*. *Cell*, **78**, 303.

128. Hartwell, L. H. (1976) Sequential function of gene products relative to DNA synthesis in the yeast cell cycle. *J. Mol. Biol.*, **104**, 803.

129. Hogan, E. and Koshland, D. (1992) Addition of extra origins of replication to a minichromosome suppresses its mitotic loss in *cdc6* and *cdc14* mutants of *Saccharomyces cerevisiae*. *Proc. Natl. Acad. Sci. USA*, **89**, 3098.

130. Li, J. J. and Herskowitz, I. (1993) Isolation of ORC6, a component of the yeast origin recognition complex by a one-hybrid system. *Science*, **262**, 1870.

131. Bueno, A. and Russell, P. (1992) Dual functions of CDC6: A yeast protein required for DNA replication also inhibits nuclear division. *EMBO J.*, **11**, 2167.

132. Kelly, T. J., Martin, G. S., Forsburg, S. L., Stephen, R. J., Russo, A., and Nurse, P. (1993) The fission yeast *cdc18*[+] gene product couples S phase to START and mitosis. *Cell*, **74**, 371.

133. Zwerschke, W., Rottjakob, H.-W., and Küntzel, H. (1994) The *Saccharomyces cerevisiae* CDC6 gene is transcribed at late mitosis and encodes a ATP/GTPase controlling S phase initiation. *J. Biol. Chem.*, **269**, 23 351.

134. Zhou, C. and Jong, A. (1990) CDC6 mRNA fluctuates periodically in the yeast cell cycle. *J. Biol. Chem.*, **265**, 19 904.

135. Tye, B.-K. (1994) The MCM2-3-5 proteins: are they replication licensing factors? *Trends Cell Biol.*, **4**, 160.

136. Moir, D., Stewart, S. E., Osmond, B. C., and Botstein, D. (1982) Cold-sensitive cell-division-cycle mutants of yeast: Isolation, properties, and pseudo-reversion studies. *Genetics*, **100**, 547.

137. Hennessy, K. M., Lee, A., Chen, E., and Botstein, D. (1991) A group of interacting yeast DNA replication genes. *Genes Dev.*, **5**, 958.

138. Maine, G. T., Sinha, P., and Tye, B.-K. (1984) Mutants of *S. cerevisiae* defective in the maintenance of minichromosomes. *Genetics*, **106**, 365.

139. Yan, H., Merchant, M., and Tye, B.-K. (1993) Cell cycle-regulated nuclear localization of MCM2 and MCM3, which are required for the initiation of DNA synthesis at chromosomal replication origins in yeast. *Genes Dev.*, **7**, 2149.

140. Blow, J. J. and Laskey, R. A. (1988) A role for nuclear envelope in controlling DNA replication within the cell cycle. *Nature*, **332**, 546.

141. Moll, T., Tebb, G., Surana, U., Robitsch, H., and Nasmyth, K. (1991) The role of phosphorylation and the CDC28 protein kinase in cell cycle-regulated nuclear import of the *S. cerevisiae* transcription factor SWI5. *Cell*, **66**, 743.

142. Thommes, P., Fett, R., Schray, B., Burkhart, R., Barnes, M., Kennedy, C., Brown, N., and Knippers, R. (1992) Properties of the nuclear P1 protein, a mammalian homologue of the yeast Mcm3 replication protein. *Nucleic Acids Res.*, **20**, 1069.

143. Adachi, Y. and Laemmli, U. K. (1994) Study of the cell cycle-dependent assembly of the DNA pre-replication centres in *Xenopus* egg extracts. *EMBO J.*, **13**, 4153.

144. Brill, S. J. and Stillman, B. (1991) Replication factor-A from *Saccharomyces cerevisiae* is encoded by three essential genes coordinately expressed at S phase. *Genes Dev.*, **5**, 1589.

145. Din, S., Brill, S. J., Fairman, M. P., and Stillman, B. (1990) Cell cycle-regulated phosphorylation of DNA replication factor A from human and yeast cells. *Genes Dev.*, **4**, 968.

146. Dutta, A. and Stillman, B. (1992) cdc2 family kinases phosphorylate a human cell DNA replication factor, RPA, and activate DNA replication. *EMBO J.*, **11**, 2189.

147. Henricksen, L. A. and Wolds, M. S. (1994) Replication protein A mutants lacking phosphorylation sites for p34[cdc2] kinase support DNA replication. *J. Biol. Chem.*, **269**, 24 203.

148. Smith, A. V. and Orr-Weaver, T. L. (1991) The regulation of the cell cycle during *Drosophila* embryogenesis: the transition to polyteny. *Development*, **112**, 997.

149. Coverley, D. and Laskey, R. A. (1994) Regulation of eukaryotic DNA replication. *Annu. Rev. Biochem.*, **63**, 745.

150. Broek, D., Bartlett, R., Crawford, K., and Nurse, P. (1991) Involvement of p34^{cdc2} in establishing the dependency of S phase on mitosis. *Nature*, **349**, 388.

151. Hayles, J., Fisher, D., Woollard, A., and Nurse, P. (1994) Temporal order of S phase and mitosis in fission yeast is determined by the state of the p34^{cdc2}-mitotic B cyclin complex. *Cell*, **78**, 813.

152. Connolly, T. and Beach, D. (1994) Interaction between the Cig1 and Cig2 B-type cyclins in the fission yeast cell cycle. *Mol. Cell. Biol.*, **14**, 768.

153. Moreno, S. and Nurse, P. (1994) Regulation of progression through the G1 phase of the cell cycle by the *rum1+* gene *Nature*, **367**, 236.

154. Li, J. J. and Deshaies, R. J. (1993) Exercising self-restraint: Discouraging illicit acts of S and M in eukaryotes. *Cell*, **74**, 223.

155. Saka, Y. and Yanagida, M. (1993) Fission yeast *cut5+*, required for S phase onset and M phase restraint, is identical to the radiation-damage repair gene *rad4+*. *Cell*, **74**, 383.

156. Cvrckova, F. and Nasmyth, K. (1993) Yeast G1 cyclins *CLN1* and *CLN2* and a GAP-like protein have a role in bud formation. *EMBO J.*, **12**, 5277.

157. Piatti, S., Lengauer, C., and Nasmyth, K. (1995) Cdc6 is an unstable protein whose *de novo* synthesis in G1 is important for the onset of S phase and for preventing a 'reductional' anaphase in the budding yeast *Saccharomyces cerevisiae*. *EMBO J.*, **14**, 3788.

158. Dirick, C., Böhm, T., and Nasmyth, K. (1995) Roles and regulation of Cln-Cdc28 kinases at the start of the cell cycle of *Saccharomyces cerevisiae*. *EMBO J.*, **14**, 4803.

159. Dahmann, C., Difley, J. F. X., and Nasmyth, K. (1995) S-phase-promoting cyclin-dependent kinases prevent re-replication by inhibiting the transition of replication origins to a pre-replicative state. *Curr. Biol.*, **5**, 1257.

8 | Regulation of S phase in the fission yeast *Schizosaccharomyces pombe*

SUSAN L. FORSBURG

1. Introduction

The fission yeast *Schizosaccharomyces pombe* provides an excellent model system for studies of cell cycle regulation (1, 2). Its cell cycle has a short G1 phase and is largely regulated at the G2/M transition in rapidly growing cells, in contrast to the cell cycle of the very distantly related budding yeast *Saccharomyces cerevisiae*. This has made study of the regulation of the G1/S phase transition in fission yeast technically difficult. However, considerable advances have been made recently, informed in part by comparisons with the models established in *S. cerevisiae* (3, 4) (see Chapter 7). Just as analysis of the G2/M phase transition shows important differences as well as broad similarities in the two yeasts, work on the control of the G1/S phase transition already has identified conserved as well as variable regulatory elements. Such comparisons between different cell types promise to extend our understanding of the control of cell cycle entry and DNA replication in all eukaryotes. The web of interactions in *S. pombe* that connects genetically defined regulators to biochemically characterized components of the replication apparatus is still being dissected. In this review, I will follow three strands of this web, which form major areas of current research:

- START, cell cycle commitment, and regulation of cell cycle exit
- the components of S phase initiation and DNA replication
- maintaining the order of cell cycle events and dependency in the cell cycle

Inevitably, there will be cross-referencing between these topics. In order to assist the reader, Fig. 1 diagrams the life cycle of fission yeast, and Table 1 lists known G1/S phase genes, their products (if known), and a brief description of their role. In keeping with standard fission yeast nomenclature, gene names are italicized in the lower case, with an appended '+' for wild-type genes (e.g., *cdc10+*); protein names, which are not italicized, are appended with 'p' (e.g., Cdc10p). Budding yeast gene names are capitalized for wild type (e.g., *MCM2*) and in lower case for mutants. Owing to space constraints, some parts of the field cannot be covered in detail, and

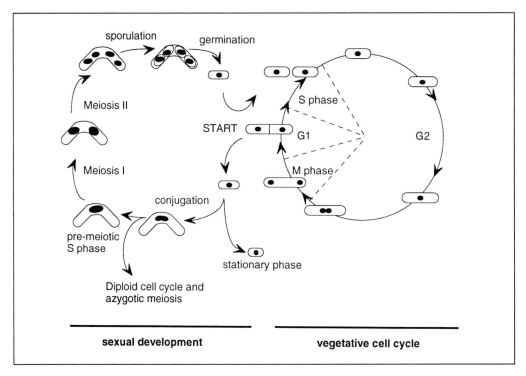

Fig. 1 The life cycle of fission yeast. The phases of the vegetative cell cycle are indicated on the right. If nitrogen starvation occurs cells exit the cycle at G1 and enter stationary phase or, if a partner of opposite mating type is present, initiate sexual development. Normally, fission yeast cells follow conjugation immediately with zygotic meiosis, but it is possible under some conditions to recover a diploid and maintain it vegetatively, or induce an azygotic meiosis (1,2,14).

so the reader is directed to recent reviews on the yeast cell cycle in general and S phase and checkpoint control in particular (2, 5–11).

2. START and the G1/S transition

2.1 The concept of START

Entry into the mitotic cell cycle occurs at START, classically defined as a point of commitment to the cell cycle (12, 13). Before START, cells are able to embark upon alternative developmental paths. After START, the cell typically must complete a cycle of division before it has the same alternatives. A fission yeast cell has only two options outside of the mitotic cell cycle: first, it can enter stationary phase, in which the cells become dormant in response to nutrient limitation, and second, it can initiate the pathway of sexual development in which conjugation and meiosis occur (see Fig. 1) (1, 2, 14). The signal to exit the cell cycle at START is conferred by nitrogen starvation. It is important to note that passage of START is defined not by what the cell can do (enter the cell cycle), but rather as what it can no longer do (initiate sexual

Table 1 G1 and S phase genes in the fission yeast *S. pombe*

Class	Gene	Product (alternative names)	Arrests	Cell cycle phenotype	Reference
START	cdc2⁺	p34 kinase	G1 and G2	cdc	(17, 18, 23)
	cdc10⁺	Transcription factor	G1/START	cdc	(17, 18, 54)
	cdc30⁺	? not cloned	G1/START	cdc	a
	res1⁺	Transcription factor; sct1⁺	G1/START	cdc	(19, 20, 70)
	res2⁺	Transcription factor; pct1⁺		NE	(68, 71)
late G1/early S	cdc18⁺	? required for DNA replication	G1/S	cut (null) cdc (ts)	(6, 77)
	cdc20⁺	? not cloned	G1/S	cdc	(77)
	cdc22⁺	Ribonucleotide reductase, large subunit	G1/S	cdc	(57, 58)
	cdt1⁺	? required for DNA replication	G1/S	cut	(56)
	rad4⁺	? required for DNA replication; cut5⁺	G1/S	cut	(76, 80–82)
	suc22⁺	Ribonucleotide reductase, small subunit	G1/S	cdc	(58)
S phase	cdc19⁺	*MCM2* homologue; nda1⁺	late S/G2	cdc	(77, 100–102)
	cdc21⁺	Novel MCM protein	late S/G2	cdc	(77, 91)
	cdc23⁺	? not cloned	S phase	cdc	(77)
	cdc24⁺	? not cloned	S phase	cdc	(77)
	mis5⁺	novel MCM protein	S phase	cdc	(103)
	nda4⁺	*MCM5/CDC46* homologue	G1/S	cdc	(100, 102)
Replication machinery	polα⁺	Polymerase α; pol1⁺; swi7⁺	G1/S/G2	cut	(84, 124, 125)ᵇ
	polδ⁺	Polymerase δ; pol3⁺; cdc6⁺	S/G2	cdc	(84, 124, 130)ᶜ
	pcn1⁺	PCNA (polδ auxiliary factor)	S/G2	cdc	(78)
	orf2⁺	ORC2 homologue	?	?	d
Cell cycle exit	cig2⁺	B type cyclin, affects exit		NE	(48–50)
	pat1⁺	Kinase, required to repress meiosis	Meiosis	Lethal	(30, 155–157)
	puc1⁺	G1 type cyclin, affects exit		NE	(38, 41)
	rep1⁺	Transcription factor; required for pre-meiotic S phase	Premeioitic S phase	NE	(75)
	res2⁺	Transcription factor; required for meiosis		NE	(68, 71)
Other	rum1⁺	Inhibits p34 activity at G2/M overproduction causes re-replication		NE	(154)ᵉ
Checkpoint	chk1⁺	Kinase; required for damage checkpoint rad27⁺		NE	(133, 148)
	hus1⁺	? required for checkpoints		NE	(143)
	rad1⁺	? required for checkpoints		NE	(141, 143–145)
	rad3⁺	? required for checkpoints		NE	(143, 144, 146)
	rad9⁺	? required for checkpoints		NE	(142, 144)
	rad17⁺	? RFC homologue required for		NE	(144)ᶠ
	rad24⁺	14-3-3 protein; required for checkpoints			(133, 158)
	rad25⁺	14-3-3 protein; required for checkpoints			(133, 158)
	rad26⁺	? required for checkpoints		NE	(133)

ᵃ B. Grallert and P. Nurse, personal communication.
ᵇ G. D'Urso and P. Nurse, personal communication.
ᶜ Y. Iino and M. Yamamoto, personal communication.
ᵈ J. Bushman and P. Russell, personal communication.
ᵉ S. Moreno, personal communication.
ᶠ A. M. Carr, personal communication
? Molecular function unknown
NE, not essential.

development: see 13). Thus START most likely represents a window of opportunity for the cell, rather than a single point of decision. This formal definition, if abstract, is useful for dividing G1 into a committed versus an uncommitted state.

In order to pass START and become committed to the cell cycle and S phase, several requirements must be met. First, the cell must have attained a sufficient size; if it does not, the G1 period will be extended to allow growth to the minimum size (15, 16). Exponentially growing fission yeast cells have passed this size control at birth, so the G1 phase is of minimal duration. Second, environmental signals such as nutrient availability and presence or absence of peptide mating pheromones must be appropriate for cell cycle progression, rather than cell cycle exit. Finally, the proper dependency relationships must be met so that the cell passes START and initiates replication once in each cell cycle (see Section 4) (7, 9).

2.2 START gene functions

The original cell cycle mutant screens in fission yeast were based upon the assumption that cells specifically deficient in regulation of cell cycle progression would be able to continue growth without dividing and become strikingly elongated as a result (Fig. 2) (17). Conditional mutants were isolated that were viable at low temperature but elongated and unable to divide at high temperature. Genes required specifically for passage of START were identified because the corresponding mutants were able to exit the cell cycle and conjugate when cell cycle progression was blocked. By these criteria, the original panel of cell cycle mutants in fission yeast identified two genes, *cdc2*⁺ and *cdc10*⁺ (17, 18). In addition, mutants in *res1* block at START in some conditions (see below) (19, 20). A new mutant, designated

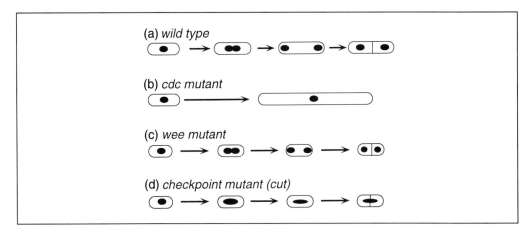

Fig. 2 A schematic describing the cell cycle phenotypes of wild-type and mutant fission yeast cells. (a) Wild-type cells grow in length and divide by medial fission. (b) Cell division cycle, or cdc mutant cells can continue growth and macromolecular synthesis, but are defective for division. The cells become highly elongated. (c) Wee mutants are accelerated through the cell cycle and divide at a smaller size than wild type, although they are viable. (d) Checkpoint deficient cells fail to restrain mitosis even though S phase is not complete. They often 'cut' or pull apart the nucleus (2,9).

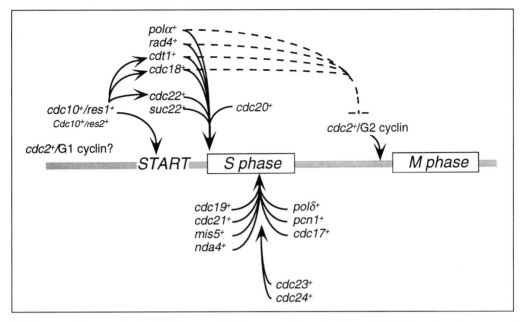

Fig. 3 A speculative model for G1/S control in fission yeast. The role of the p34^{cdc2} kinase is not known, and no G1 cyclins have been identified. The Cdc10p transcription factor activates expression of *cdc18+*, *cdc22+*, and *cdt1+*, and is required for passage of START (**Section 2.2.2**). *cdc18+*, *cdc20+*, *cdc22+*, *cdt1+*, *rad4+*, *suc22+*, and probably *polα+* are all required for initiation of S phase (**Section 3.1.3**). *cdc18+*, *cdt1+*, *rad4+*, and probably *polα+* are also required for the checkpoint that restrains M phase until S phase is complete (**Section 4.1.2**). Other genes that affect S-phase progression include the MCM proteins *cdc19+*, *cdc21+*, *mis5+*, and *nda4+* **Section 3.1.2**), or replication factors including *polδ+* *pcn1+* or *cdc17+*, as well as the still uncloned *cdc23+* and *cdc24+*. References are given in the text and in Table 1.

cdc30, also shows a START-specific block and is currently under investigation (B. Grallert and P. Nurse, personal communication). The relationship between many of the gene products involved in the G1/S transition is shown in Fig. 3, to which I will refer throughout the review. The biochemical details are largely unknown, and so the indicated interactions are not necessarily meant to infer molecular mechanisms, but rather to provide a model framework.

2.2.1 *cdc2+*, G1 cyclins, and cyclin-dependent kinases

The cyclin-dependent protein kinase p34^{cdc2} is a central cell cycle regulator at the G2/M transition in all eukaryotes examined. Regulation of the kinase is provided at several levels, including a positively acting phosphorylation of Thr167, a negatively acting phosphorylation of Tyr15, and association with a positively acting regulatory subunit called a cyclin (21–23). Kinase activity is lost when the mitotic cyclin protein is destroyed as the cells traverse M phase; this periodic degradation gives cyclins their name (23–25). This interaction has provided a paradigm for the function of kinase/cyclin pairs in the cell cycle. In metazoans, other closely related protein kinases called CDKs (cyclin-dependent kinases) are required at different points in the cell cycle, which thus resembles a linked series of cyclin/kinase cascades (3, 26). However, in fission yeast (as in budding yeast), the same p34^{cdc2} kinase acts in G1 and in G2, its altered specificity presumably being conferred by its association with different regulatory cyclin proteins (3, 4, 18). Although the regulation of the p34^{cdc2} kinase at the G2/M transition in fission yeast is quite well understood, its G1 role is still murky, and efforts to identify G1-specific cyclins in *S. pombe* have also been unsuccessful (see below). In addition, cyclin/kinase pairs are probably not limited to the cell cycle in fission yeast. In budding yeast, different cyclin/kinases have been recently found to affect phosphate regulation (27). Although no other CDK proteins have yet been identified in fission yeast, it seems likely that cyclin/kinase pairs will be a general regulatory paradigm found broadly in the cell.

cdc2^{+}. Numerous dominant and recessive mutants of the fission yeast *cdc2*$^{+}$ gene are known that block at both G1 and G2, or at G2 only; no alleles that block the cells solely in the G1 phase of the cell cycle have been identified. In fact, demonstrating kinase activity in G1 of the fission yeast cell cycle has been difficult, which may be due to the short period of G1, or weak activity of the G1 form of the kinase against the G2/M defined substrate histone H1 (28, 29). That p34^{cdc2} is required at the G1 point in the fission yeast cell cycle is genetically proven, because nitrogen-starved cells which are returned to the mitotic cell cycle are unable to enter S phase without *cdc2*$^{+}$ function (18). Still, its biochemical role and potential substrates are unknown in the G1 phase of the cell cycle. *cdc2*$^{+}$ apparently is not required for premeiotic S phase, which has led to the conclusion that the G1 function of p34^{cdc2} in the mitotic cell cycle does not reflect an actual role in initiation of DNA replication, but rather a regulatory function reflecting specific control of M phase exit or entry into the vegetative cycle (Section 2.3.2) (30–32).

The search for G1 cyclins. Cyclins are a large family of proteins with numerous sequence subclasses (33, 34). They share a core of homology in a block of 100–150 amino acids. The distinctive G1 cyclin sequence class was first identified in budding yeast, and has thus far only been found in yeasts (35–38). In contrast, the B-type cyclins have been found in all eukaryotes. The first cyclin identified in fission yeast was a B-type cyclin, encoded by the *cdc13*$^{+}$ gene, which is essential for the G2/M transition (39, 40). Several other fission yeast cyclins have been identified, but it has not been possible to assign any of these a role as positive regulators of the G1/S phase transition and, in fact, their precise molecular role in fission yeast remains unclear.

There is one fission yeast cyclin with a high sequence similarity to the G1 cyclins of budding yeast, encoded by *puc1*$^{+}$ (38). Not surprisingly, given this sequence relatedness, the *puc1*$^{+}$ gene is able to rescue a budding yeast cell that is deleted for all its own G1 cyclins (38). However, there is no evidence that Puc1p functions in the fission yeast as a genuine G1 cyclin. Although overexpression of the *puc1*$^{+}$ gene causes a G2 delay, possibly by competing with the Cdc13p cyclin for binding to Cdc2p, this is likely to be an artefact of high expression levels (38, 41). Normally the *puc1*$^{+}$ gene is expressed at very low levels in cycling cells. Its expression is induced as cells exit the cell cycle in response to nutrient limitation (41). Deletion of *puc1+*, a nonessential gene, modestly accelerates the rate of cell cycle exit (41). These results suggest that Puc1p functions in modulating the transition from cycling to non-cycling cells, rather than in the entry into a cycling state. Although *puc1+* is not demonstrably required for normal G1 progression, in contrast to its homologues in budding yeast, there is one intriguing similarity. As well as being required for the G1/S cell cycle transition, the G1 cyclins in *S. cerevisiae* act to couple mating pheromone signaling to the cell cycle (36, 42–44). In wild-type *S. pombe* cells, mating pheromone has no direct effects upon the cell cycle. However, it is possible that Puc1p is acting to modulate the signals to exit the cell cycle and begin differentiation in a way at least loosely analogous to this signal transduction function of the *S. cerevisiae* proteins.

The *mcs2+* gene encodes a fission yeast cyclin with a low level of sequence identity to the C-type cyclins of metazoa (45). *mcs2+* is an essential gene; cells lacking *mcs2+* largely arrest as binucleate, septated cells, but their DNA content is unknown (45). Although there is some genetic evidence that Mcs2p interacts with p34^{cdc2} (46), biochemical analysis suggests that Mcs2p-associated kinase activity is due to a distinct protein kinase (45). None of the experiments link it particularly with G1.

Two fission yeast B-type cyclins with rather ambiguous roles have been identified, encoded by *cig1+* and *cig2+* (also known as *cyc17+*) (47–50). As is the case with *puc1+*, *cig1+* and *cig2+* are able to rescue the lethality of a budding yeast strain that is deficient in its own G1 cyclins (49). Deletion of *cig1+* has no obvious phenotype and its role remains mysterious (41, 49–51). Deletion of *cig2+* does not affect cycling cells, but affects the rate of their exit from the cell cycle, similar to the phenotype seen when *puc1+* is deleted (41, 49, 50). However, there is no obvious synthetic phenotype in a double mutant *Δcig2 Δpuc1* (S. L. Forsburg, unpublished; 50). Expression of *cig2+* varies in the cell cycle, peaking near G1/S, and is induced under conditions that favor cell cycle exit (49, 50). It has also been shown that *cig2+* expression requires Res1p and Res2p under different conditions (see Section 2.2.2) (50). These results suggest that Cig2p functions both in mitosis and sexual development. One possibility is that the Puc1p and Cig2p proteins are involved, perhaps independently, in coupling the signal transduction pathways that monitor nutrient and pheromone signals to the cell cycle, but their ambiguous phenotypes and expression patterns make it difficult at this point to arrive at a plausible model that explains their role. It is important to stress that there is no evidence that they function as true G1 cyclins.

Thus, in formal terms there is no evidence that p34^{cdc2} protein kinase requires a cyclin or indeed functions as a kinase in the G1 phase of the fission yeast cell cycle. Yet as will be discussed in Section 4, there is considerable evidence that the status of p34^{cdc2} is important for regulation of S-phase entry, and all known models of p34 kinase function require cyclin subunits. So where are the G1 cyclins? One possibility is that they are heavily redundant, as is the case in *S. cerevisiae*. This would make it difficult to isolate them on the basis of their mutant phenotype. The *S. cerevisiae* cyclins *CLN1* and *CLN2* were identified as dosage suppressors of a G1-specific mutation in the p34^{CDC28} kinase (37). In fission yeast, p34^{cdc2} mutants primarily affect G2 and dosage suppression of these mutants has largely identified G2 factors (e.g., 39, 52). Another possibility is that putative *S. pombe* G1 cyclins are a distinctive and highly degenerate sequence class, which are unable to function in a heterologous system and cannot be identified by structure-based assays such as degenerate PCR and low stringency hybridization. A third possibility is that the G1 cyclins have been identified, and exist in the gene products just discussed, but their function in the fission yeast may be distinct from the budding yeast-derived model. This is probably true in any case, as the regulation of the G1/S phase transition in *S. pombe* appears to be quite different from that in *S. cerevisiae*.

2.2.2 *cdc10⁺* and transcriptional control

cdc10⁺. The *cdc10⁺* gene encodes the second essential START function in fission yeast (53). In contrast to *cdc2* mutants, *cdc10* mutant cells block only at START in the cell cycle; the DNA of the mutant cells is unreplicated and they are capable of mating (17, 53). Cdc10p is part of a transcriptional activation complex (19, 54). Various screens have identified some of the targets of this complex which are expressed periodically in the cell cycle, with expression peaking near the G1/S trans-ition (Fig. 3, see below) (54–57). Their promoters contain the short sequence ACGCGT, called the MCB box (54–58). The same cell cycle-regulated *cis*-acting pro-moter element (the MCB box) was first identified as a regulatory element in *S. cere-visiae*, in which system it also confers periodic transcription of S-phase genes (59–61).

The sequence of Cdc10p is related to that of the budding yeast protein SWI6p with a putative protein interaction domain at the C-terminus and a set of repeated sequences in the middle domain of the protein, variously referred to as cdc10, SWI6, or ankyrin repeats (Fig. 4) (20, 62, 63). In budding yeast, SWI6p binds the MCB promoter element together with a second protein called MBP1p; SWI6p uses a different partner, SWI4p, to bind a separate cell cycle-regulated sequence called the SCB box. The MCB and SCB responsive systems function at the same time in the S.

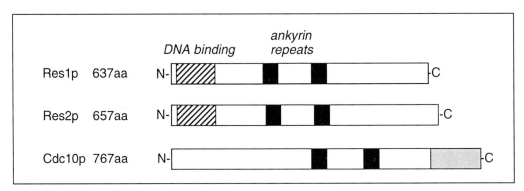

Fig. 4 Cdc10p and its partners share domains of repeated structure called Cdc10p, or ankyrin repeats (dark boxes). Res1p and Res2p share putative DNA-binding domains at the N-terminus (hatched boxes). Cdc10p has a C-terminal domain that is related to the C-terminus of SWI6, its cognate in budding yeast (stippled box) (20, 62, 66, 68, 71, 160).

cerevisiae cell cycle (see Chapter 7) (59–61). The SCB promoter element has not been identified in fission yeast.

Despite these broad similarities, the transcriptional systems in the two yeasts have some important differences. *SWI6* is not itself an essential gene in budding yeast, and although a double deletion of *SWI6* and *SWI4*, or *SWI6* and *MBP1*, is synthetically lethal, such a mutant does not have a G1 specific phenotype (64–66). In contrast, mutations of fission yeast *cdc10⁺* show a cell cycle phenotype, and this gene product is strictly required for passage of START (17, 18). This is a curious result, because although mutations in the known target genes of *cdc10⁺* arrest prior to S phase, they arrest after START (see Section 3.1.1) (54–57). This implies that Cdc10p has additional targets, or additional roles at START beyond transcriptional activation of S-phase genes. One possibility is that a START-specific target of Cdc10p might be the G1 cyclin partner for p34^{cdc2}, as is the case for the *CLN* genes in *S. cerevisiae* (59–61). The activity of Cdc10p may be influenced by the p34^{cdc2} kinase (67), but it is not clear whether this is an important G1 role for *cdc2⁺*. Cdc10p is also required for S-phase initiation in the meiotic cell cycle, presumably because expression of S-phase genes depends upon Cdc10p activity; Cdc10p also may affect later meiotic events (30, 68, 69).

res1⁺ and res2⁺. Proteins that associate with Cdc10p have been identified using a variety of screens. The *res1⁺* gene, also called *sct1⁺*, was first identified as a chromosomal suppressor of a temperature-sensitive *cdc10* allele and subsequently identified independently as a dosage suppressor of *cdc10* and *pat1* mutants (19, 20, 70). Gratifyingly, Res1p is similar in overall structure to SWI4p and MBP1p proteins from *S. cerevisiae*, the known partners of the SWI6p transcription factor (20). Res1p has a putative DNA binding domain at the N-terminus, with the Cdc10p repeated elements in the middle (see Fig. 4) and it associates with Cdc10p (19). *res1* mutants show a START-specific cdc phenotype similar to *cdc10* mutants, although some conditions where the gene is not essential have been reported (19, 20). This suggests that there may be additional partners of Cdc10p that can function in the G1/S phase transition in rare instances (see below). Res1p can also substitute for Cdc10p if produced on a high copy plasmid, or if present as a dominant mutant allele (19, 20, 70).

A second partner of Cdc10p is the *res2⁺* gene product, also called *pct1⁺*. *res2⁺* was cloned as a suppressor of a *res1* mutant and independently by a screen for its ability to activate a Cdc10p-responsive promoter when coexpressed with Cdc10p in budding yeast (68, 71). In contrast to the deletion of *res1*, vegetatively growing *Δres2* cells are not sick, and only modestly slowed in growth; however, they are unable to initiate premeiotic S phase or complete meiosis (68, 71). Genetic experiments suggest that there are interactions between Res2p and Cdc10p: *res2⁺* expression cannot rescue a *cdc10* null mutant, but it can rescue the conditional temperature-sensitive *cdc10-129*, and the *Δres2* strain is synthetically lethal in a *cdc10-129* background (68). Expression of *res2⁺* is sharply induced under conditions that favor conjugation (68). Res2p is structurally very similar to Res1p, so some overlap of function in rapidly growing cells is not surprising (see Fig. 4) (68, 71).

Although these results suggest that Cdc10p and Res1p are required for expression of genes for S phase of the mitotic cell cycle, and Cdc10p and Res2p are required for expression of genes for premeiotic S phase, such a model is too simple to explain all the data (reviewed in 72). There is clearly functional overlap between Res1p and Res2p, so that depending upon levels of expression and growth phase, the genes can partially substitute for one another (68, 71). *res1* mutants are not only defective in mitosis, but also have some meiotic phenotypes, and several lines of genetic and biochemical evidence suggest that *res2*[+] functions in mitotically cycling cells (68, 71). *res2* mutants also show meiotic defects outside of premeiotic S phase. Finally, the double mutant *Δres1 Δres2* is lethal under conditions where each single mutant is viable (68). Thus, although there is some division of labor between these two proteins, there also appears to be considerable overlap. This seems to be particularly true of rapidly growing cells that experience sudden shifts in growth conditions. In cells that are starved for extended periods, on the other hand, the division of roles between the Res1p and Res2p complexes appears more pronounced (68). Interestingly, deletion of either *res1*[+] or *res2*[+] accelerates cells out of the cell cycle, as seen for mutants in *puc1*[+] or *cig2*[+] (Section 2.1.2) (50).

Targets of Cdc10p. Three targets of the Cdc10p transcription complex have been identified thus far: the *cdc18*[+], *cdc22*[+], and *cdt1*[+] genes (54–58). These genes are expressed in a Cdc10p-dependent manner that is periodic in the cell cycle, with the peak of expression occurring at G1/S. All of these genes contain the MCB box element in their promoter region. Although there are probably additional Cdc10p target genes, they have not been identified.

The *cdc18*[+] gene was cloned by its ability to partially suppress a temperature-sensitive mutant of *cdc10-129* when overexpressed under a heterologous promoter (55). One explanation for this phenomenon is as follows. The *cdc10-129* mutation is known to be somewhat leaky. That is, with time the arrested cells will begin to replicate their DNA. There are multiple targets of the Cdc10p transcription complex, and some of these may have a more stringent requirement for Cdc10p activity than others. If *cdc18*[+] expression has a particularly stringent requirement for Cdc10p, it may be a rate-limiting target for S-phase progression, and thus by supplying *cdc18*[+] independently, the partial expression of the other targets is sufficient to allow the cells to proceed through S phase, albeit slowly. Ectopic expression of *cdc18*[+] is unable to rescue a null mutation in *cdc10*[+], so *cdc18*[+] cannot bypass a requirement for Cdc10p (55). The *cdt1*[+] gene was cloned using a similar screen for rescue of the *cdc10-129* mutation (56). It is not known whether ectopic expression of the third known target gene, *cdc22*[+], can also suppress the *cdc10-129* allele. Each of these targets are required for DNA replication, as will be discussed in Section 3.1.1. *cdc18*[+] and *cdt1*[+] are also apparently required for checkpoint control (see Section 4.1).

2.3 Cell cycle exit and sexual development

Cells exit the cell cycle at START if nitrogen becomes limiting. As they exit, they have two options (Fig. 1). One is to enter stationary phase, in which the cells become dormant: they cease growing, are metabolically inactive and highly resistant to heat shock (15, 73). The second alternative is to initiate sexual development in response to mating pheromones, during which cells conjugate with partners of the opposite mating type, undergo karyogamy, and immediately enter meiosis to form four spores (14, 74).

The process of meiosis is a specialized cell cycle. Premeiotic S phase does not require $cdc2^+$, but does require $cdc10^+$ (30, 31, 69). This suggests that p34^{cdc2} is not required specifically for the process of replication. Rather, its role appears regulatory and it is required later in the meiotic cell cycle for the meiotic divisions (30, 31, 69). In contrast, the requirement for $cdc10^+$ may reflect its regulation of the expression of targets such as $cdc18^+$, $cdc22^+$, and $cdt1^+$ (30, 31, 69). Curiously, results with the $res2$ mutant suggest that the Cdc10p/Res2p complex is also necessary for later stages in meiosis (68, 71). This may be analogous to the requirement for Cdc10p at START of the vegetative cell cycle: there may be additional targets for this transcriptional activation complex at different points in meiosis. Genetic analysis of various cdc mutants suggests that meiosis is blocked in $cdc10$, $cdc20$, $cdc21$, $cdc23$, and $cdc24$ mutants, which is consistent with these genes playing a role in actual DNA replication (69).

The $rep1^+$ gene product appears to be specifically required for premeiotic S phase (75). This gene, encoding a novel transcription factor, is induced in conditions that favor conjugation, and is required for exit-induced expression of the Cdc10p partner, $res2^+$. In the absence of $rep1^+$, premeiotic S phase is almost absent. In addition, ectopic expression of $rep1^+$ is able to suppress a mutation in $res1$ (75). Once again, a striking overlap of function between meiotic and mitotic regulators is apparent. One possibility is that just as Res1p and Res2p are overlapping but nonidentical factors with mitotic and meiotic functions, Rep1p will have a mitotic counterpart which could regulate $res1^+$. There may be a transcriptional cascade regulating G1 control; perhaps Cdc10p functions with rep gene products to regulate Res/Cdc10p activity. This could be another possible START function for Cdc10p.

3. S phase

Fewer genes affecting S-phase entry have been isolated in S. *pombe* than have been isolated in budding yeast, in part because of the short G1 phase in fission yeast. Also, because mating pheromone in fission yeast does not arrest the cell cycle in wild-type cells, many of the approaches to identify G1 factors used successfully in budding yeast have not been possible. Those genes affecting S phase that have been identified can be divided into two broad classes, based on genetic characterization and molecular cloning. First, there are genes required specifically for S-phase initiation, in the absence of which no DNA synthesis takes place. Second, there are genes that affect S-phase progression. These can be further subdivided into those affecting regulation, and those actually part of the replication apparatus and DNA synthesis machinery. These groups are shown on Fig. 3.

3.1 Genes required for S-phase initiation

3.1.1 Mutants that block prior to S phase

A subset of mutants that arrest after START completely block the initiation of DNA replication (Fig. 3). *cdc18+*, *cdc22+*, and *cdt1+* are targets of the Cdc10p transcription factor (see Section 2.2.2). *cdc18+*, *cdt1+* and *rad4+* (*cut5+*) are all part of the replication checkpoint control as well as required for S phase (Section 4.1.2) (55, 56, 76). Mutants lacking *cdc20+* also reportedly block in G1; however, *cdc20+* has not been cloned so its molecular function is still unknown (77).

cdc18+, *cdt1+* and *rad4+*: *replication and checkpoint mutants.* The *cdc18+* gene product was cloned by suppression of a *cdc10* mutant and is similar to the *CDC6* gene from budding yeast (55). The temperature-sensitive allele *cdc18-K46* (*cdc18ts*) was originally shown to arrest cells with a G2 DNA content and the typical elongated cdc phenotype (77). The chromosome structure of the *cdc18ts* strain at the restrictive temperature is abnormal, as the chromosomes cannot enter a pulsed field gel; this phenotype is also seen for chromosomes from cells in which DNA replication is blocked due to treatment with hydroxyurea and has been interpreted to indicate incomplete replication and the presence of replication fork structures (55, 78, 79).

Deletion of *cdc18* has a very different phenotype from that of the temperature sensitive allele of the gene. Instead of arresting in interphase with an elongated cdc phenotype, the *Δcdc18* cells are small, and show signs of mitosis (55). Also in contrast to the original conditional mutant, the deletion strain has an unreplicated DNA content. Thus, *Δcdc18* cells appear to be undergoing mitosis in the absence of DNA replication and the temperature-sensitive allele is therefore leaky for *cdc18+* function (55). The simplest interpretation of these data is that Cdc18p is required for DNA replication and also for the checkpoint control that restrains mitosis until S phase is complete (which will be further discussed in Section 4). Because *cdc18+* is also dependent upon *cdc10+* for expression, it couples START to S phase as well as S phase to M phase through its checkpoint control (6, 55).

The *cdt1+* gene is also required for S phase: cells deleted for *cdt1* appear to undergo cell division in the absence of replication, although they may leak through more quickly than *Δcdc18* (56). *cdt1+*, like *cdc18+*, is a target of the Cdc10p transcription complex and thus couples S phase to START and mitosis (Section 2.2.2) (19, 56). It has no significant homology to known proteins (56). The *rad4+* (or *cut5+*) gene has a similar mutant phenotype of unreplicated DNA but inappropriate mitosis (76, 80, 81). However, *rad4+* expression is constitutive, and so *rad4+* does not link S to START (82). The *rad4+* gene product is related to the human XRCC1 protein (76, 82, 83). The molecular roles of Cdc18p, Cdt1p, and Rad4p are all unknown. One possibility is that they interact directly with the replication complex as structural or catalytic components. Alternatively, they could play a regulatory role, and affect the timing of S phase.

The results with *cdc18+*, *rad4+*, and *cdt1+* suggest that some fraction, perhaps very large, of putative S-phase mutants will not have the elongated cdc phenotype, but the small, 'cut' mitotic phenotype (see Section 4.1.2 and Fig. 2). Deletion of poly-merase α has a morphological arrest phenotype indistinguishable from that of the *cdc18* deletion, suggesting that it too plays a role in checkpoint control (84). Recent evidence suggests that this is the case (G. D'Urso and P. Nurse, personal communi-cation). Over 19 cut mutants are now known, and they range in function from motor proteins to replication mutants, so the class is large and diverse (76, 81, 85, 86). Therefore, other cut mutants may identify additional early S-phase genes.

cdc22⁺ and suc22⁺: ribonucleotide reductase. cdc22⁺ encodes the large subunit of ribonucleotide reductase, required for nucleotide precursor synthesis (57, 58, 87). A mutation in *cdc22⁺* blocks the cells with an unreplicated DNA content (77). The other subunit of ribonucleotide reductase, encoded by *suc22⁺*, is also essential, but is not regulated by *cdc10⁺* (58). Extensive analysis of the promoter of *cdc22⁺* has shown that the MCB box, the target of Res1p/Cdc10p, is responsible for its regulated expression (19, 54, 58). Both *cdc22* and *suc22* mutants have a typical elongated cdc arrest without any phenotypes indicative of checkpoint deficiency (58, 77).

3.1.2 Mutants that block in S/G2

A number of mutants block the cell cycle after START and S-phase initiation but prior to M phase. Many have been classified as G2 mutants because they block with an apparently replicated DNA content. However, it is not possible to distinguish a late S-phase block from a G2 block under most conditions that have been examined. The *cdc23⁺* and *cdc24⁺* genes have been characterized as probable S-phase genes, but they are not yet cloned, so their molecular function is unknown (77). Other putative G2 genes have been discovered to encode replication factors, such as *cdc6⁺* (poly-merase δ; see Section 3.2; Y. Iino and M. Yamamoto, personal communication). Genetic experiments to characterize the execution point of various mutants in the cell cycle, which is the point at which the gene products are required, as distinct from the point at which the mutants block cell cycle progression, have been helpful in identifying such S/G2 mutants. A distinct subset of S/G2 mutants with G1 execution points include the members of the conserved MCM protein family.

MCM genes. The MCM genes are members of a family that is conserved in all eukaryotes, and are compelling candidates for S-phase regulators (88). The gene family was first identified in *S. cerevisiae* in a screen for mutants that affect minichromosome maintenance, and there are at least six members of the family (79, 89–94). Three of the *S. cerevisiae* members, *MCM2*, *MCM3*, and *MCM5* (*CDC46*), have been closely examined in molecular terms. They are closely related to one another in sequence, but each gene is essential (79, 93, 94). These gene products all undergo periodic nuclear translocation near the G1/S boundary; once inside the nucleus, they appear to associate with the chromatin and they disappear from the nucleus at S phase (93, 95–97). They have structural features suggestive of DNA-dependent ATPases (98). These characteristics have all led to speculation that these proteins may play a role in the initiation of replication, and their regulated nuclear

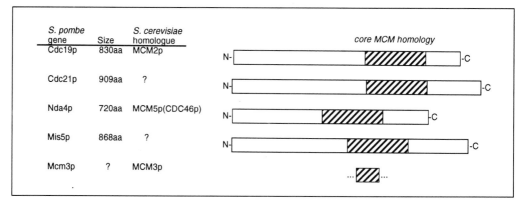

Fig. 5 The MCM proteins of fission yeast share a central core sequence in common with other MCM proteins that has homology to a degenerate class of DNA-dependent ATPases (hatched boxes) (98). There are five members known so far in *S. pombe*, three of which have known cognates in budding yeast *S. cerevisiae* (91, 101–103). Outside the central homology domain there is more variation between these proteins.

entry may provide a clock to help maintain the timing of S phase in the cell cycle (88). This model is intriguing because the behavior of the *S. cerevisiae* MCM proteins suggests licensing factor as posited by Blow and Laskey to explain how cells replicate their DNA once and only once in the cell cycle (see Chapter 6) (88, 96, 99).

In *S. pombe*, four members of the MCM family have been identified: *cdc19+* (also known as *nda1+*), *cdc21+*, *nda4+*, and *mis5+*. A fragment of another member, an MCM3p homologue, has been identified by PCR (91). A cartoon comparing the structure of the different members of the family is presented in Fig. 5. Both *cdc19* and *cdc21* temperature-sensitive mutants were isolated in the original panel of cdc mutants thought likely to affect S phase (77). The *cdc19* mutant was shown genetically to have an early execution point, in G1, although it blocked with a replicated DNA content (77). Subsequently, *cdc19+* was shown to be allelic with the *nda1+* gene, which was isolated as a 'nuclear division arrest' mutant (100, 101). Cdc19p is most closely related to MCM2p of budding yeast and is essential for viability (101, 102). Deletion of the gene shows that although it is essential for viability, it is not required for DNA synthesis, although S phase is delayed in Δ*cdc19* cells (101, 102). An intact replication checkpoint is required for *cdc19ts* arrest (see Section 4) (101).

cdc21+ encodes a novel member of the MCM family; a *cdc21ts* mutant also delays, but does not prevent DNA synthesis (77, 91). In addition, *cdc21ts* has an origin-specific defect in plasmid maintenance, reminiscent of the original screens for MCM proteins in budding yeast (90, 91). It is the only member of the family in fission yeast for which nuclear localization is known, and is found constitutively in the nucleus (D. Maiorano and S. E. Kearsey, personal communication). It is not yet clear whether the regulation of nuclear entry or nuclear retention is a conserved method of regulation between the yeasts, both of which undergo closed mitoses. Formally, only one or a subset of members of a complex need to be regulated to confer regulation to the complex; no homologue of Cdc21p has been studied in other systems, so it will be of interest to see whether those family members with known budding yeast homologues (Cdc19p, Nda4p) are also constitutively nuclear in the fission yeast cell cycle.

The *nda4⁺* gene encodes a homologue of MCM5p (CDC46p, 102). *nda4cs* mutants have a more striking defect in replication than either *cdc19ts* or *cdc21ts* mutants, and puzzlingly their cold sensitivity can be suppressed by exogenous calcium in the medium (102). Recently, a screen for mutants defective in maintaining minichromosomes in fission yeast has identified a variety of new genes, one of which, *mis5⁺*, is another novel member of the MCM family (103).

The genetic data suggest that the *S. pombe* MCM genes are each essential, and operate at an early stage of replication, although the biochemical evidence suggests that DNA synthesis can in most cases initiate without them (77, 91, 101–103). This argues against a simplistic model in which these proteins are obligate activators of S phase, or initiator proteins. They could be involved in origin choice or usage, regulation of the replication complex, processivity of replication, or structural modification of the chromatin or replication factors. The early execution point of mutants of *cdc19* and *cdc21* (77) suggests that they are required for an initiation event, even though they are not required for DNA synthesis (91, 101, 104). It is possible that all the MCM proteins interact together in a complex that retains some function in the absence of any one of the group, although the details of their individual regulation or localization may differ. Initiation of DNA synthesis might be abolished if more than one of the family were deleted, although each gene is itself essential. In budding yeast, there is genetic evidence that these proteins interact with the origin recognition complex (105). Whether the MCM proteins function as initiation factors themselves, or modify the process of replication, their conservation amongst eukaryotes suggests that they will play a conserved role of fundamental importance to S-phase progression, although their contribution to its regulation may differ in different systems (91, 106–109).

3.2   DNA replication machinery

The actual process of DNA replication, as opposed to its regulation, requires the same essential components in all systems: origins of replication, and the enzymes and associated proteins of the replication machinery that allow origin recognition, unwinding, initiation, and elongation of the DNA chain. Efforts to isolate and characterize these components have been based upon identification by function or by homology.

3.2.1   Origins

Autonomous replicating sequence, or *ars*, elements in fission yeast have been isolated by their ability to confer upon plasmids the ability to replicate autonomously (110–116). It has been shown for several *ars* sequences that they are indeed origins of plasmid replication in fission yeast, which validates the assay (112, 114, 117). A number of elements have been sequenced (111, 118). The elements characterized by

Maundrell *et al.* (111) varied from 800 to 1800 base pairs, and were in the order of 70% AT content, which appears typical (118). Sequence analysis of two of the three *ura4+ ars* elements shows they fit this pattern, and suggests that the sequences are likely to be easily unwound (118). Although a short sequence in common between all these *ars* elements was identified, mutational analysis failed to define any role for it in the plasmid assay (111, 115). In contrast, in budding yeast, a small well-defined consensus sequence is required for *ARS* function (reviewed in 5). Fine structure studies have recently been undertaken on fission yeast *ars1* using a series of deletions, and these experiments suggest that the functional elements are redundant (R. Clyne and T. J. Kelly, personal communication). Of a series of 50 base pair deletions constructed across the element, only two reduced efficiency of transformation; however, deletion constructs from the ends to define the boundaries of the element suggest that additional sequences are required for efficient transformation in the *ars* plasmid assay (R. Clyne and T. J. Kelly, personal communication). This again suggests that the functional unit is extended over some distance. There is evidence that an *S. pombe ars* cannot function efficiently in *S. cerevisiae* transformation assays (110; R. Clyne and T. J. Kelly, personal communication), arguing that the sequences required and the factors that bind them are likely to be different in the two yeasts.

Because *ars* elements are identified by a functional assay upon an episome, their identity as actual chromosomal origins is assumed in most cases. However, based on these cloning strategies and on this assumption, it has been estimated that chromosomal origins in fission yeast occur every 20 to 50 kb (110–112). Curiously, there is evidence that *S. pombe* chromosomal origins are clustered (112, 117). At first, it appeared that the origin was very broad, but now several examples exist where the origins are clearly adjacent (112, 113, 117). There appears to be interference between them, and one model suggests that firing one origin inactivates its neighbors (117). Although the reason one origin is weaker than another is unclear, the clustering of these elements might be a fail-safe device to guarantee that initiation takes place at regular intervals in the chromosome (see Chapter 4).

3.2.2 Replication proteins

Origin binding. A protein complex called origin recognition complex (ORC) binds the *ARS* elements in the chromosome in *S. cerevisiae* throughout the cell cycle, although its footprint changes through the cycle and in different growth conditions (119–121) (Chapter 3). Components of budding yeast ORC have been cloned either by biochemical purification, or by molecular genetic selection (109, 122, 123). No analogous experiments have yet been carried out in fission yeast, and no ORC homologues exist amongst the known S-phase mutants. Intriguingly, however, a homologue of ORC2p was recently identified in a two hybrid screen by its interaction with p34^{cdc2}. The fission yeast *orp2+* gene product is essential and 22% identical overall to budding yeast ORC2p, with an identity of 33% between the C-terminal third of the proteins (J. Bushman and P. Russell, personal communication).

Polymerases. DNA polymerase α (*pol1+*; *polα+*) was cloned by low stringency hybridization to the budding yeast *POL1* and is 39% identical to the *S. cerevisiae* gene (124, 125). Expression of *polα⁺* is constitutive across the cell cycle, and the protein also appears to be constitutive (124, 126, 127). However, as assayed by electrophoretic mobility, there appears to be a cell cycle-specific mobility shift, with the smaller form appearing at the G1/S boundary (126). This could indicate cell cycle-specific modifications of the protein. *polα⁺* is an essential gene (84, 128). When deleted, cells germinate and divide without striking elongation, morphologically similar to the *cdc18* mutant (84). There is some evidence for DNA synthesis, although the division appears to be abnormal with a typical cut phenotype (G. D'Urso and P. Nurse, personal communication). This suggests that *polα⁺* may have a checkpoint role, as will be discussed in Section 4.1.

Curiously, *polα⁺* was also cloned independently as a mating type switching gene and is allelic to *swi7⁺*. Fission yeast cells switch their mating type, using information donated by silent loci to an expression cassette. Replication appears to be important for the switching event, which is thought to occur in S phase or G2 (129). In addition, silenced loci in fission yeast have neighboring *ars* elements (116). The *swi7* mutation affects a nonconserved region of the protein suggesting that there may be an interaction between replication factors and switching factors in fission yeast (128), and it will be interesting to see if other essential DNA replication mutants also affect the switching process.

Polymerase δ (*pol3⁺*, *polδ⁺*) was cloned by low stringency hybridization to the budding yeast *POL3* and is 55% identical to it (124, 130). It is also able to complement a budding yeast *pol3* mutant (84). Its expression is also constitutive in the cell cycle (124, 126). The gene is essential, but in contrast to a polymerase α mutation, polymerase δ mutants elongate with a typical cdc phenotype and a DNA content between 1C and 2C, suggesting an S-phase block (84). This phenotype also suggests that the checkpoint response is intact (see Section 4). The *polδ⁺* gene is allelic to *cdc6⁺*, which was originally identified as a G2 mutant (Y. Iino and M. Yamamoto, personal communication; 17).

Other factors. Fission yeast ligase is encoded by *cdc17⁺*, an essential gene (104, 131). Unlike its homologue in *S. cerevisiae*, the *S. pombe* ligase is expressed constitutively in the cell cycle (132). The cells arrest with replicated but fragmented DNA, and the mutant is often used to generate DNA damage to examine checkpoint phenotypes (see Section 4) (104, 131, 133). The three subunits of *S. pombe* RP-A have been identified biochemically and recently cloned. Their sequence structure is not yet known although it is anticipated that they will be similar to the RPA subunits from other organisms (A. Parker and T. J. Kelly, personal communication). The polymerase δ associated factor PCNA, encoded by *pcn1⁺*, was cloned by using antisera to a conserved epitope (78). The predicted protein is 52% identical to human PCNA, which can complement the pombe disruption. The gene is expressed constitutively in the cell cycle. It is an essential gene; most *Δpcn1* cells manage one division, blocking with a delayed S phase and approximately G2 DNA content and elongated morphology (78). The RF-C genes have not yet been identified in fission yeast. The checkpoint mutant *rad17+* has some limited homology to RF-C, but it is not an essential gene and deletion does not apparently affect DNA replication, so this similarity may indicate overlapping interactions with the replication complex (A. M. Carr, personal communication). Two topoisomerases have been identified in fission yeast. A type I toposiomerase, encoded by *top1+*, is a nonessential gene (134, 135). The type II form, encoded by *top2+*, is an essential gene; cells lacking *top2+* die in mitosis unable to segregate their DNA (136–138). Cells lacking both topoisomerases die rapidly throughout the cycle, suggesting that *top2+* overlaps with *top1+* for some functions (134, 136).

4. Checkpoints and dependency

As well as machinery to couple the entry into S phase to external signals, the cells requires machinery to maintain the order of events in the cycle in normal conditions and when the cycle is perturbed. A replication checkpoint monitors initiation of DNA synthesis and a damage checkpoint responds to insults such as radiation damage. The cell also has systems enabling arrest by other means elsewhere in the cycle. A comprehensive analysis of *S. pombe* checkpoints is beyond the scope of this chapter, and the reader is directed to recent reviews on the topic (7–9, 11, 139). There are two general questions affecting the regulation and feedback control of S phase that I will address here. One asks how the cell monitors the completion of S phase to maintain the dependence of M on S. The second asks how the cell monitors the completion of M phase, and prevents reinitiation or rereplication, to maintain the dependency of S on M.

4.1 Replication checkpoint: dependence of M on S

Once the cell has entered S phase, M phase onset must be prevented until replication is complete. It would be disastrous to attempt chromosome segregation if replication were still in progress. The simplest way to do this is temporal: it takes longer to initiate M phase than to complete S phase. Such a simple clock could work by monitoring cell growth or accumulation of a marker protein or modification. This would be adequate unless S phase were delayed or perturbed (Fig. 6). If the clock were unable to monitor S phase progression directly under such conditions, then M phase would proceed inappropriately with lethal consequences. However, this is clearly not the case in wild-type cells.

Previously it has been shown that inhibiting DNA replication with hydroxyurea generates a signal that ultimately affects the activity of the p34^{cdc2} kinase at M phase (140). This demonstrated that a mechanism must exist in fission yeast to monitor S-phase progression so that if S phase is delayed, there is a corresponding delay of M phase: such a mechanism defines a checkpoint. Genetic analysis has identified mutants selected on the basis of their inability to arrest properly when challenged by hydroxyurea, which shows that there are nonidentical but overlapping mechanisms that sense incomplete DNA replication and DNA damage. Amongst these,

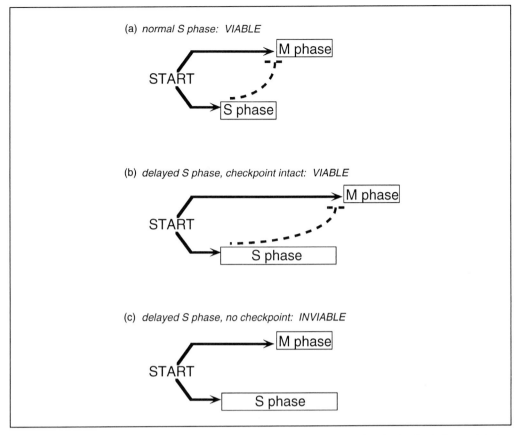

(a) *normal S phase: VIABLE*

(b) *delayed S phase, checkpoint intact: VIABLE*

(c) *delayed S phase, no checkpoint: INVIABLE*

Fig. 6 How checkpoints prevent premature mitosis. (a) Normally, although the checkpoint is intact, it takes longer to initiate M phase than to complete S phase, so cells are perfectly viable even if the checkpoint is missing. (b) If S phase is delayed or extended, the checkpoint delays the onset of M phase until S phase is completed to maintain viability. (c) In the absence of the checkpoint in cells with delayed S phase, M phase occurs with wild-type timing while S phase is still in progress. This is a lethal event (6–9, 11).

rad1⁺, *rad3⁺*, *rad9⁺*, *rad17⁺*, *rad26⁺*, and *hus1⁺* are all involved in both damage responsive and replication responsive checkpoints, as measured by their sensitivity to hydroxyurea, which blocks replication, and to radiation, which causes damage (133, 141–146).

The cell cycle arrest of a variety of S-phase mutants depends upon an intact checkpoint control (101, 140). If the checkpoint is abrogated, then M phase occurs prematurely with lethal consequences. However, a different checkpoint is required for cell cycle arrest pre-START. Passage of START is presumably required to enable both DNA replication and M phase, so cells blocked prior to START have other means of preventing inappropriate mitosis (see Section 4.2.2). Thus, there is a difference in the response of committed and uncommitted cells to the absence of replication. A summary of these results is presented in Table 2.

As discussed in Section 3.1.1, a number of genes are apparently required specifi-

Table 2 Summary of checkpoint mutant phenotypes

Checkpoint mutant	Cell cycle perturbation	Phenotype	Relevant checkpoint	Reference
None	Hydroxyurea	G1/early S phase block	Intact	(140)
None	cdc10	START block	Intact	(17, 18)
None	cdc17	Mid-S phase block	Intact	(104, 131)
cdc2-3w	None	Viable	–	(140, 159)
cdc2-3w	Hydroxyurea	Cut	Missing	(140)
cdc2-3w	cdc10	START block	Intact	(140)
cdc2-3w	cdc17	Cut	Missing	(140)
Δchk1	None	Viable, radiation-sensitive	–	(133, 148)
Δchk1	cdc10	Cut	Missing	(9)
Δchk1	cdc17	Cut	Missing	(133, 148)
Δchk1	Hydroxyurea	G1/early phase block	Intact	(133, 148)
Δrum1	None	Viable, sterile, not radiation sensitive	–	(154)[a]
Δrum1	Hydroxyyrea	G1/early phase block	Intact	(154)
Δrum1	cdc10	Cut	Missing	(154)

[a] A. M. Carr, personal communication

cally for checkpoint response as well as S-phase onset. One possible model is that the proper assembly of a replication complex is monitored in some way by the checkpoint apparatus (6, 7, 9–11, 143, 147). The $cdc18^+$, $cdt1^+$, $pol\alpha^+$, and $rad4^+$ gene products may be components or modifiers of this complex and thus contribute to a signal that is monitored by the identified checkpoint genes. These checkpoint gene products might associate with the replication complex or be modified during its function. If replication is not initiated, then no checkpoint is activated; the unreplicated DNA cannot be distinguished from replicated DNA, and mitosis proceeds (Figs. 6 and 7). This provides one relatively simple model for how a replication checkpoint might work.

After S-phase initiation, checkpoint signals must still be present, because mutants blocked in late gene functions such as $pol\delta$ or $cdc17$ are able to arrest their cell cycles without entering mitosis (77, 84). These mutants block the cell cycle after some DNA synthesis has occurred, so it seems likely that the signal provided by $cdc18^+$, $cdt1^+$, $rad4^+$, and $pol\alpha^+$ is intact in these cells. The continued presence of some replication structures may lead to the S-phase delay. Alternatively, later progression through S phase may be monitored via the damage checkpoint. Cells blocked late in S probably look similar to cells insulted by DNA damage, with incomplete or damaged DNA, or stalled replication complexes. That the later S-phase checkpoint may be distinct from the replication checkpoint is suggested by phenotype of the $chk1$ ($rad27$) mutant (133, 146). Mutants lacking $chk1$ are not hydroxyurea sensitive, so the replication checkpoint is apparently intact and does not require $chk1^+$ to function. However, $chk1$ mutants are radiation sensitive, so the damage checkpoint is missing (133, 148). The $chk1$ mutant phenotype distinguishes two overlapping checkpoint pathways from one another.

In the absence of a checkpoint system, cells are viable as long as there are no per-

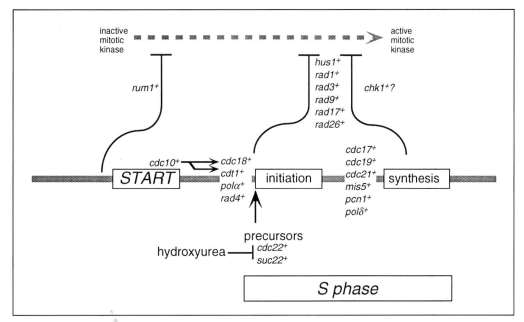

Fig. 7 A speculative, simplified model of checkpoint control. If there is any delay prior to passing START, the mitotic kinase is restrained by inhibition by Rum1p. At START, Cdc10p activates expression of *cdc18+* and *cdt1+*. These gene products, as well as *rad4+* and *polα+*, are required for initiation of S phase, perhaps by assembling into a replication complex. If initiation does not occur, then no checkpoint signal is generated and the p34*cdc2* kinase is activated in its G2 form. If the complex assembles, a checkpoint signal is generated, so if the supply of precursors is removed because of hydroxyurea or mutation in *cdc22+*, a checkpoint signal is still generated. If cell cycle arrest occurs after initiation, a damage checkpoint senses the partially synthesized DNA and causes arrest. Some DNA synthesis occurs even in *cdc17, cdc19, cdc21, mis5, polδ*, or *pcn1* mutants. If the components of this checkpoint control are missing, then no checkpoint signal is registered and the initiation checkpoint is abrogated. Both sides of the pathway must be intact for checkpoint control to respond normally. Note that no particular molecular mechanism is suggested by the relationships proposed here. References are in the text and in Table 1.

turbations of DNA replication. The overlapping checkpoint pathways provide a safety net, by monitoring S-phase progression and adjusting the cell cycle accordingly. This rather abstract concept will become more complete as the biochemical function of the checkpoint proteins are determined.

4.2 Initiation control: dependence of S on M

A central question in studies of DNA replication asks how cells replicate their DNA once and only once in the cell cycle. This can be further divided to ask how DNA is replicated completely in a single S phase, and how S phase occurs only once in a given cell cycle.

4.2.1 Using origins once

How cells replicate their genome exactly in a given S phase without reinitiation is still a mystery. As suggested by the licensing factor model of Blow and Laskey (99), some modifications must either enable replication of a given sequence or block it once it has been initially replicated (Chapter 6). As discussed in Section 3.1.2, the MCM proteins are candidates for such a licensing factor, by extrapolating from their behavior in budding yeast. One model suggests they bind origins and activate them (88). However, while compelling, this model is still unproven especially because DNA synthesis still occurs in *cdc19*, *cdc21*, or *mis5* mutants (77, 91, 101–103). Another possibility is that MCM proteins and other replication factors may be targets of an unidentified licensing factor, such as a protein kinase, and that their role in replication is not dependent upon their regulated localization (149). Certainly, observed interference between adjacent origins supports the model that some replication-dependent event blocks reinitiation but the molecular mechanism remains unknown (112, 117).

4.2.2 Preventing rereplication

The second question, how cells ensure that S phase itself occurs only once per cell cycle, has been investigated extensively in *S. pombe*, with results suggesting that the regulation of G1/S may be strikingly different than that in budding yeast.

By exploiting the genetics of the fission yeast system, mutants able to repeat S phase have been isolated. A bank of temperature-sensitive mutants was treated with a pulse of high temperature and enriched for clones that became diploid (150, 151). When analyzed in detail, several of the mutants were shown to be alleles of *cdc2⁺*, the p34^{cdc2} protein kinase, and of *cdc13⁺*, the mitotic cyclin partner of p34^{cdc2} (150, 151). Upon inactivation by high temperature, these alleles caused complete rereplication of the genome without intervening M phase, by initiating multiple rounds of sequential S phase. A similar phenotype is apparent in spores deleted for *cdc13*, which although inviable are able to overreplicate, again leading to a periodic increase of DNA in genome equivalents (151). In both these cases, the mitotic form of p34^{cdc2} is presumed to be disrupted, either by destroying a temperature-labile protein kinase itself, or by destroying its M-phase cyclin partner (150, 151). In the absence of Cdc13p cyclin partner, p34^{cdc2} is able to fulfill its still-mysterious G1 role even from G2 of the cycle, and can activate the G1/S transition: it can be reset from G2 to G1. If specific assembly of the mitotic form of the kinase is necessary to convey to the cell that it is in G2, then the opposite may also be true: p34^{cdc2} may be reset from G1 to G2 by forcing its premature association with the Cdc13p cyclin. Hayles *et al.* (151) overexpressed both *cdc2⁺* and *cdc13⁺* in cells blocked at START by the *cdc10-129* allele and induced a premature mitosis in the absence of replication. This demonstrates that providing an inappropriately active mitotic cyclin/kinase complex is sufficient to induce M phase in cells pre-START. This is not the case in *S. cerevisiae* where such premature cyclin/kinase activation does not induce mitosis, and where deletion of mitotic cyclins does not induce an additional S phase (152, 153) showing a significant difference in regulation of these events in the two yeasts.

These results suggest that under normal conditions, a mechanism exists to prevent mitotic kinase activation when cells are blocked in the uncommitted stage before START. As discussed previously, the replication and damage-responsive checkpoints are activated after START, so a separate control is necessary. Some insight into its nature is provided by another over-replicating strain. An independent screen for *S. pombe* cDNAs that cause rereplication when overproduced identified the *rum1*+ gene (154). A small (ca. 25 kD) protein with no significant sequence homology to known proteins, Rum1p when overproduced induces regular rounds of S phase in a Cdc10p-dependent fashion (154). That is, cells must pass the *cdc10*+-defined START for each round to occur and *rum1*+ overproduction somehow resets the cell cycle clock to the uncommitted G1 state. Overproduction of Rum1p thus results in a phenotype very similar to the rereplication of the *cdc2* or *cdc13* mutants. In agreement with this, biochemical experiments show that Rum1p is an inhibitor of the p34^{cdc2} mitotic kinase (S. Moreno, J. Correa, personal communication). Obviously Rum1p activity does not inhibit the G1 function of the kinase, since overproduction of Rum1p allows multiple rounds of S phase in a *cdc2*+ and *cdc10*+-dependent fashion (154). The *rum1*+ gene is not essential, but cells lacking *rum1*+ are unable to arrest their cell cycle before START; the cells are also sterile because arrest in G1 is required for mating (154). In addition, a double mutant Δ*rum1 cdc10ts* enters mitosis at the restrictive temperature (154). Δ*rum1* mutant cells are therefore unable to prevent mitotic activation during the pre-START window of the cell cycle. Curiously, Δ*chk1 cdc10-129* mutants have the same phenotype (9). Δ*rum1* cells respond like wild type to hydroxyurea and to radiation, so the post-START DNA checkpoints function normally (154, A. M. Carr, personal communication). Rum1p therefore appears to define a specialized role, distinct from known checkpoints, that prevents activation of the mitotic kinase prior to passage of START. In normally cycling cells, the mitotic kinase does not become prematurely active even in the absence of Rum1p, perhaps because its association with Cdc13p cyclin is rate limiting in these conditions, or because it is blocked by associations with a G1 cyclin. However, in cells blocked at START, the assembly of the mitotic kinase can occur leading to an active G2 kinase and resetting the cells to G2. One testable prediction of such a model is that it should be possible to titrate levels of Rum1p to prevent the premature mitosis induced by overproduction of Cdc2p and Cdc13p in a *cdc10ts* background.

4.3 A speculative checkpoint model

Taken together, these results all suggest that *S. pombe* has multiple control mechanisms to prevent inappropriate activation of the mitotic p34^{cdc2}. If the mitotic form of the kinase is not assembled and activated, p34^{cdc2} is constitutively able to perform its START function and initiate replication. One can speculate in the complete absence of any data that an intriguing role for Cdc10p in START would be to relieve the Rum1p inhibition of the activation of the mitotic form of p34^{cdc2} kinase. The default state of the kinase appears to be G1 competent. After passage of START, the initiation of replication events carries cells directly into S phase and into replication checkpoints. Once through S phase, the G2 regulatory pathways carry the cycle forward. All these events impart a momentum and directionality to the cell cycle, which normally prevents resetting of the kinase from G2 to G1. Checkpoints therefore operate as fail-safe devices to help maintain this directionality.

The molecular details of how the cells monitor S-phase events remain to be determined. One possibility is that some components of checkpoint signaling, such as Cdc18p, assemble with the replication complex, generating a signal that affects activation of mitotic p34^{cdc2} (6, 8, 9, 143, 147). If the replication complex fails to assemble, then it appears to the cell that replication is complete. Mutation of *cdc22*, or hydroxyurea treatment, blocks the cells between the initiation and synthesis stages, so that the complex is assembled and monitored. The process of replication and synthesis of DNA changes the signal, either modifying some associated factor or otherwise activating the damage checkpoint. The signal imparted by simple assembly is no longer sufficient to maintain the checkpoint signal, or is modified in some way. Now it is the ongoing synthesis of DNA or later replication complexes that are monitored by the checkpoint transducing genes. Only when replication is complete and the now modified complex is disassembled can the cell cycle proceed to activate the M-phase kinase. This model posits at least three distinct ways of preventing the activation of the mitotic kinase: pre-START, early S, and late S. It remains to be seen whether it will fit with experimental detail.

5. Conclusions and perspectives

Informed by new results on control of cell cycle entry and DNA replication in other systems, our understanding of the molecular control of *S. pombe* S phase is proceeding rapidly. The more abstract genetic descriptions are beginning to be replaced by increasingly concrete biochemical models, although the broad picture is still diffuse. The short, lightly regulated fission yeast S phase may prove to be a model system for similar cycles in other cell types, such as embryonic cell cycles. In general, as is the case elsewhere in the cell cycle, fission yeast differs from budding yeast in many subtle and not-so-subtle ways, proving again the utility of studying such a complicated problem in two divergent genetic systems. As we identify more of the genes involved in DNA replication in *S. pombe*, we acquire the genetic tools that complement the work in other systems. Just as a dramatic synergy between different systems and approaches led to a revolution in our understanding of the control of G2/M transition in the late 1980s, we can anticipate a burst of new data in the next five years from a variety of sources that will with luck fuse into a cohesive model of DNA replication for all eukaryotic cells.

Acknowledgements

Thanks are due to Giuseppe Baldacci, Janet Bushman, Michelle Calos, Tony Carr, Rosemary Clyne, Jaime Correa, Gennaro D'Urso, Stefana Francesconi, Beata Grallert, Joel Huberman, Yuichi Iino, Stephen Kearsey, Kohta Takahashi, Tom Kelly, Domenico Maiorano, Sergio Moreno, Paul Nurse, Andy Parker, Paul Russell, Viesturs Simanis, Masayuki Yamamoto, and Mitsuhiro Yanagida, for communication of unpublished results. I am grateful to Julian Blow, Tony Carr, Tamar Enoch, Bea Grallert, Tom Kelly, Olaf Nielsen, and the members of my lab for many helpful discussions, and to Gordy Hering, Tom Kelly, and members of the Forsburg and Kelly labs for critical comments on the manuscript.

References

1. Hayles, J. and Nurse, P. (1992) Genetics of the fission yeast *Schizosaccharomyces pombe*. *Annu. Rev. Genet.*, **26**, 373.
2. Forsburg, S. L. and Nurse, P. (1991) Cell cycle regulation in the yeasts *Saccharomyces cerevisiae* and *Schizosaccharomyces pombe*. *Annu. Rev. Cell Biol.*, **7**, 227.
3. Nasmyth, K. (1993) Control of the yeast cell cycle by the Cdc28 protein kinase. *Curr. Opin. Cell Biol.*, **5**, 166.
4. Reed, S. I. (1992) The role of p34 kinases in the G1 to S phase transition. *Annu. Rev. Cell Biol.*, **8**, 529.
5. Campbell, J. L. and Newlon, C. S. (1991) Chromosomal DNA replication. In *The molecular and cellular biology of the yeast Saccharomyces: Genome dynamics, protein synthesis, and energetics*. Broach, J., Jones, E., and Pringle, J. (eds.). Cold Spring Harbor Laboratory Press, Cold Spring Harbor, NY, vol. 1, p. 41.
6. Kelly, T. J., Nurse, P., and Forsburg, S. L. (1993) Coupling DNA replication to the cell cycle. *Cold Spring Harbor Symp. Quant. Biol.*, **58**, 637.
7. Enoch, T. and Nurse, P. (1991) Coupling M phase and S phase: controls maintaining the dependence of mitosis on chromosome replication. *Cell*, **65**, 921.
8. Carr, A. M. (1994) Cell cycle coordination after too much rum. *Bioessays*, **16**, 309.
9. Sheldrick, K. S. and Carr, A. M. (1993) Feedback controls and G2 checkpoints: fission yeast as a model system. *Bioessays*, **15**, 775.
10. Murray, A. W. (1993) Cell cycle: sunburnt fission yeast. *Nature*, **363**, 302.
11. Murray, A. W. (1992) Creative blocks: cell cycle checkpoints and feedback controls. *Nature*, **359**, 599.
12. Hartwell, L. H. (1974) *Saccharomyces cerevisiae* cell cycle. *Bacteriol. Rev.*, **38**, 164.
13. Nurse, P. (1981) Genetic control of the yeast cell cycle: a reappraisal of 'START'. In *The fungal nucleus*. Gull, K. and Oliver, S. (eds.). Cambridge University Press, Cambridge, p. 331.
14. Egel, R., Nielsen, O., and Weilguny, D. (1990) Sexual differentiation in fission yeast. *Trends Genet.*, **6**, 369.
15. Nurse, P. and Thuriaux, P. (1977) Controls over the timing of DNA replication during the cell cycle of fission yeast. *Exp. Cell. Res.*, **107**, 365.
16. Nasmyth, K. (1969) A control acting over the initiation of DNA replication in the yeast *Schizosaccharomyces pombe*. *J. Cell Sci.*, **36**, 215.
17. Nurse, P., Thuriaux, P., and Nasmyth, K. (1976) Genetic control of the cell division cycle in the fission yeast *Schizosaccharomyces pombe*. *Mol. Gen. Genet.*, **146**, 167.
18. Nurse, P. and Bissett, Y. (1981) Gene required in G1 for commitment to cell cycle and in G2 for control of mitosis in fission yeast. *Nature*, **292**, 558.
19. Caligiuri, M. and Beach, D. (1993) Sct1 functions in partnership with Cdc10 in a transcription complex that activates cell cycle START and inhibits differentiation. *Cell*, **72**, 607.
20. Tanaka, K., Okazaki, K., Okazaki, N., Ueda, T., Sugiyama, A., Nojima, H., and

Okayama, H. (1992) A new cdc gene required for S phase entry of *Schizosaccharomyces pombe* encodes a protein similar to the *cdc10+* and *SWI4* gene products. *EMBO J.*, **11**, 4923.

21. Gould, K. L. and Nurse, P. (1989) Tyrosine phosphorylation of the fission yeast *cdc2+* protein kinase regulates entry into mitosis. *Nature*, **343**, 39.
22. Gould, K., Moreno, S., Owen, D., Sazer, S., and Nurse, P. (1991) Phosphorylation at Thr 167 is required for fission yeast p34^{cdc2} function. *EMBO J.*, **10**, 3297.
23. Nurse, P. (1990) Universal control mechanism regulating onset of M phase. *Nature*, **344**, 503.
24. Pines, J. (1992) Cell proliferation and control. *Curr. Opin. Cell Biol.*, **4**, 144.
25. Hunter, T. and Pines, J. (1991) Cyclins and cancer. *Cell*, **66**, 1071.
26. Pines, J. (1993) Cyclin-dependent kinases: clear as crystal. *Curr. Biol.*, **3**, 544.
27. Kaffman, A., Herskowitz, I., Tjian, R., and O'shea, E. K. (1994) Phosphorylation of the transcription factor PHO4 by a cyclin-cdk complex, PHO80-PHO85. *Science*, **263**, 1153.
28. Moreno, S., Hayles, J., and Nurse, P. (1989) Regulation of p34^{cdc2} protein kinase during mitosis. *Cell*, **58**, 361.
29. Creanor, J. and Mitchison, J. M. (1994) The kinetics of H1 histone kinase activation during the cell cycle of wild type and wee mutants of the fission yeast *Schizosaccharomyces pombe*. *J. Cell. Sci.*, **107**, 1197.
30. Beach, D., Rodgers, L., and Gould, J. (1985) RAN1+ controls the transition from mitotic division to meiosis in fission yeast. *Curr. Genet.*, **10**, 297.
31. Grallert, B. and Sipiczki, M. (1990) Dissociation of meiotic and mitotic roles of the fission yeast cdc2 gene. *Mol. Gen. Genet.*, **222**, 473.
32. Grallert, B. and Sipiczki, M. (1989) Initiation of the second meiotic division in *Schizosaccharomyces pombe* shares common functions with that of meiosis. *Curr. Genet.*, **15**, 231.
33. Lew, D. J. and Reed, S. I. (1992) A proliferation of cyclins. *Trends Cell Biol.*, **2**, 77.
34. Xiong, Y. and Beach, D. (1991) Population explosion in the cyclin family. *Curr. Biol.*, **1**, 362.
35. Nash, R., Tokiwa, G., Anand, S., Erickson, K., and Futcher, A. B. (1988) The *WHI1+* gene of *S. cerevisiae* tethers cell division to cell size and is a cyclin homologue. *EMBO J.*, **7**, 4335.
36. Cross, F. (1988) *DAF1*, a mutant gene affecting size control, pheromone arrest and cell cycle kinetics of *Saccharomyces cerevisiae*. *Mol. Cell. Biol.*, **8**, 4675.
37. Richardson, H. E., Wittenberg, C., Cross, F., and Reed, S. I. (1989) An essential G1 function for cyclin-like proteins in yeast. *Cell*, **59**, 1127.
38. Forsburg, S. L. and Nurse, P. (1991) Identification of a G1-type cyclin *puc1+* in the fission yeast *Schizosaccharomyces pombe*. *Nature*, **351**, 245.
39. Booher, R. and Beach, D. (1987) Interaction between *cdc13+* and *cdc2+* in the control of mitosis in fission yeast: dissociation of the G1 and G2 roles of the *cdc2+* protein kinase. *EMBO J.*, **6**, 3441.
40. Hagan, I., Hayles, J., and Nurse, P. (1988) Cloning and sequencing of the cyclin-related *cdc13+* gene and a cytological study of its role in fission yeast mitosis. *J. Cell Sci.*, **91**, 587.
41. Forsburg, S. L. and Nurse, P. (1994) Analysis of the *Schizosaccharomyces pombe* cyclin puc1: evidence for a role in cell cycle exit. *J. Cell Sci.*, **107**, 601.
42. Nasmyth, K. A. (1990) FAR-reaching discoveries about the regulation of START. *Cell*, **63**, 1117.
43. Peter, M., Gartner, A., Horecka, J., Ammerer, G., and Herskowitz, I. (1993) Far1 links the signal transduction pathway to the cell cycle machinery in yeast. *Cell*, **73**, 747.

44. Tyers, M. and Futcher, B. (1993) FAR1 and FUS3 link the mating pheromone signal-transduction pathway to three G1-phase CDC28 kinase complexes. *Mol. Cell. Biol.*, **13**, 5659.

45. Molz, L. and Beach, D. (1993) Characterization of the fission yeast *mcs2* cyclin and its associated protein kinase activity. *EMBO J.*, **12**, 1723.

46. Molz, L., Booher, R., Young, P., and Beach, D. (1989) *cdc2* and the regulation of mitosis: six interacting *mcs* genes. *Genetics*, **122**, 773.

47. Bueno, A., Richardson, H., Reed, S. I., and Russell, P. (1991) A fission yeast B-type cyclin functioning early in the cell cycle. *Cell*, **66**, 149.

48. Bueno, A. and Russell, P. (1993) Two fission yeast B-type cyclins, Cig2 and Cdc13, have different functions in mitosis. *Mol. Cell. Biol.*, **13**, 2286.

49. Connolly, T. and Beach, D. (1994) Interaction between the Cig1 and Cig2 B type cyclins in the fission yeast cell cycle. *Mol. Cell. Biol.*, **14**, 768.

50. Obara-Ishihara, T. and Okayama, H. (1994) A B-type cyclin negatively regulates conjugation via interacting with cell cycle 'start' genes in fission yeast. *EMBO J.*, **13**, 1863.

51. Bueno, A., Richardson, H., Reed, S. I., and Russell, P. (1993) Erratum. *Cell*, **73**, 1050.

52. Hayles, J., Beach, D., Dukacz, B., and Nurse, P. (1985) The fission yeast cell cycle control gene *cdc2*: isolation of a sequence *suc1* that suppresses *cdc2* mutant function. *Mol. Gen. Genet.*, **202**, 291.

53. Aves, S., Durkacz, B., Carr, T., and Nurse, P. (1985) Cloning, sequencing and transcriptional control of the *Schizosaccharomyces pombe cdc10* 'start' gene. *EMBO J.*, **4**, 457.

54. Lowndes, N. F., McInerny, C. J., Johnson, A. L., Fantes, P. A., and Johnston, L. H. (1992) Control of DNA synthesis genes in fission yeast by the cell cycle gene *cdc10*⁺. *Nature*, **355**, 449.

55. Kelly, T. J., Martin, G. S., Forsburg, S. L., Stephen, R. J., Russo, A., and Nurse, P. (1993) The fission yeast *cdc18*⁺ gene product couples S phase to START and mitosis. *Cell*, **74**, 371.

56. Hofmann, J. F. X. and Beach, D. (1994) *cdt1*⁺ is an essential target of the Cdc10/Sct1 transcription factor: requirement for DNA replication and inhibition of mitosis. *EMBO J.*, **13**, 425.

57. Gordon, C. and Fantes, P. (1986) The *cdc22* gene of *Schizosaccharomyces pombe* encodes a cell cycle regulated transcript. *EMBO J.*, **5**, 2981.

58. Fernandez Sarabia, M.-J., McInerny, C., Harris, P., Gordon, C., and Fantes, P. (1993) The cell cycle genes *cdc22*⁺ and *suc22*⁺ of the fission yeast *Schizosaccharomyces pombe* encode the large and small subunits of ribonucleotide reductase. *Mol. Gen. Genet.*, **238**, 241.

59. Koch, C. and Nasmyth, K. (1994) Cell cycle regulated transcription in yeast. *Curr. Opin. Cell Biol.*, **6**, 451.

60. Merrill, G. F., Morgan, B. A., Lowndes, N. F., and Johnston, L. H. (1993) DNA synthesis control in yeast: an evolutionarily conserved mechanism for regulating DNA synthesis genes? *BioEssays*, **14**, 823.

61. Johnston, L. H. and Lowndes, N. F. (1992) Cell cycle control of DNA synthesis in budding yeast. *Nucl. Acids Res.*, **20**, 2403.

62. Breeden, L. and Nasmyth, K. (1987) Similarity between cell cycle genes of budding and fission yeast and the *notch* gene of *Drosophila*. *Nature*, **329**, 651.

63. Primig, M., Sockanathan, S., Auer, H., and Nasmyth, K. (1992) Anatomy of transcription factor important for the cell cycle in *Saccharomyces cerevisiae*. *Nature*, **358**, 593.

64. Breeden, L. and Nasmyth, K. (1987) Cell cycle control of the yeast HO gene: *cis* and *trans* activating regulators. *Cell*, **48**, 389.

65. Ogas, J., Andrews, B. J., and Herskowitz, I. (1991) Transcriptional activation of *CLN1*, *CLN2* and a putative new G1 cyclin (*HCS26*) by *SWI4*, a positive regulator of G1-specific transcription. *Cell*, **66**, 1015.

66. Koch, C., Moll, T., Neuberg, M., Ahorn, H., and Nasmyth, K. (1993) A role for the transcription factors Mbp1 and Swi4 in the progression from G1 to S phase. *Science*, **261**, 1551.

67. Reymond, A., Marks, J., and Simanis, V. (1993) The activity of *S. pombe* DSC-1-like factor is cell cycle regulated and dependent on the activity of p34cdc2. *EMBO J.*, **12**, 4325.

68. Miyamoto, M., Tanaka, K. and Okayama, H. (1994) *res2+* a new member of the *cdc10+/SWI4* family, controls the START of mitotic and meiotic cycles in fission yeast. *EMBO J.*, **13**, 1873.

69. Grallert, B. and Sipiczki, M. (1991) Common genes and pathways in the regulation of the mitotic and meiotic cell cycles of *Schizosaccharomyces pombe*. *Curr. Genet.*, **20**, 199.

70. Marks, J., Fankhauser, C., Reymond, A., and Simanis, V. (1992) Cytoskeletal and DNA structure abnormalities result from bypass of requirement for the *cdc10* start gene in the fission yeast *Schizosaccharomyces pombe*. *J. Cell Sci.*, **101**, 517.

71. Zhu, Y., Takeda, T., Nasmyth, K. and Jones, N. (1994) *pct1+*, which encodes a new DNA-binding partner of p85^{cdc10}, is required for meiosis in the fission yeast *Schizosaccharomyces pombe*. *Genes Dev.*, **8**, 885.

72. Forsburg, S. L. (1994) In and out of the cell cycle. *Curr. Biol.*, **4**, 828.

73. Mitchison, J. M., Creanor, J., and Novak, B. (1991) Coordination of growth and division during the cell cycle of fission yeast. *Cold Spring Harbor Symp. Quant. Biol.*, **56**, 557.

74. Nielsen, O. (1993) Signal transduction during mating and meiosis in *S. pombe*. *Trends Cell Biol.*, **3**, 60.

75. Sugiyama, A., Tanaka, K., Okazaki, K., Nohima, H., and Okayama, H. (1994) A zinc finger protein controls the onset of premeiotic DNA synthesis of fission yeast in a mei2-independent cascade. *EMBO J.*, **13**, 1881.

76. Saka, Y. and Yanagida, M. (1993) Fission yeast *cut5+*, required for S phase onset and M phase restraint, is identical to the radiation-damage repair gene *rad4+*. *Cell*, **74**, 383.

77. Nasmyth, K. and Nurse, P. (1981) Cell division cycle mutants altered in DNA replication and mitosis in the fission yeast *Schizosaccharomyces pombe*. *Mol. Gen. Genet.*, **182**, 119.

78. Waseem, N. H., Labib, K., Nurse, P., and Lane, D. P. (1992) Isolation and analysis of the fission yeast gene encoding polymerase δ accessory protein PCNA. *EMBO J.*, **11**, 5111.

79. Hennessy, K. M., Lee, A., Chen, E., and Botstein, D. (1991) A group of interacting yeast DNA replication genes. *Genes Dev.*, **5**, 958.

80. Duck, P., Nasim, A., and James, A. P. (1976) Temperature sensitive mutant of *Schizosaccharomyces pombe* exhibiting enhanced radiation sensitivity. *J. Bacteriol.*, **128**, 536.

81. Hirano, T., Funahashi, S., Uemura, T., and Yanagida, M. (1986) Isolation and characterization of *Schizosaccharomyces pombe cut* mutants that block nuclear division but not cytokinesis. *EMBO J.*, **5**, 2973.

82. Fenech, M., Carr, A. M., Murray, J., Watts, F. Z., and Lehmann, A. R. (1991) Cloning and characterization of the *rad4* gene of *Schizosaccharomyces pombe*: a gene showing short regions of sequence similarity to the human *XRCC1* gene. *Nucl. Acids Res.*, **19**, 6737.

83. Lehmann, A. R. (1993) Duplicated region of sequence similarity to the human XRCC1 DNA repair gene in the *Schizosaccharomyces pombe rad4/cut5* gene. *Nucleic Acids Res.*, **21**, 5274.

84. Francesconi, S., Park, H., and Wang, T. S. F. (1993) Fission yeast with DNA polymerase delta temperature sensitive alleles exhibits cell division cycle phenotype. *Nucleic Acids Res.*, **21,** 3821.

85. Samejima, I., Matsumoto, T., Nakaseko, Y., Beach, D., and Yanagida, M. (1993) Identification of 7 new cut genes involved in *Schizosaccharomyces pombe* mitosis. *J. Cell Sci.*, **105,** 135.

86. Hagan, I. and Yanagida, M. (1990) Novel potential mitotic motor protein encoded by the fission yeast *cut7+* gene. *Nature*, **347,** 563.

87. Dickson, D. and Nurse, P. (1981) The *cdc22* mutation in *Schizosaccharomyces pombe* is a temperature sensitive defect in nucleoside diphosphokinase. *Eur. J. Biochem.*, **119,** 341.

88. Tye, B.-K. (1994) The MCM2-3-5 proteins: are they replication licensing factors? *Trends Cell Biol.*, **4,** 160.

89. Sinha, P., Chang, V., and Tye, B.-K. (1986) A mutant that affects the function of autonomously replicating sequences in yeast. *J. Mol. Biol.*, **192,** 805.

90. Maine, G. T., Subgam O., and Tye, B.-K. (1984) Mutants of *S. cerevisiae* defective in the maintenance of minichromosomes. *Genetics*, **106,** 365.

91. Coxon, A., Maundrell, K., and Kearsey, S. E. (1992) Fission yeast *cdc21+* belongs to a family of proteins involved in an early step of chromosome replication. *Nucl. Acids Res.*, **20,** 5571.

92. Bussereau, F., Mallet, L., Gaillon, L., and Jacquet, M. (1993) A 12.8 kb segment on the right arm of chrmosome II from *Saccharomyces cerevisiae* including part of the *DUR1, 2* gene contains five putative new genes. *Yeast*, **9,** 767.

93. Gibson, S. I., Surosky, R. T., and Tye, B.-K (1990) The phenotype of the minichromosome maintenance mutant mcm3 is characteristic of mutants defective in DNA replication. *Mol. Cell. Biol.*, **10,** 5707.

94. Yan, H., Gibson, S., and Tye, B. K. (1991) Mcm2 and Mcm3, two proteins important for ARS activity, are related in structure and function. *Genes Dev.*, **4,** 968.

95. Yan, H., Merchant, A. M., and Tye, B. K. (1993) Cell cycle-regulated nuclear localization of MCM2 and MCM3, which are required for the initiation of DNA synthesis at chromosomal replication origins in yeast. *Genes Dev.*, **7,** 2149.

96. Hennessy, K. M., Clark, C. D., and Botstein, D. (1990) Subcellular localization of yeast CDC46 varies with the cell cycle. *Genes Dev.*, **4,** 2252.

97. Chen, Y. R., Hennessy, K. M., Botstein, D., and Tye, B. K. (1992) Cdc46/Mcm5, a yeast protein whose subcellular localization is cell cycle-regulated, is involved in DNA replication at autonomously replicating sequences. *Proc. Natl. Acad. Sci. USA*, **89,** 10459.

98. Koonin, E. V. (1993) A common set of conserved motifs in a vast variety of putative nucleic acid-dependent ATPases including MCM proteins involved in the initiation of ukaryotic DNA replication. *Nucl. Acids Res.*, **21,** 2541.

99. Blow, J. J. and Laskey, R. A. (1988) A role for the nuclear envelope in controlling DNA replication within the cell cycle. *Nature*, **332,** 546.

100. Toda, T., Umesono, K., Hirata, A., and Yanagida, M. (1983) Cold sensitive nuclear division arrest mutants of the fission yeast *Schizosaccharomyces pombe*. *J. Mol. Biol.*, **168,** 251.

101. Forsburg, S. L. and Nurse, P. (1994) The fission yeast *cdc19+* gene encodes a member of the MCM family of replication proteins. *J. Cell Sci.*, **107,** 2779.

102. Miyake, S., Okishio, N., Samejima, I., Hiraoka, Y., Toda, T., Saitoh, I., and Yanagida, M. (1993) Fission yeast genes *nda1+* and *nda4+*, mutations of which lead to S-phase block, chromatin alteration and Ca^{2+} suppression, are members of the *CDC46/MCM2* family. *Mol. Biol. Cell*, **4,** 1003.

103. Takahashi, K., Yamada, H., and Yanagida, M. (1994) Fission yeast minichromosome loss mutants *mis* cause lethal aneuploidy and replication abnormality. *Mol. Biol. Cell*, **5,** 1145.

104. Nasmyth, K. (1977) Temperature-sensitive lethal mutants in the structural gene for DNA ligase in the yeast *Schizosaccharomyces pombe*. *Cell*, **12,** 1109.

105. Li, R. and Murray, A. W. (1991) Feedback control of mitosis in budding yeast. *Cell*, **66,** 519.

106. Thömmes, P., Fett, R., Schray, B., Burkhart, R., Barnes, M., Kennedy, C., Brown, N. C., and Knippers, R. (1992) Properties of the nuclear P1 protein, a mammalian homologue of the yeast Mcm3 replication protein. *Nucl. Acids Res.*, **20,** 1069.

107. Hu, B., Burkhart, R., Schulte, D., Musahl, C., and Knippers, R. (1993) The P1 family – a new class of nuclear mammalian proteins related to the yeast MCM replication proteins. *Nucleic Acids Res.*, **21,** 5289.

108. Todorov, I. T., Pepperkok, R., Philipova, R., Kearsey, S. E., Ansorge, W., and Werner, D. (1994) A human nuclear protein with sequence homology to a family of early S phase proteins is required for entry into S phase and for cell division. *J. Cell Sci.*, **107,** 253.

109. Li, J. J. and Herskowitz, I. (1993) Isolation of ORC6, a component of the yeast origin recognition complex by a one-hybrid system. *Science*, **262,** 1870.

110. Maundrell, K., Wright, A. P., and Shall, S. (1985) Evaluation of heterologous ARS activity in *S. cerevisiae* using cloned DNA from *S. pombe*. *Nucleic Acids Res.*, **13,** 3711.

111. Maundrell, K., Hutchison, A., and Shall, S. (1988) Sequence analysis of ARS elements in fission yeast. *EMBO J.*, **7,** 2203.

112. Wohlgemuth, J. G., Bulboaca, G. H., Moghadam, M., Caddle, M. S., and Calos, M. P. (1994) Physical mapping of origins of replication in the fission yeast *Schizosaccharomyces pombe*. *Mol. Biol. Cell*, **5,** 839.

113. Zhu, J., Brun, C., Kurooka, H., Yanagida, M., and Huberman, J. A. (1992) Identification and characterization of a complex chromosomal replication origin in *Schizosaccharomyces pombe*. *Chromosoma*, **102,** S7.

114. Caddle, M. S. and Calos, M. P. (1994) Specific initiation at an origin of replication from *Schizosaccharomyces pombe*. *Mol. Cell. Biol.*, **14,** 1796.

115. Johnston, L. H. and Barker, D. G. (1987) Characterisation of an autonomously replicating sequence from the fission yeast *Schizosaccharomyces pombe*. *Mol. Gen. Genet.*, **207,** 161.

116. Olsson, T., Ekwall, K., and Ruusala, T. (1993) The silent P mating type locus in fission yeast contains two autonomously replicating sequences. *Nucleic Acids Res.*, **21,** 855.

117. Dubey, D. D., Zhu, J., Carlson, D. L., Sharma, K., and Huberman, J. A. (1994) Three ARS elements contribute to the *ura4* replication origin in the fission yeast *Schizosaccharomyces pombe*. *EMBO J.*, **13,** 3638.

118. Zhu, J., Carlson, D. L., Dubey, D. D., Sharma, K., and Huberman, J. A. (1994) Comparison of two major ARS elements of the *ura4* replication origin region with other ARS elements in the fission yeast *Schizosaccharomyces pombe*. *Chromosoma* **103,** 414.

119. Diffley, J. F. X. and Cocker, J. H. (1992) Protein–DNA interactions at a yeast replication origin. *Nature*, **357,** 169.

120. Diffley, J. F. X., Cocker, J. H., Dowell, S. J., and Rowley, A. (1994) Two steps in the assembly of complexes at yeast replication origins in vivo. *Cell*, **78,** 303.

121. Bell, S. P. and Stillman, B. (1992) ATP-dependent recognition of eukaryotic origins of DNA replication by a multiprotein complex. *Nature*, **357,** 128.

122. Bell, S. P., Kobayashi, R., and Stillman, B. (1993) Yeast origin recognition complex functions in transcription silencing the DNA replication. *Science*, **262,** 1844.

123. Foss, M., McNally, F. J., Laurenson, P., and Rine, J. (1993) Origin Recognition Complex (ORC) in transcriptional silencing and DNA replication in *S. cerevisiae. Science*, **262**, 1838.

124. Park, H., Francesconi, S., and Wang, T. (1993) Cell cycle expression of two replicative DNA polymerases α and δ from Schizosaccharomyces pombe. *Mol. Biol. Cell*, **4**, 145.

125. Damagnez, V., Tillit, J., Recondo, A.-M., and Baldacci, G. (1991) The POL1 gene from fission yeast, *Schizosaccharomyces pombe*, shows conserved amino acid blocks specific for eukaryotic DNA polymerase alpha. *Mol. Gen. Genet.*, **226**, 182.

126. Pignede, G., Moussy, G., Bouvier, D., Tillit, J., de Recondo, A. M., and Baldacci, G. (1992) Expression of the catalytic subunits of pol alpha and pol delta from fission yeast *Schizosaccharomyces pombe. Chromosoma*, **102**, S128.

127. Bouvier, D., Pignede, G., Damagnez, V., Tillit, J., De Recondo, A.-M., and Baldacci, G. (1992) DNA polymerase α in the fission yeast *Schizosaccharomyces pombe*: identification and tracing of the catalytic subunit during the cell cycle. *Exp. Cell Res.*, **198**, 183.

128. Singh, J. and Klar, A. J. S. (1993) DNA polymerase alpha is essential for mating type switching in fission yeast. *Nature*, **361**, 271.

129. Klar, A. J. S. (1992) Molecular genetics of fission yeast cell type: mating type and mating type interconversion. In *The molecular and cellular biology of the yeast Saccharomyces: Gene expression*. Jones, E., Pringle, J. and Broach, J. (eds.). Cold Spring Harbor Laboratory Press, Cold Spring Harbor, NY, vol. 2, p. 745.

130. Pigneded, G., Bouvier, D., Derecondo, A. M., and Baldacci, G. (1991) Characterization of the pol3 gene product from *Schizosaccharomyces pombe* indicates interspecies conservation of the catalytic subunit of DNA polymerase delta. *J. Mol. Biol.*, **222**, 209.

131. Johnston, L., Barker, D., and Nurse, P. (1986) Molecular cloning of the *Schizosaccharomyces pombe* DNA ligase gene *cdc17* and an associated sequence promoting high frequency plasmid transformation. *Gene*, **41**, 321.

132. White, J., Barker, D., Nurse, P., and Johnston, L. (1986) Periodic transcription as a means of regulating gene expression during the cell cycle: contrasting modes of expression of DNA ligase genes in budding and fission yeast. *EMBO J.*, **5**, 1705.

133. Al-Khodairy, F., Fotou, E., Sheldrick, K. S., Griffiths, D. J. F., Lehmann, A. R., and Carr, A. M. (1994) Identification and characterization of new elements involved in checkpoint and feedback controls in fission yeast. *Mol. Biol. Cell*, **5**, 147.

134. Uemura, T. and Yanagida, M. (1984) Isolation of type I and type II DNA topoisomerase mutants from fission yeast: single and double mutants show different phenotypes in cell growth and chromatin organization. *EMBO J.*, **3**, 1737.

135. Uemura, T., Morino, K., Uzawa, S., Shiozaki, K., and Yanagida, M. (1987) Cloning and sequencing of *Schizosaccharomyces pombe* DNA topoisomerase I gene, and effect of gene disruption. *Nucleic Acids Res.*, **15**, 9727.

136. Uemura, T. and Yanagida, M. (1986) Mitotic spindle pulls but fails to separate chromosomes in type II DNA topoisomerase mutants: uncoordinated mitosis. *EMBO J.*, **5**, 1003.

137. Uemura, T., Morikawa, K., and Yanagida, M. (1986) The nucleotide sequence of the fission yeast DNA toposiomerase II gene: structural and functional relationships to other DNA topoisomerases. *EMBO J.*, **5**, 2355.

138. Uemura, T., Ohkura, H., Adachi, Y., Morino, K., Shiozaki, K., and Yanagida, M. (1987) DNA toposiomerase II is required for condensation and separation of mitotic chromosomes in *S. pombe. Cell*, **50**, 917.

139. Murray, A. W. (1994) Cell cycle: rum tale of replication. *Nature*, **367**, 219.

140. Enoch, T. and Nurse, P. (1990) Mutation of fission yeast cell cycle control genes abolishes dependence of mitosis on DNA replication. *Cell*, **60**, 665.

141. Sunnerhagen, P., Seaton, B. L., Nasim, A., and Subramani, S. (1990) Cloning and analysis of a gene involved in DNA repair and recombination, the *rad1* gene of *Schizosaccharomyces pombe*. *Mol. Cell. Biol.*, **10**, 3750.

142. Murray, J. M., Carr, A. M., Lehmann, A. R., and Watts, F. Z. (1991) Cloning and characterization of the DNA repair gene *rad9* from *Schizosaccharomyces pombe*. *Nucleic Acids Res.*, **19**, 3525.

143. Enoch, T., Carr, A. M., and Nurse, P. (1992) Fission yeast genes involved in coupling mitosis to completion of DNA replication. *Genes Dev.*, **6**, 2035.

144. Al-Khodairy, F. and Carr, A. M. (1992) DNA repair mutants defining G2 checkpoint pathways in *Schizosaccharomyces pombe*. *EMBO J.*, **11**, 1343.

145. Rowley, R., Subramani, S., and Young, P. G. (1992) Checkpoint controls in *Schizosaccharomyces pombe: rad1*. *EMBO J.*, **11**, 1335.

146. Seaton, B. L., Yucel, J., Sunnerhagen, P., and Subramani, S. (1990) Isolation and characterisation of the *Schizosaccharomyces pombe rad3+* which is involved in the DNA damage and DNA synthesis checkpoints. *Gene*, **119**, 83.

147. Li, J. J. and Deshaies, R. J. (1993) Exercising self-restraint: discouraging illicit acts of S and M in eukaryotes. *Cell*, **74**, 223.

148. Walworth, N., Davey, S., and Beach, D. (1993) Fission yeast *chk1* protein kinase links the rad checkpoint pathway to *cdc2*. *Nature*, **363**, 368.

149. Kimura, H., Nozaki, N., and Sugimoto, K. (1994) DNA polymerase alpha associated protein P1, a murine homolog of yeast MCM3, changes its intranuclear distribution during the DNA synthetic period. *EMBO J.*, **13**, 4311.

150. Broek, D., Bartlett, R., Crawford, K., and Nurse, P. (1991) Involvement of p34^{cdc2} in establishing the dependency of S phase on mitosis. *Nature*, **349**, 388.

151. Hayles, J., Fisher, D., Woollard, A., and Nurse, P. (1994) Temporal order of S phase and mitosis in fission yeast is determined by the state of the p34^{cdc2} mitotic B cyclin complex. *Cell*, **78**, 813.

152. Amon, A., Irniger, S., and Nasmyth, K. (1994) Closing the cell-cycle circle in yeast: G2 cyclin proteolysis initiated at mitosis persists until the activation of G1 cyclins in the next cycle. *Cell*, **77**, 1037.

153. Fitch, I., Dahmann, C., Surana, U., Amon, A., Nasmyth, K., Goetsch, L., Byers, B., and Futcher, B. (1992) Characterisation of four B-type cyclin genes of the budding yeast *Saccharomyces cerevisiae*. *Mol. Biol. Cell*, **3**, 805.

154. Moreno, S. and Nurse, P. (1994) Regulation of progression through the G1 phase of the cell cycle by the *rum1+* gene. *Nature*, **367**, 236.

155. Iino, Y. and Yamamoto, M. (1985) Negative control for the initiation of meiosis in *Schizosaccharomyces pombe*. *Proc. Natl. Acad. Sci. USA*, **82**, 2447.

156. Nurse, P. (1985) Mutants of the fission yeast *Schizosaccharomyces pombe* which alter the shift between cell proliferation and sporulation. *Mol. Gen. Genet.*, **198**, 497.

157. McLeod, M. and Beach, D. (1986) Homology between the *ran1+* gene of fission yeast and protein kinases. *EMBO J.*, **5**, 3665.

158. Ford, J. C., Al-khodairy, F., Fotou, E., Sheldrick, K. S., Griffiths, D. J. F., and Carr, A. M. (1994) 14-3-3-protein homologs required for the DNA-damage checkpoint in fission yeast. *Science*, **265**, 533.

159. Nurse, P. (1975) Genetic control of cell size at cell division in yeast. *Nature*, **256**, 547.

160. Andrews, B. J. and Herskowitz, I. (1989) The yeast SWI4 protein contains a motif present in development regulators and is part of a complex involved in cell cycle dependent transcription. *Nature*, **342**, 830.

Index

Italic numbers denote reference to illustrations
Genes and gene products from *S. cerevisiae* are uppercase, and those from *S. pombe* are lower case.